PROTEINS: FORM AND FUNCTION

Edited by

RALPH A. BRADSHAW AND MARY PURTON

1990

ELSEVIER TRENDS JOURNALS

Cambridge

ISBN: 1 85166 512 9

Published by:

Elsevier Trends Journals
68 Hills Road
Cambridge
CB2 1LA

Sole distributors worldwide except for the USA and Canada

Elsevier Science Publishers Ltd
Crown House
Linton Road
Barking
Essex IG11 8JU
UK

Sole distributors for the USA and Canada

Elsevier Science Publishing Company Inc.
655 Avenue of the Americas
New York
NY 10010
USA

Cover illustration: Computer graphic representation of the structure of the aspartic proteinase endothiapepsin, courtesy of Lawrence Pearl and Tom Blundell

Design by In-Toto Services
Typeset by Goodfellow & Egan Ltd, Cambridge
Printed in England by Black Bear Press, Cambridge

Contents

PREFACE

Protein chemistry is the term generally used to describe studies of the structure and function of proteins. In many respects, however, this term is too limiting. Clearly, the successful study of proteins requires the knowledge and techniques of many fields of science ranging from physics to biology: it is important not only to understand the three-dimensional structure of proteins in atomic detail but also to understand the manner in which they function within cells, organs and whole animals. Biophysics, biochemistry and biology are all equally important in this task.

This volume comprises a series of articles and short reviews that covers this broad spectrum of protein science. Many of the chapters first appeared in a special issue devoted to proteins which appeared as the July 1989 issue of *Trends in Biochemical Sciences*. That issue proved to be highly popular, particularly for senior undergraduate and graduate teaching and has prompted us to reprint and expand this collection, including additional articles that have or will shortly appear in *TIBS* or were specially commissioned for it. In this manner, greater balance and some additional depth in certain areas has been achieved. However, comprehensive coverage in any volume dealing with such a diverse topic is, like beauty, 'in the eye of the beholder'. Clearly, it is not possible to treat every topic in the detail that is required to render the reader a complete understanding. Rather, like the journal from which the articles are taken, the chapters are meant to provide a reasonable introduction, and to provide appropriate leads and direction for further reading to the level desired. Its suitability as instructional material will depend upon the course, the instructor, and the level of expertise to be achieved.

To facilitate its use, the volume is divided into sections that deal with both structure and function. The sections on structure are heavily methodological, focusing on how data are collected and analysed, but are supplemented with articles describing the application of these methods. The sections of the book devoted to function emphasize how structure dictates and controls activity. Considerable attention is devoted to the modification of proteins, both those generated by the powerful arsenal of techniques derived from molecular biological manipulations and those which occur naturally. These latter alterations in protein structure begin at the time of synthesis as co-translational events and ultimately terminate in the proteolytic degradation of the molecule. In between, a large variety of covalent modifications of both a reversible and irreversible character are utilized by cells to ensure their correct translocation, the regulation of their activity, and ultimately their turnover.

It is important to point out that certain functional properties are not developed in great detail. For example, there are not chapters on the properties of enzymes, since many excellent volumes already exist dealing with this material. Similarly specialized structures such as membrane-bound receptors are not addressed directly. Their omission is not meant to minimize their importance; only to indicate the limitations in space and size of a volume of this type.

Having stressed these shortcomings, it is worth pointing out a few strengths. Importantly the articles are quite recent and therefore represent some of the most up-to-date information on the subject. Some of the material covered is indeed already well known, but most of it focuses on the cutting edge of protein science, as we enter the last decade of this century. It emphasizes the importance of bringing together all of the germane techniques from the aforementioned branches of science if the study of proteins is to advance at the speed and intensity that one expects will characterize the 1990s. We hope that this volume will prove a useful tool for both researchers and students.

We would like to thank the many authors whose original articles appeared in *TIBS*, and those who have provided either updated articles or new contributions for this book. Their willingness to participate in this venture is greatly appreciated.

RALPH A. BRADSHAW AND MARY PURTON

HIGHLIGHTS OF PROTEIN STRUCTURAL ANALYSIS

JOHN E. SHIVELY, RAYMOND J. PAXTON AND TERRY D. LEE

This article describes highlights of the state of the art in protein structural analysis, and comments on the current trends toward increased sensitivity and integrated isolation–structure methodologies.

The biological activity of a protein depends on its three-dimensional structure and its interaction with other molecules. In order to understand the fundamental relationship of a protein's structure to its function, one must start at the beginning, namely, the determination of its structure. Since proteins comprise linear polymers of amino acids, structural analysis of proteins begins with amino acid sequence determination. With the advent of molecular cloning of genes encoding for proteins, and rapid DNA sequencing methods, it has been possible to shorten, or in the opinion of some, to bypass protein sequence analysis by simply deducing amino acid sequence information from nucleotide sequences. However, proteins often undergo extensive post-translational modifications which are not predicted from the gene sequence and so a strong argument can be made for extensive protein structural analysis. In addition, preliminary protein structural information is often used as the basis for gene identification, or for generating oligonucleotides which can be used to clone the gene of interest. In this respect, the emphasis is usually on 'rare' proteins. Recombinant DNA technology allows the production of recombinant proteins which makes rare proteins available in large amounts, or allows the production of site specific modifications in proteins. With recombinant proteins, it is important to confirm that the predicted structure conforms to the expressed product. In fact, it can be said that the recent growth of modern protein structural analysis and molecular biology have been complementary, and that molecular biologists and protein chemists have become dependent on each other for the isolation and characterization of rare proteins and the confirmation of the structure of recombinant proteins.

The science of protein structural analysis is considerably complicated by the need to integrate protein and peptide isolation techniques with amino acid sequence determination. One of the most common complaints is that the sample was lost between the isolation and sequence determination steps. Another problem is concerned with the sequence analysis itself. With 20 naturally occurring amino acids, and hundreds possible by post-translational modifications, the challenge inherent in protein sequence determination is obvious. Nonetheless, the problems are solvable and the sensitivity and speed of the approaches are impressive.

A comprehensive overview of modern protein sequence analysis is beyond the scope of this article. Books have been written on the subject, and contain valuable practical hints, but are usually out of date before they appear in print[1–5]. A recent review by Simpson and co-workers[6] can be recommended for the integration of isolation and structural determination methods. This article will cover the broad scope of the field with an emphasis towards expected new developments over the next few years.

STRATEGY

The strategy begins with assurance of a steady supply of the 'purified' protein. As analysis proceeds, the results may dictate further purification. The chief criteria of purity include single bands or peaks using *several* chromatographic and/or electrophoretic methods (these are discussed by Ken Wilson, see p. 15). Researchers who wish to sequence proteins directly from one- or two-dimensional gel electrophoretograms, should elute the sample and verify its homogeneity by an additional technique. Proteins which are composed of subunits may exhibit several bands or peaks in the pure state, but will be evident by their reproducible stoichiometry and specific biological activity.

Paramount to the opening gambit is the knowledge of how much sample is present. Although proteins can be detected on electrophoresis gels by various staining techniques, they cannot be quantitated unless they are standards of already known structure. In all cases the sample should be eluted or directly hydrolysed and quantitated by amino acid compositional analysis. This composition also provides a valuable piece of data which should ultimately match that predicted for the total structure. It can be used as a criteria of purity if it is reproducible and unique. From the quantitation and molecular weight (most commonly obtained from SDS gel electrophoresis), one can calculate molar amounts. Although there are occasional examples of obtaining structural information from subpicomole amounts of proteins, it is fair to say that current technology requires at least 10 pmol to begin the analysis. The actual amount used at each step will also depend on the skill and resources of the protein chemist.

Assuming that sample quantitation is known and supply lines are assured, one can plan strategies aimed at three levels of increasing involvement:

(1) 10–100 pmol for several N-terminal sequence and amino acid analyses.
(2) 100–200 pmol for single peptide maps, sequence and mass spectrometric analyses. Important information gained on post-translational modifications.
(3) 100–1000 pmol for complete (or >90%) sequence analysis. Involves multiple peptide maps and complete analysis of post-translational modifications.

If the sample is so rare and difficult to prepare that you can only hope for an N-terminal sequence analysis, then the method of choice may be to sequence the protein following SDS gel electrophoresis and transfer-ence to a polyvinylidiene (PVDF) membrane, or microbore (1 mm ID) HPLC purification. Even in desperate circumstances, the sample mass should be quantitated by amino acid analysis, by splitting the sample (best), or by running a parallel sample (second best). If no N-terminal sequence is obtained, then the quantitative analysis will allow one to conclude if it is blocked or simply below the level of analysis. If a low yield of the N-terminal group relative to the mass analysis is obtained, it is possible that only a trace impurity was sequenced. The latter possibility may have dire consequences, for many an impurity has been mistaken for the real sample. Thus, even at level (1) above, there is no excuse for omitting quantitative amino acid analysis.

Approaches (2) or (3) are the most desirable, since they will provide enough information to relate the structure to other proteins, reveal post-translational modifications, and with reasonable luck (and skill) provide the entire structure. In these approaches, the key is to obtain peptide maps of the protein by digestion with chemicals such as cyanogen bromide, or proteases such as trypsin. The peptide maps (usually obtained by reversed phase HPLC) provide multiple chances to obtain amino acid sequence information (unlike the N-terminal sequence of the intact protein), but ultimately have to be aligned with either overlapping peptides or the cDNA deduced sequence information. Thus, level (2) assumes a cooperative approach with the molecular biologist, and level (3) offers the chance to 'go it alone'. The speed of peptide mapping and sequence analysis is fast: 20–30 amino acids per day can be sequenced in the average lab. However, it is a general rule that some portion of the protein structure will be refractory, and one usually obtains 90% of the structure within the first few weeks or months, with the remaining 10% requiring an indefinite period of time.

Structural approaches must be integrated with protein-purification strategies in at least three areas. The first is that most analytical methods are adversely affected by salts and require the use of volatile buffers. Thus, the sample may have to be desalted by reversed phase HPLC or precipitation prior to sequence analysis. The second is that the structural methods require concentrated samples (0.1–10 $\mu g\ \mu l^{-1}$), thus the last step of purification may also involve concentration. These strategies are discussed by Ken Wilson (see p. 15). The third is often the trickiest, namely, that the sample may precipitate or be adsorbed to a surface just prior to analysis. Problems with protein precipitation or adsorption can be prevented: (1) by collecting or immediately transferring the sample to a membrane which will be directly analysed; (2) by adding a carrier or substance which prevents protein adsorption to surfaces; or (3) by integrating protein purification and sample analysis. Usually, when these problems become evident, researchers are content to 'scale up', concentrate their samples, or wash the sample tube with solvents of extremes of polarities or pH. Super solvents such as formic acid or hexafluoroacetone trihydrate are recommended for recovery of adsorbed samples. The addition of carriers is still an art, and since proteins differ widely in their adsorptive properties, no universal carrier has been devised. Organic solvents such as propanol have become popular for preventing protein adsorption to polyethylene or polypropylene tubes. Integrated approaches, for the most part, are still in the future, but the PVDF electrotransfer technology is a first approach.

AMINO ACID ANALYSIS

This all important analytical tool has been recently re-examined in terms of sensitivity and sample preparation. Nonetheless, the analysis of submicrogram amounts of proteins is a daunting problem because even traces of dust or contamination on hydrolysis tubes, for example, will lead to backgrounds greater than the sample mass itself. A recent review on the very popular PTC (phenylthiocarbamyl) method is recommended as a starting point[7]. Amino acid analysis has always been a critical tool in protein chemistry, but has only recently been re-examined in terms of sensitivity and sample handling. We can predict that integrated methods for sample handling, hydrolysis and analysis will ultimately solve these challenging problems.

REDUCTION AND ALKYLATION OF CYSTEINE RESIDUES

Prior to sequence analysis or peptide mapping the sample may require reduction and *S*-alkylation of cysteine residues. If the protein has cysteines involved in disulfide bonds, it will be necessary to establish their location by peptide mapping of the intact protein (usually involving more aggressive proteases). Standard denaturing conditions for proteins are dissolution in 6 M guanidinium HCl (best) or 8 M urea (requires careful deionization and minimization of sample exposure to prevent blockage of the N-terminal group). The reduction step is performed under nitrogen or argon with DTT (dithiothreitol), mercaptoethanol or tributyl phosphine. Tributyl phosphine is popular because it is volatile and may be used directly on sequencer supports[8]. Popular alkylating agents include iodoacetic acid which can be obtained radiolabeled to trace *S*-carboxymethyl cysteine derivatives[9], and vinyl pyridine which imparts a unique UV absorbance to the *S*-ethylpyridyl cysteine derivatives[10] and is volatile[8]. The references cited demonstrate the methods and discuss strategies for removing the denaturing agent prior to peptide mapping. Denatured proteins are even trickier to handle than the native proteins, often precipitating during the desalting steps. In cases where low microgram amounts of sample are handled, the precipitates are invisible. Care must be taken to follow the sample quantitatively, never discarding tubes in which the sample has resided. Strategies to retain protein solubility include the use of organic solvents such as propanol or detergents such as SDS; however, these additives will frequently affect subsequent mapping and chromatographic steps.

PEPTIDE MAPPING

Two major options are open for peptide mapping. One is treatment with cyanogen bromide which cleaves at the C-terminal side of methionine residues, and usually results in very large, relatively insoluble peptides. Before choosing this option, ascertain that the protein has methionine residues by amino acid analysis. If the sample has already precipitated, it can be redissolved in 70–98% formic acid (10–100 μl) and treated with cyanogen bromide[11]. The sample can be dried, redissolved in formic acid for reversed phase HPLC analysis, or in SDS for gel electrophoresis. Remembering that large peptides often aggregate and elute together in low yields on reversed phase HPLC, a safer approach may be to separate the peptides by SDS gel electrophoresis adapted for the resolution of low molecular weight peptides[12]. The peptides can be electrotransferred to PVDF, stained and sequenced. For preservation of disulfide bonds, 6 M guanidinium HCl can be used as the denaturant, and the cyanogen bromide treated samples directly analysed by reversed phase HPLC[13]. If poor yields or poorly resolved peaks are observed for reversed phase HPLC, the sample can be dissolved in 40–60% acetic acid and subjected to gel permeation chromatography. Acidic conditions will promote the formation of homoserine lactone at the C-terminus of the methionyl cleaved peptides. The C-terminus of the protein will be the only peptide not ending in homoserine (lactone) unless nonspecific peptide bond cleavages occur.

The second, more widely used approach, is peptide mapping with proteases. The most useful proteases are trypsin (cleavage at C-terminal side of Lys or Arg), *S. aureus* V8 protease (cleavage at C-terminal side of Glu and sometimes Asp), and Asp-N-protease (cleavage at N-terminal side of Asp, cysteic acid and sometimes Glu). The samples are usually digested in ammonium bicarbonate at pH 8, and the resulting peptides resolved by reversed phase HPLC. Many proteins produce an insoluble peptide core which has to be solubilized and redigested with another protease. Trypsin or Asp-N-protease are good choices for first peptide maps. *S. aureus* V8 protease is a good second choice for overlapping peptides or for digesting peptides in the presence of SDS. Since proteases are inhibited by organic solvents to varying degrees, it is advisable to perform test runs on standards before settling on optimal conditions for maintaining protein solubility and proteolytic activity[13]. In addition to reversed phase HPLC, excellent peptide maps can also be obtained on ion exchange columns such as the sulfoethyl aspartamide column[14,15]. This recent work is of interest because the samples were further analysed without the need for desalting.

Proteins in the size range of 10–50 000 daltons will produce a moderate number of peptides (10–40) which can be resolved by a single HPLC run plus a few selected rechromatographic runs (usually on the same column). The peptides can be preselected by FAB/MS prior to sequence analysis to analyse for purity and molecular weight. Larger proteins give complicated peptide maps requiring extensive rechromatography, but with the existing narrow bore

(2 mm ID) and microbore (1 mm ID) columns, the yields and sensitivities are equal to the task.

Edman Chemistry

N-terminal sequence analysis involves the reaction of a free N-terminal amino acid with phenylisothiocyanate (PITC) in base to form a phenythiocarbamyl (PTC) derivative which under acidic conditions forms the cyclic anilinothiazolinone (ATZ) amino acid derivative and a new free N-terminal residue. The unstable ATZ is usually converted by acid hydrolysis to the more stable phenylthiohydantoin (PTH) derivative which is identified by reversed phase HPLC. The chemistry remains unchanged from the original formulation by Edman in 1950[16]. Major improvements include automation[17], and the use of Polybrene to immobilize the sample, especially small peptides[18,19]. The modern automated instrument is similar to that described by Hewick *et al.* in 1981[20], and has a sensitivity in the range of 10–100 pmol when connected on-line to a narrow-bore HPLC. The limit of sensitivity of analysis depends on the detection of PTH amino acid derivatives (currently about 0.1 pmol), initial yields (~30–80%), and repetitive yields (~92–95%). Thus, starting with 10 pmol of sample and (assuming initial and repetitive yields of 50% and 95%, respectively, and no injection losses) the amount of PTH derivative detected after 20 cycles would be about 1.8 pmol. Practically speaking, less than 100% of the sample is analysed, repetitive yields are lower for real samples (compared to standards), and some PTH derivatives are recovered in low yields (e.g. Ser, Thr, His and Arg are detected at about 30–50% the level of the other derivatives). In order to obtain reasonably reliable data with the current methods available, one should analyse about 10 pmol of sample. Since the confidence of assigning amino acid derivatives falls with increasing cycle number, the method has inherent limitations. A determination of the amino acid composition or molecular weight of the analysed peptide is an important check on sequence data. Given the need for high sensitivity, mass spectrometry is the current method of choice for providing molecular weight information. If sequence data and molecular weight data agree, there is an extremely high likelihood that the sequence is correct. If the data don't agree, either an error has been made in the sequence, or a modified amino acid is present.

Electroblotting Techniques

Since SDS gel electrophoresis is the mainstay of protein molecular weight analysis, it is not surprising that many researchers have directed their efforts towards the direct sequencing of proteins from one- or two-dimensional gels. The early work of Hunkapiller and co-workers[21] which describes the elution and sequence analysis of stained protein bands from gels was fraught with problems of recovery and background noise. This approach has now been updated and automated with a continuous elution system commercially available from Applied Biosystems. Highly sensitive electroblotting techniques have now become the mainstay in the field. The initial approach reported by Abersold and co-workers[22], who performed electrotransfer of proteins from SDS gels to derivatized glass fiber paper, later gave way to the more reproducible method of Matsudaira[23] in which proteins were electrotransferred to PVDF membranes. Importantly, the transferred proteins could be stained with Coomassie blue and sequenced without interfering background peaks on HPLC analysis of the PTH derivatives. There has been a flurry of papers describing improvements on the method, including the use of Polybrene[24], reducing agents[25], and detergents[6] to improve transfer and sequencer recoveries. It is also possible to elute proteins from PVDF and recover enzyme activity[26]. Other advantages of the PVDF membrane include its hydrophobic, protein-adsorbing properties, and its relative chemical inertness towards strong acid and base, and organic solvents. For example, samples can be subjected to amino acid analysis or cyanogen bromide cleavage on PVDF membranes without prior elution[6]. PVDF can be chemically derivatized with a wide variety of functional groups[27,28] leading to the possibility of covalent immobilization of peptides and proteins. This concept has revitalized interest in solid-phase Edman chemistry originally described by Laursen in 1971[29], and has recently been adapted to PVDF as a solid phase support[30]. It is too early to evaluate the realized advantages of this technology, but in theory it promises increased sensitivity and cycle number.

Membrane proteins pose special problems for protein solubilization, recovery, and sequence analysis. A recent approach describes solubilization of samples in neat trifluoroacetic acid, disaggregation in SDS, and separation by SDS gel electrophoresis[31]. For complex mixtures, a single dimension separation is inadequate. Successful approaches include two-dimensional gel electrophoresis (isoelectric focusing in one dimension followed by SDS electrophoresis in the second). In this approach sample sizes of about 0.5–1 μg can be electrophoresed, electrotransferred, and sequenced[32]. The obvious problem is sensitivity: 0.5 μg of a 50 kDa protein is only 10 pmol, which is at the limit of sensitivity of current sequencing methods (especially considering transfer and sequencing yields). A second approach is tandem gel electrophoresis or tandem HPLC/SDS gel electrophoresis. One example uses first isoelectric focusing and then acetic acid/triton/urea gels[33], and a second uses first microbore HPLC and then SDS gel electrophoresis[34]. These methods are limited to N-terminal sequence information only. In order to obtain internal sequence information, the sample must be eluted and chemically or proteolytically digested. One approach is described by Abersold *et al.*[35] who transferred gel electrophoresis samples to nitrocellulose and digested reversibly stained proteins directly on the membrane. Due to limitations of sensitivity, the approach often requires multiple samples, especially for samples separated by two-dimensional gel electrophoresis. These techniques point the way to integrated isolation/sequence analysis methods, and demand a further increase in the sensitivity of sequence analysis.

MASS SPECTROMETRY

New developments in mass spectrometry have revolutionized peptide analysis in the last ten years. Among the most widely used is fast atom bombardment mass spectrometry (FAB/MS), in which underivatized peptides are placed in a liquid matrix and desorbed and ionized with a neutral atomic beam (e.g. xenon) or metal ion beam (e.g. cesium), giving rise to high yields of the protonated molecular ion. The high field magnetic instruments are capable of scanning to m/z of 10 000 at full ionizing potential, thus allowing analysis of the majority of peptide fragments obtained from peptide maps. The resolution of these instruments is sufficiently high to give molecular formulae below mass 1000 and are accurate to within one mass until even up to mass 10 000. Thus, one can distinguish amides from free carboxyl derivatives in peptides. Although both negative and positive ions can be detected, greater sensitivity is usually observed in the positive ion mode. However, negative FAB/MS may be the method of choice for phosphorylated peptides. The sensitivity of the analysis is in the low picomole range, competing well with microsequence and amino acid analysis. The results are based on direct mass measurements and are not complicated by the identification of chemical derivatives or measuring retention times. Knowledge of the molecular weight of a peptide allows rapid confirmation of structure or provides evidence for a post-translational modification. Many post-translational modifications have known masses, thus guiding the researcher towards the proper type of analysis. In the case of complex glycosyl units, the majority have well-defined structures, and thus can be predicted with high confidence from mass alone.

An example of the analysis of a glycopeptide is shown in Figs 1 and 2. When a tryptic fragment of the heavy chain (50 kDa) of an immunoglobulin was subjected to microsequence analysis (Fig. 1), clear signals were seen for PTH amino acids at each cycle except cycle 5, suggesting a post-translational modification. The suspicion was further strengthened when FAB/MS analysis revealed a protonated molecular ion of 2603 (monoisotopic mass of 2602.00). This mass matches the structure shown in Fig. 2. Fragmentation ions are consistent with much of the proposed glycosyl structure. This analysis cannot be considered a proof of structure, but based on our knowledge of immunoglobulin structure including glycosylation sites and structure, it is a highly likely structure. Further structural analysis could be initiated at this point if necessary. When mixtures of peptides are analysed the protonated molecular ion intensities vary and usually do not give an indication of the relative amounts present. In some cases, a major species is suppressed, with only the minor species observed. In the case of glycopeptides, because of their low intensities and competition for the matrix surface, it is a good idea to purify them first[36] or convert them to more hydrophobic derivatives[37].

Mass spectrometric methods also provide structural information for peptides. Sample ionization techniques cause fragmentation of the peptide backbone and can provide significant if not complete sequence information. The low mass region of FAB/MS spectra are usually dominated by ions due to the matrix and sample fragments. Thus, sequence-related fragments may be obscured by spectral background for sample amounts less than a few nanomoles. In ideal cases, considerable sequence information can be obtained from FAB/MS spectra for samples in the 100 pmol range. An example is provided in Fig. 3. The sample was a tryptic peptide obtained from a branched peptide provided to the Protein Society as an unknown in 1987. The presence of the branch point complicated structural analysis, providing a unique opportunity for solving the structure by FAB/MS. The spectrum provides sequence information from both the N and C termini.

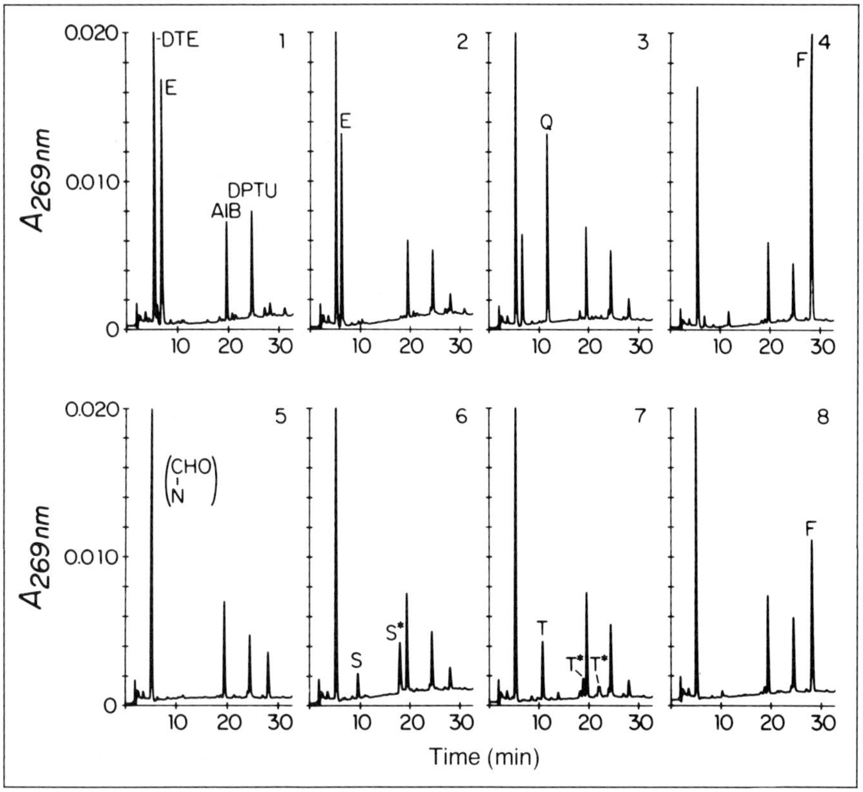

FIGURE 1

Microsequence analysis of a glycosylated peptide. Purified heavy chain from an immunoglobulin was digested with trypsin and mapped by reversed phase HPLC (not shown.) Peptide T13 was sequenced on PVDF using a continuous flow reactor on a gas phase sequencer. The sequence was NH_2-Glu-Glu-Gln-Phe-?-Ser-Thr-Phe-Arg-COOH. (The last cycle is not shown.) Approximately 100 pmol of sample were analysed. Cycle 5, which gave no signal, was suspected to be a glycosylated asparagine residue (N-CHO) by the presence of the sequence (Asn)-XXX-Ser/Thr.

Improvements on sample introduction and sensitivity include the technique of 'flow FAB'[38], in which the sample can be introduced as a continuous stream of liquid. In addition to providing a means of directly coupling FAB/MS to HPLC[39] and capillary electrophoresis, the use of a more volatile flowing matrix greatly reduces the contribution of interfering matrix ions to the mass spectrum. The end result is an order of magnitude increase in the sensitivity of the analysis. A second approach is the use of a focused liquid metal ion gun[40] which gives a similar increase in sensitivity.

Tandem mass spectrometry has the capability to obtain significant structural information on peptides without ancillary analyses. In a tandem experiment, an ion of interest in the spectrum is isolated from all the rest and then fragmented either by collision-activated decomposition or laser photodissociation. The resulting daughter ion spectrum is generally free from interfering ions from the matrix or other peptides present in the sample. Consequently, it is possible to sequence a mixture of peptides without prior separation. All two-sector magnetic instruments have limited tandem MS capabilities by using B/E linked

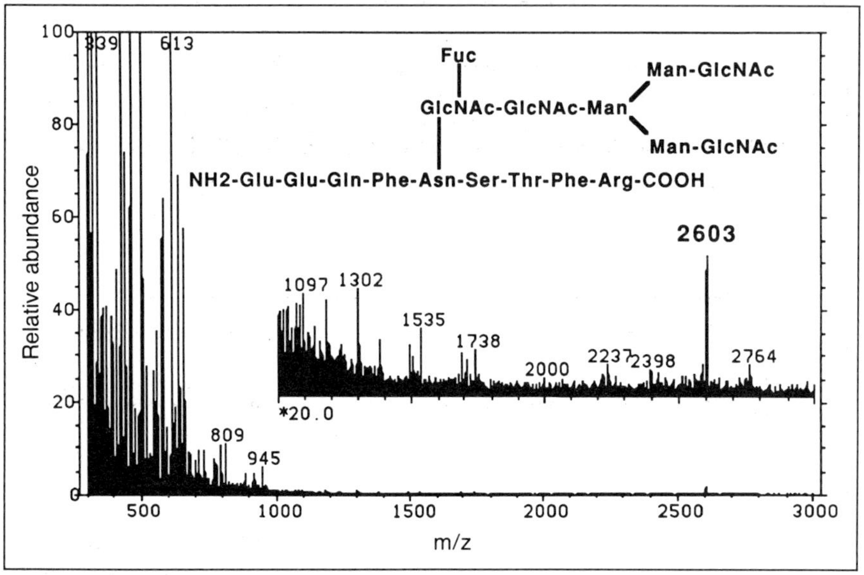

FIGURE 2

Fast atom bombardment mass spectrometric analysis of a glycosylated peptide. Peptide T13 was subjected to FAB/MS analysis (approximately 200 pmol). The observed protonated molecular ion (2602) agrees well with mass 2602 predicted for the parent ion. Fragmentation of the parent ion results in a series that is consistent with the proposed structure, including loss of GlcNAc (2398), loss of Man-GlcNAc (2237), a cluster corresponding to loss of Man-(Man-GlcNAc)$_2$ (1738), and a cluster corresponding to loss of GlcNAc-Man-(Man-GlcNAc)$_2$ (1535). The intense ion at 1302 corresponds to the doubly protonated parent ion.

scans. However, four sectors are needed in order to get high resolution of both the parent and daughter ion spectra. An example of the power of this technique is shown in the work of Biemann and co-workers[41]. The limit of sensitivity for this approach is currently about 1 nmol. The sensitivity of four-sector systems can be further improved with the addition of array detectors which allows a percentage of the mass range to be detected simultaneously. Unfortunately, the cost of such instruments is high (>\$1 million).

Although magnetic sector instruments have been the work horses for protein structural analysis, other types of mass spectrometers, including time of flight (TOF), quadrupole, and fourier transform (FT), are attractive alternatives to magnetic sector systems for many applications. TOF instruments determine the mass of an ion by measuring the time it takes to cover the distance between the ion source and the detector. In this respect, they require a pulsed source. Most common are the plasma desorption (PDMS) instruments which utilize the radioactive isotope ^{252}Cf. The decay of a ^{252}Cf nucleus yields two high energy (MeV) particles traveling in opposite directions. One passes through the sample matrix and the other is used to start the timing pulse for the TOF measurement. In the past these instruments provided the only means to mass analyse proteins of 10–50 kDa[42]. More recently, laser desorption TOF instruments have become available which have successfully analysed ions in excess of 100 kDa[43]. The PDMS instrument may be appealing because of its cost (\$200 000), while the laser desorption instruments are more expensive (>\$300 000). Problems with resolution and very limited capabilities for tandem mass spectral analysis may limit the utility of the TOF instruments.

Quadrupole instruments have several advantages including low cost, ease of operation, and facile coupling to various separation techniques (including HPLC and capillary electrophoresis). The mass range and resolution of the quadrupole instruments is not as great as the large magnetic sector instruments, thus making them less popular for FAB/MS of peptides. However, Hunt and co-workers[44] have shown that triple-quad instruments have great utility in the sequence analysis of peptide mixtures. Recently, Fenn and co-workers[45] have coupled electrospray ionization to a quadrupole analyser and obtained spectra of high molecular weight peptides and proteins. Electrospray produces multiply charged ions for large molecules, thus the mass to charge ratio remains within the range of the quadrupole mass spectrometer. Simple computer analysis of the spectrum yields values for the mass of the protein molecule that are accurate to within 0.1%. The technique is very sensitive, requiring only a few picomoles of sample. Commercial systems are now available, and should become widely used in protein labs.

A fourth type of instrument is the fourier transform (FT) mass spectrometer, which holds the record for sensitivity of sequence determination. In this instrument, the sample ion is trapped in an extremely high magnetic field (6–8 Tesla, produced by a superconducting magnet), synchronized into a stable cyclotron orbit, and analysed by its characteristic RF frequency corresponding to its mass (cyclotron frequency). Through the use of fourier transformations frequencies for multiple ions can be determined simultaneously (similar to NMR spectroscopy). The ions have very long half-times, allowing high sensitivities of detection, and the ability to perform studies on them because they are not destroyed by the detection process. Although these instruments have been available for over ten years for the analysis of small molecules, their use for peptides is rather recent[46,47]. In the instrument developed by Don Hunt at the University of Virginia, the sample is ionized by a FAB gun, and transmitted to the ion trap by a quadrupole sector. Once trapped and analysed, peptide ions can be fragmented by a laser, giving substantial structural information. The sensitivity of this method is already at the low picomole range, a range still not achieved for the other instruments.

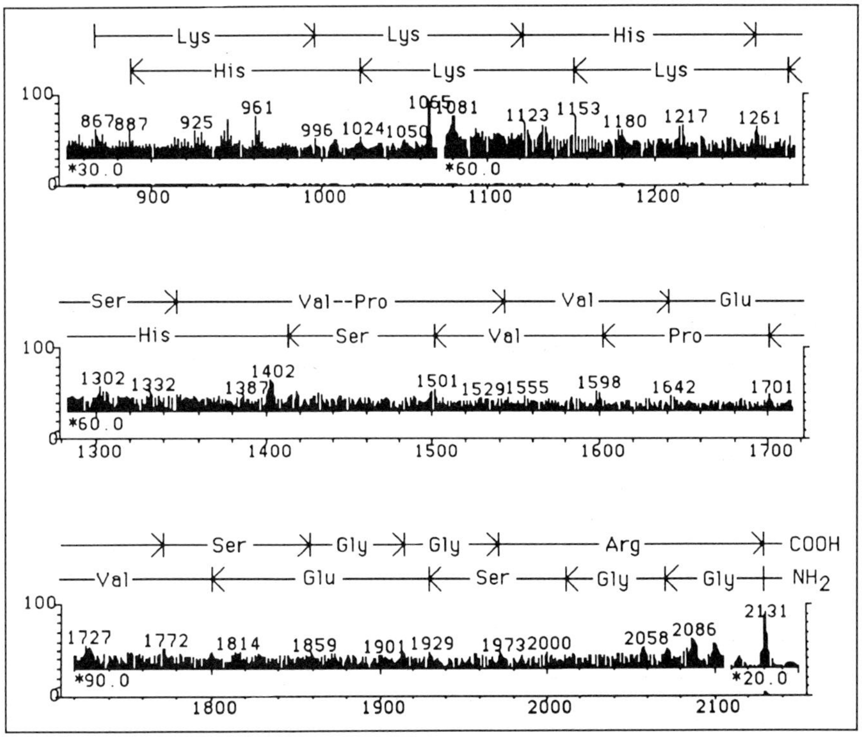

FIGURE 3

FAB/MS fragment ion series for a branched peptide. The structure of the peptide is NH_2-Gly-Gly-Ser-Glu-Val-Pro-Val-Ser-His-Lys-(ε-NH_2)Lys-His-Ser-Val-Pro-Val-Glu-Ser-Gly-Gly-Arg-COOH. It was obtained from a larger branched peptide (M+H)=5445.1 after tryptic digestion and separation by reversed phase HPLC. The tryptic peptide contains the branch point (the C^{α} group of Lys^{10} is connected to the NH_2^{ε} group of Lys11) and has a (M+H)=2130.1. The fragmentation series for the peptide shows the complete N- and C-terminal series as indicated. Approximately 100 pmol was analysed. The assignments were further confirmed by amino acid and microsequence analyses.

DATA ANALYSIS

The large amounts of data generated in amino acid sequence analysis present another problem, namely data handling. In the case of N-terminal sequence analysis, data from the on-line HPLC system can be compiled, including peak assignment, background subtraction and determination of the sequence within reasonable confidence levels. This software is available on the Applied Biosystems sequencer. In sequence analysis by mass spectrometry, the entire process can be automated by software developed by Biemann and others. Peptide mapping predictions, fragment mass prediction and peptide alignment software are becoming increasingly available. Protein data bases allow comparisons to published gene and protein sequences. Eventually, integrated software will be available which will eliminate the need to reenter and recompile data from one type of analysis to another.

NEW DEVELOPMENTS

Carboxy-terminal sequence analysis would be an obvious complement to Edman chemistry. Since proteins are linear polymers, access to N- and C-terminal analyses would help define proteins better, and eliminate guesswork about the processing of nascent proteins to their mature forms. One promising approach, analogous to Edman chemistry, is the thiocyanate degradation method dating back to 1926[48]. Proteins are activated at their C-termini with acetic anhydride, reacted with a thiocyanate reagent to form a peptidylthiohydantoin, and hydrolysed to release an amino acid thiohydantoin product and the shortened peptide. The reagent, reaction conditions, and sensitivity have been recently reinvestigated[49,50], and there is considerable optimism that this method can be eventually used with the same proficiency as Edman chemistry. Enzymatic approaches such as carboxypeptidase based methods also hold some promise, but are limited by their accessibility to the free C^{α} group, differential rates of peptide bond hydrolysis, and the need for kinetic analyses (limited cycle numbers). In this approach the released amino acids can be analysed with high sensitivity methods such as PTC, OPA, dansyl or FMOC.

Further increases in the sensitivity of Edman chemistry are predictable. In order to analyse nanogram amounts of sample, a 10–1000-fold increase in sensitivity is required. Fluorescent derivatives and narrowbore column (1 mm ID) or capillary (100 microns or less) based separation systems will be required. Although many fluorescent Edman reagents have been explored, most alter the reaction conditions substantially posing many new problems for optimization of the Edman degradation scheme. Alternative choices include the conversion of the ATZ derivatives to fluorescent products, which can be separated and detected in the attomole to femtomole range by capillary electrophoresis. A recent report demonstrates subattomole detection of FTC (fluorescein thiocarbamyl) amino acid derivatives[51]. In spite of the promise of these methods, many technical problems must be solved first, including interfacing Edman sequencers to capillary HPLC or electrophoresis separation systems.

In order to match sample isolation techniques with sequence analysis without sample loss, more integrated systems must be developed. One can envision automated sample analysis and collection from capillary HPLC or electrophoresis onto membranes. The membranes can be analysed directly by amino acid, sequence and mass spectrometric analyses. In this respect, capillary electrophoresis has already been interfaced to mass spectrometry[52]. Sample purification by tandem methods, such as HPLC and electrophoresis, can be fully automated, including structural analysis. Each step can be monitored by the researcher, and aborted if necessary. The emphasis will be on speed, resolution and sensitivity. Mass spectrometric methods will be of special importance because of their mass analysis capabilities. UV and fluorescent analytical methods will remain important because of their quantitative aspects.

One daunting aspect of the trend towards sophisticated, integrated instrumentation is cost. Expensive equipment tends to limit use to core facilities where standardized analytical and volume practices are fostered. The opportunities for individual researchers, especially postdoctoral fellows and graduate students, experimenting with methods, strategies, and equipment will be limited. Thus, there will always be room for simpler, manual approaches. In this regard it is still encouraging to see reports of manual Edman chemistry[53,54]. Properly done, manual sequencing can handle multiple samples with the same sensitivity as automated sequencing. Over the next few years, an even bigger challenge will be to provide low cost alternatives for micromethods in protein chemistry, and to promote education and training in this important field.

References

1 Walker, J. M. (ed.) (1988) *New Protein Techniques*, Humana Press

2 Shively, J. E. (ed.) (1986) *Methods of Protein Microcharacterization,* Humana Press

3 Schlesinger, D. H. (ed.) (1988) *Macromolecular Sequencing and Synthesis,* Alan R. Liss

4 Brown, A. S. (ed.) (1988) *Protein/peptide Sequence Analysis: Current Methodologies,* CRC Press

5 Walsh, K. A. (ed.) (1986) *Methods in Protein Sequence Analysis,* Humana Press

6 Simpson, R. J., Moritz, R. L., Begg, G. S., Rubira, M. R. and Nice, E. (1989) *Anal. Biochem.* 177, 221–236

7 Cohen, S. A. and Strydom, D. J. (1988) *Anal. Biochem.* 174, 1–16

8 Andrews, P. C. and Dixon, J. E. (1978) *Anal. Biochem.* 161, 524–528

9 Pan, Y-C. E., Widerman, J., Blacher, R., Chang, M. and Stein, S. (1984) *J. Chromatogr.* 297, 13–19

10 Inglis, A. S. (1983) *Methods Enzymol.* 91, 26–30

11 Gross, E. and Witkop, B. J. (1962) *J. Biol. Chem.* 237, 1856–1860

12 Schagger, H. and von Jagow, G. (1987) *Anal. Biochem.* 166, 368–379

13 Welinder, K. G. (1988) *Anal. Biochem.* 174, 54–64

14 Crimmins, D. L., Thoma, R. S., McCourt, D. W. and Schwartz, B. D. (1989) *Anal. Biochem.* 176, 255–260

15 Alpert, A. J. and Andrews, P. C. (1988) *J. Chromatogr.* 443, 85–96

16 Edman, P. (1950) *Acta Chem. Scand.* 4, 283–293

17 Edman, P. and Begg, G. (1967) *Eur. J. Biochem.* 1, 80–91

18 Tarr, G. E., Beecher, J. F., Bell, M. and McKean, D. J. (1978) *Anal. Biochem.* 84, 622–627

19 Klapper, D. G., Wilde, C. E. and Capra, J. D. (1978) *Anal. Biochem.* 85, 126–131

20 Hewick, R. M., Hunkapiller, M. W., Hood, L. E. and Dreyer, W. J. (1981) *J. Biol. Chem.* 256, 7990–7997
21 Hunkapiller, M. W., Lujan, E., Ostrander, F. and Hood, L. E. (1983) *Methods Enzymol.* 91, 227–236
22 Aebersold, R. H., Teplow, D. B., Hood, L. E. and Ken, S. B. H. (1986) *J. Biol. Chem.* 261, 4229–4238
23 Matsudaira, P. (1987) *J. Biol. Chem.* 262, 10035–10038
24 Xu, Q-Y. and Shively, J. E. (1988) *Anal. Biochem.* 170, 19–30
25 Moos, M., Jr, Nguyen, N. Y. and Liu, T-Y. (1988) *J. Biol. Chem.* 263, 6005–6008
26 Szewczyk, B. and Summers, D. F. (1988) *Anal. Biochem.* 168, 48–53
27 Dias, A. J. and McCarthy, T. J. (1987) *Macromolecules* 20, 2068–2076
28 Lee, K-W. and McCarthy, T. J. (1987) *Macromolecules* 20, 1439–1440
29 Laursen, R. A. (1971) *Eur. J. Biochem.* 20, 89–102
30 Laursen, R. A., Dixon, J. D., Liang, S-P., Nguyen, D. M., Kelcourse, T., Udell, L. and Pappin, D. (1988) Abstract in *Methods in Protein Sequence Analysis, 7th International Conference, Berlin*
31 Hennessey, J. P., Jr and Scarborough, G. A. (1989) *Anal. Biochem.* 176, 284–289
32 Bauw, G., De Loose, M., Inze, D., Van Montagu, M. and Vaderkerckhove, J. (1987) *Proc. Natl Acad. Sci. USA* 84, 4806–4810
33 Vanfleteren, J. R. (1989) *Anal. Biochem.* 177, 388–392
34 Yuen, S., Hunkapiller, M. W., Wilson, K. J. and Yuan, P. M. (1988) *Anal. Biochem.* 168, 5–15
35 Aebersold, R. H., Leavitt, J., Saavedra, R. A., Hood, L. E. and Kent, S. B. H. (1987) *Proc. Natl Acad. Sci. USA* 84, 6970–6974
36 Carr, S. A. and Roberts, G. D. (1986) *Anal. Biochem.* 157, 396–406
37 Boulenguer, P., Leroy, Y., Alonso, J. M., Montreuil, J., Ricart, G., Colbert, C., Duquet, D., Dewaele, C. and Fournet, B. (1988) *Anal. Biochem.* 168, 164–170
38 Caprioli, R. M., Moore, W. T., Da Gue, B. and Martin, M. (1988) *J. Chromatogr.* 443, 355–362
39 Caprioli, R. M., Da Gue, B. and Wilson, K. (1988) *J. Chromatogr. Sci.* 26, 640–644
40 Jiang, L. F., Barofsky, E. and Barofsky, D. F. (1988) *Proc. 36th ASMA Conf. Mass Spec.* pp. 1209–1210
41 Martin, S. A., Rosenthal, R. S. and Biemann, K. (1987) *J. Biol. Chem.* 262, 7514–7522
42 Tsarbopoulos, A., Becker, G. W., Occolowitz, J. L. and Jardine, I. (1988) *Anal. Biochem.* 171, 113–123
43 Karas, M. and Hillenkamp, F. (1988) *Anal. Chem.* 60, 2301–2303
44 Hunt, D. F., Yates, J. R., Shabanowitz, J., Winston, S. and Hauer, C. R. (1986) *Proc. Natl Acad. Sci. USA* 83, 6233–6237
45 Whitehouse, C. M., Dreyer, R. N., Yamashita, M. and Fenn, J. B. (1988) *Anal. Chem.* 57, 675–679
46 Hunt, D. F., Shabanowitz, J., Yates, J. R., III, Zhu, N-Z., Russell, D. H. and Castro, M. E. (1987) *Proc. Natl Acad. Sci. USA* 84, 620–623
47 Brinegar, A. C., Cooper, G., Stevens, A., Hauer, C. R., Shabanowitz, J., Hunt, D. F. and Fox, J. E. (1988) *Proc. Natl Acad. Sci. USA* 85, 5927–5931
48 Schlack, P. and Kumpf, W. (1926) *Hoppe-Seyler's Z. Physiol. Chem.* 154, 125–170
49 Meuth, J. L., Harris, D. E., Dwulet, F. E., Crowl-Powers, M. L. and Gurd, F. R. N. (1982) *Biochemistry* 16, 3750–3757
50 Hawke, D. H., Lahm, H-W., Shively, J. E. and Todd, C. W. (1987) *Anal. Biochem.* 166, 298–307
51 Cheng, Y-F. and Dovichi, N. (1988) *Science* 242, 562–564
52 Smith, R. D., Barinaga, C. J. and Udseth, H. R. (1988) *Anal. Chem.* 60, 1948–1952
53 Fischer, P. M. and Howden, M. E. H. (1989) *Anal. Biochem.* 177, 46–49
54 Haniu, M. and Shively, J. E. (1988) *Anal. Biochem.* 173, 296–306

J. E. Shively, R. J. Paxton and T. D. Lee are at the Division of Immunology, Beckman Research Institute of the City of Hope, 1450 East Duarte Road, Duarte, CA 91010-0269, USA.

MICRO-LEVEL PROTEIN AND PEPTIDE SEPARATIONS

KENNETH J. WILSON

The current major limitation in protein micro-structural and -functional studies is our inability to purify and manipulate microgram or smaller quantities of material. Both high performance liquid chromatography (HPLC) and polyacrylamide gel electrophoresis (PAGE) have been widely used in attempts to address such needs. This review summarizes the current status of preparative and analytical applications for both methods, and suggests which directions the micro-separations field will take within the next few years.

The structural–functional roles of many biologically active polypeptides have been studied and elucidated using various separation techniques. For example, the minimum molecular weight of a protein can be determined electrophoretically under denaturing conditions; the presence or absence of post-translational modifications chromatographically detected following their removal by hydrolysis; and the native polymeric structures investigated using either electrophoresis or chromatography under non-denaturing conditions.

As the available quantities of 'interesting' proteins have decreased over the years, there has been a concomitant requirement for increased instrument sensitivity, faster throughput, reliability, simplicity, higher recoveries and biocompatibility. The separation techniques most frequently used today, electrophoresis and chromatography, are employed not only to isolate components, but also to evaluate homogeneity analytically at each step of an isolation protocol.

The trends in biochemical separations have been toward smaller sample volumes, rapid purification and analysis, high resolution of complex mixtures, mild conditions, and high recoveries. In the area of protein primary structure determination these have resulted in the development of specialized methods using high performance liquid chromatography (HPLC) and polyacrylamide gel electrophoresis (PAGE) for both preparative and analytical purposes. The utility of these methods in sequencing is quite illustrative: either method is employed to prepare samples for structure determination and HPLC for PTH-amino acid identification and quantitation.

Over the past 10 years instrument design and chemistry optimizations have resulted in a 100–1000-fold increase in sensitivity. In fact it is generally recognized that the current limitation in protein microsequencing at the picomolar level (or below) is our inability to purify and manipulate such quantities prior to chemical degradation. Both HPLC and PAGE have been widely used to address such needs and this brief review outlines the current 'state of affairs' and, as covered in the previous article by Shively *et al.* (see p. 1) suggests where the field will develop over the next few years.

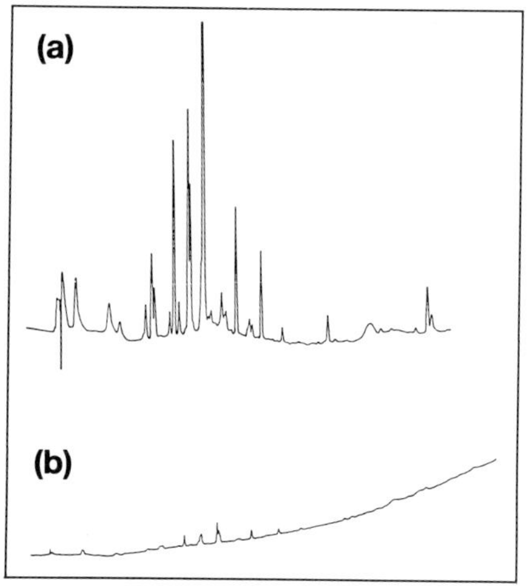

FIGURE 1

Normal and microbore peptide HPLC. Identical amounts (100 pmol) of tryptically digested apomyoglobin were chromatographed on 1 mm **(a)** and 4.6 mm **(b)** ID columns of equal length (25 cm). The same gradient conditions were used except that the smaller ID column was eluted at 50 μl min^{-1} and the larger at 1 ml min^{-1}.

HPLC

The miniaturization of chromatographic instrumentation to improve overall yield, detection sensitivity and resolution is commonly referred to as microbore HPLC. Through the judicious choice of operating conditions (≤100 μl min^{-1}) and column sizes (< 2 mm ID) it is possible to increase sensitivities 15–20-fold[1]. The separation of identical sample sizes (100 pmol) on 1 mm and 4.6 mm ID columns clearly illustrates this point for peptide mapping (Fig. 1). Concomitant with the use of smaller column diameters and reduced flow rates, is the reduction in the total volume in which each component elutes[1,2]. Smaller recovery volumes – higher concentrations – may eliminate the need for lyophilization and/or desalting steps prior to further characterization, thus reducing potential mass loss. It should be emphasized that handling microgram levels of certain peptides and proteins can be extremely difficult and that losses due to surface adsorption are often encountered.

Future developments in microbore HPLC will probably involve even further reductions in column sizes (≤ 500 μm ID), improvements in associated solvent delivery and detection hardware, and alternative means of sample collection. In fact, for many types of analyses (sequencing, amino acid analysis, peptide mapping), collection onto an appropriate surface such as glass, polyvinylidene difluoride (PVDF) or nylon can be suggested[3]. Provided that potentially interfering buffers, salts, etc. can be removed, this is an excellent means to minimize handling and transfer losses, while immobilizing the sample on a 'usable' surface.

The use of reverse phase supports in microbore applications has certainly received the most attention. With the continued development of improved ion exchange, size exclusion and hydrophobic interaction silica-based supports, as well as those on polymeric materials, we can expect to see increased usage of these as well. Similarly, further development of 'optimized' solvent systems will be aimed at improving yields, reducing or minimizing protein denaturation, and allowing for even higher detection sensitivities.

ELECTROPHORESIS

The most widely used analytical method in protein chemistry involves the electrophoretic separation of components through gels, acetate strips, or other support media. Separations using PAGE are commonly done either in the presence of a denaturant (sodium dodecyl sulfate, SDS), its absence ('native' conditions), or in an ampholite-containing gel (isoelectric focusing). The combination of these in a two-dimensional manner can be used for detecting thousands of components in cellular extracts, or investigating cellular regulation.

Traditionally, the levels of sensitivity, inherent resolution, simplicity and reliability of the methods have been compatible with their analytical uses. However, as higher and higher sensitivites were required new stains, thinner/smaller gels and improved general procedures were developed. These modifications improved detectability but there still remained major disadvantages with many of the techniques: labor-intensive solution and sample preparation, semi-quantitative staining, and poor recoveries when used preparatively. The more recently developed methods for using SDS–PAGE in microgram-level isolations have focused on minimizing amino-terminal blockage and/or amino acid side chain modification, eliminating polypeptide fragmentation during staining and destaining, and reducing the introduction of contaminates.

Electroelution / dialysis

The first generally acceptable recovery method involved electroelution of the stained protein from a gel piece, subsequent electrodialysis to lower contaminate levels, and final recovery by precipitation in an organic solvent[4]. Samples isolated using this technique are quite amenable to primary sequence determination but losses incurred with many proteins at the microgram level can be unacceptable. However, for the isolation of tens to hundreds of micrograms of a particular protein from an enriched sample this is an excellent method.

Electroblotting

The development of procedures for transferring separated species from a polyacrylamide gel onto a support surface has received considerable attention. The first materials onto which such transfers were made utilized derivatives of glass fiber sheets. These included glass surfaces that had been acid washed, derivatized to provide high positive charge densities, or chemically modified to introduce reactive moieties that would form covalent bonds with transferred proteins[5].

Probably the most commonly used and widely available blotting support today is polyvinylidene difluoride (PVDF). Its use in conjunction with electroblotting and high sensitivity sequencing has been extensively described in a number of reports[6–8]. Similarly, its utility in amino acid analysis[9] and peptide mapping[8] has received some attention. Although PDVF and some of the other supports have been used quite successfully in a large number of sequencing investigations, there are inherent drawbacks to the methodology. These include the limited binding capacities of the membranes, transfer times differing as a function of the molecular size of the solutes, slow diffusion/migration out of the blotting support ('over-transferring'), and probable mass losses between the gel and blotting surfaces.

Continuous elution

There have been numerous attempts to introduce preparative electrophoretic methods that circumvent some of the above mentioned disadvantages. Common to most has been the continuous elution concept[10,11] where solutes are electrophoretically separated, allowed to exit the gel and then swept toward a detection/collection system. In contrast to the slab gels used for preparative electroblotting purposes, these instruments typically employ a cylindrically shaped gel that is

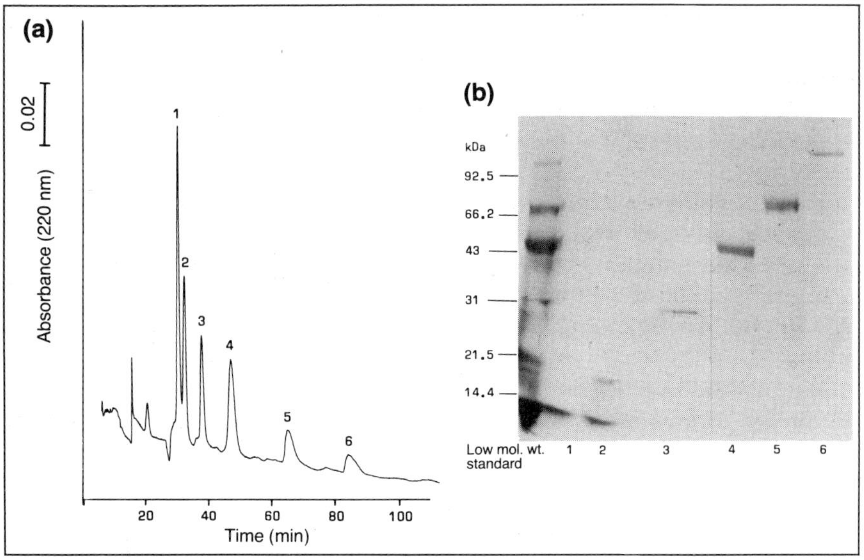

Figure 2

Preparative protein isolation by continuous elution SDS–PAGE. **(a)** A standard containing 2 μg each of lysozyme (1), soybean trypsin inhibitor (2), carbonic anhydrase (3), ovalbumin (4), bovine serum (5), and phosphorylase B (6), was separated on a 2.5×50 mm 7% acrylamide gel using an SDS-Tris/tricine buffer system at 40 μA constant current **(b)**. Aliquots of the collected fractions were analysed by slab gel electrophoresis (same percentage gel and buffer system).

cooled to minimize joule heating and thus protein denaturization and/or damage to the separation medium. One recently introduced instrument accommodates gels of various diameters and lengths, has a continuous elution rate of a few microlitres per minute, on-line spectrophotometric detection and an integrated micro-fraction collector[12]. Figure 2 illustrates the preparative isolation of a protein standard mixture on such an apparatus and the subsequent analytical analysis of the fractions. Since the separation was carried out under denaturing conditions, the solutes elute as a function of increasing molecular weights and, as with other size exclusion separations, the peaks become less pronounced due to time-dependent diffusions in the gel. Yields of 70–95% for radioactively-labeled proteins (10–100 kDa) have been determined at the microgram level[13]. Proteins fractioned on such a system can be concentrated and desalted for further characterization studies using normal[3] or reversed-gradient[14] reverse phase HPLC, or centrifugation through PVDF membrane[13]. This particular approach toward micropreparative optimization circumvents many of the difficulties involving sample recovery following either electroelution/dialysis or electroblotting; i.e. samples are recovered in an aqueous medium, detection is not dependent on staining–destaining, and microgram quantities can be readily recovered.

Capillary electrophoresis

Another electrophoretic technique creating a high level of interest for many analytical protein/peptide applications[15–17] is currently referred to as either capillary electrophoresis (CE) or high performance capillary electrophoresis (HPCE). By this method solutes are separated in narrow (25–100 μm ID) fused-quartz capillaries which allow for rapid heat dissipation at the voltage used for separation (10–30 kV). The four electrophoretic modes that have been described are (1) free-solution in buffers, (2) size exclusion in SDS-containing polyacrylamide, (3) isoelectric focusing in an ampholyte solution and (4) micellar-containing buffers. Capillary lengths are normally 50–200 cm, sample loads of $\leqslant$ 10 nl are used and separation efficiencies often exceed 300 000 theoretical plates per meter. On-line spectrophotometric, fluorescence, electrochemical, post-column derivatization and mass spectrometer detectors have been developed[18].

The screening of fractions from a micropreparative isolation of peptides for structural analysis is a typical application for this method (Fig. 3). In this example sequencing confirmed the heterogeneity of the fractions predicted by CE: fraction (a) contained three peptides, (b), one and (c), two peptides. Since only 5 nl amounts were analysed the technique is virtually non-destructive. Other studies have indicated that differences in the extent of glycosylation, and deamidation[17] as well as side chain oxidation and phosphorylation (J. C. Colburn, unpublished) can be readily detected and quantitated. In fact, the higher resolution inherent to this technique will make it extremely useful in quality control, especially with recombinant protein products[17].

Low percentages of 'contaminates' are frequently detected by CE in samples that appeared homogeneous by all other criteria. For many this presents a dilemma and will probably create the need for and the development of preparative CE instrumentation. Another direction will be to employ CE and microbore HPLC in an orthogonal manner, similar to the microbore HPLC/SDS–PAGE electroblotting combination[7], to achieve truly high sensitivities at the nanogram level with analysis times of only a few minutes.

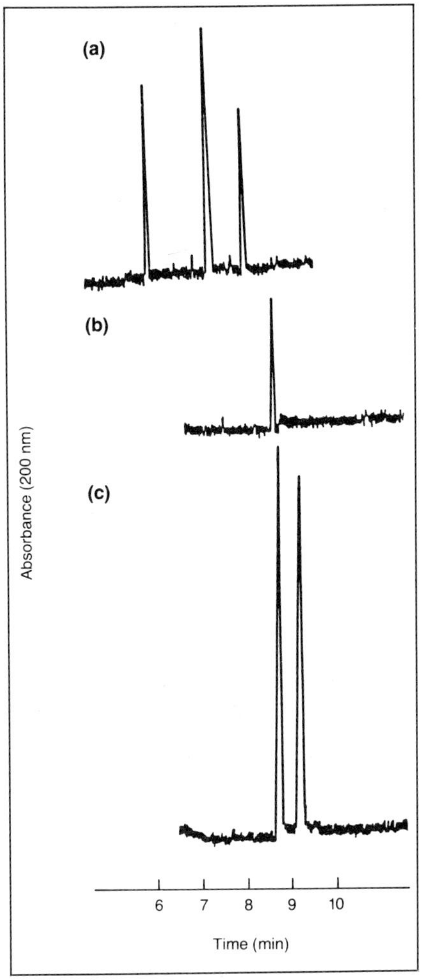

FIGURE 3

Capillary zone electrophoresis analysis of purified peptides prior to sequencing. Tryptically digested β-lactoglobulin was fractionated by microbore HPLC and individual peaks analysed by CE: 20 mm citric acid, pH 2.50, as buffer in a 50 μm ID capillary (45 cm to detector); field strength, 277 V cm^{-1}, current 24 μA at 30°C. Each electrophoretogram (**a–c**) represents an individual peak from the micropreparative HPLC separation, see Ref. 17 for details.

REFERENCES

1. Scott, R. P. W. and Kucera, P. (1979) *J. Chromatog.* 185, 27–41
2. Schlabach, T. D. and Wilson, K. J. (1987) *J. Chromatogr.* 385, 65–74
3. Wilson, K. J., Hawke, D. H. and Yuan, P. M. (1987) in *Protein Purification: Micro to Macro Vol. 68*, (Burgess, R., ed.), pp. 35–47, Alan R. Liss
4. Hunkapiller, M. W., Lujan, E., Ostrander, F. and Hood, L. E. (1983) *Methods Enzymol.* 91, 227–236
5. Aebersold, R. H., Teplow, D. B., Hood, L. E. and Kent, S. B. H. (1986) *J. Biol. Chem.* 261, 4229–4238
6. Matsudaira, P. (1987) *J. Biol. Chem.* 262, 10035–10038
7. Yuen, S., Hunkapiller, M. W., Wilson, K. J. and Yuan, P. M. (1988) *Anal. Biochem.* 168, 5–15
8. Yuen, S., Chui, A. J., Wilson, K. J. and Yuan, P. M. (1989) *BioTechniques* 7, 74–82
9. Keim, P. S., Bello, R., Bozzini, M., Chui, A. H. and Dupont, D. R. (1988) *The Protein Society 2nd Symp.* (Abstr. T926), San Diego, CA, USA
10. Gordon, A. H. and Louis, L. N. (1967) *Anal. Biochem.* 21, 190–200
11. Akaiwa, S. (1979) *Anal. Biochem.* 92, 205–212
12. Hsi, K. L., Efcavitch, J. W., Yuan, P. M. and Hunkapiller, M. W. (1988) *Electrophoresis '88* (Abstr. 37) Copenhagen, Denmark
13. Sheer, D. G. (1989) *Applied Biosystems User Bulletin, Separation Systems* 4
14. Simpson, R. J., Moritz, R. L., Nice, E. C. and Grego, B. (1987) *Eur. J. Biochem.* 65, 21–29
15. Grossman, P. D., Wilson, K. J., Petrie, G. and Lauer, H. H. (1988) *Anal. Biochem.* 173, 165–270
16. Nielsen, R. G., Sittampalam, G. S. and Richard, E. C. (1989) *Anal. Biochem.* 177, 20–26
17. Grossman, P. D., Colburn, J. C., Lauer, H. H., Nielsen, R. G., Riggin, R. M., Sittampalam, G. S. and Richard, E. C. (1989) *Anal. Chem.* 61, 1186–1194
18. Abstracts (1989) from *First International Symposium on High Performance Capillary Electrophoresis*, Boston, MA, USA

K. J. Wilson is at Applied Biosystems, Foster City, CA 94404, USA.

Site-directed Mutagenesis and Protein Engineering

Michael Smith

The systematic replacement of nucleotides in DNA fragments isolated by recombinant DNA technology can be achieved by a variety of chemical and enzymatic manipulations. Application of these procedures to DNA that encodes the amino acid sequence of a protein makes it possible to obtain specifically mutated protein when the encoding DNA is expressed in living cells. This variant of genetic engineering – protein engineering by site-directed mutagenesis – is having a major impact on protein structure–function analysis and in the production of proteins with changed or novel properties.

The understanding of biology in molecular terms generally requires an understanding of the roles of specific amino acids in the catalytic activity or the substrate/ligand binding specificity of proteins. Classically, such understanding has depended on such techniques as the observation of changes in enzyme activity resulting from specific chemical modifications of a protein[1]. The comparison of amino acid sequences of proteins with the same biological function isolated from different species can identify conserved and, presumably, essential amino acid residues: cytochrome *c* has been studied by this method[2]. The combination of specific chemical modification, comparison of amino acid sequence and the determination of protein's three-dimensional structure is particularly informative and has revealed much about the serine proteases[3]. Lysozyme[4] and DNA polymerase[5] are examples of proteins that have been studied by determination of the three-dimensional structure of an enzyme to which the substrate is bound. But the information provided by these experiments, however insightful, results in functional models that generally require experimental testing for proof of their validity. In the last decade, site-directed mutagenesis has emerged as an excellent procedure for testing these models in a systematic and detailed manner[6]. An example of this approach is provided by studies on the enzyme carboxypeptidase A. Structural studies on the enzyme–substrate complex suggested that a particular amino acid residue, Tyr248, has a role in catalysis involving the phenolic hydoxyl group[7]. Site-directed mutagenesis was used to produce a carboxypeptidase A in which residue 248 was replaced by phenylalanine. The mutant enzyme was an efficient catalyst, clearly demonstrating that the hydroxyl group of Tyr248 does not have a crucial role[8]. Beyond testing specific models of protein function, site-directed mutagenesis is being directed at more general properties of enzymes such as thermal stability[9–11]. Most excitingly, site-directed mutagenesis is now leading to mutant proteins with specifically designed catalytic functions[12] or general properties[13].

The basic principles and alternate strategies for site-directed mutagenesis have been reviewed in detail[14] and so this article, after describing the different approaches to site-directed mutagenesis, will focus on new developments. For more detailed methodology and specific protocols the reader is referred to other literature on biotechnological methods, including publications such as *Methods in Enzymology* (Academic Press), *Current Protocols in Molecular Biology* (Wiley Interscience) and *Molecular Cloning: A Laboratory Manual* (Cold Spring Harbor).

GENERAL PRINCIPLES OF SITE-DIRECTED MUTAGENESIS

Any experiment in site-directed mutagenesis involves several separate DNA manipulations. First, the DNA encoding the protein, either as an isolated fragment or incorporated in a circular plasmid or phage vector, is modified using chemical and/or enzymatic reagents. Next, the modified DNA, incorporated in a DNA vector, is used to transform bacterial cells, usually *Escherichia coli*, and individual clones are isolated. The mutated recombinant DNA is then isolated and sequenced to deduce the amino acid sequence of the mutant protein. Finally, characterized mutant genes are used to transform host cells in which the gene is transcribed and translated. In many cases the host cell will perform post-translational modification of the initial polypeptide product to produce the mature protein. Characterization of a mutant protein with regard to the properties of interest completes the experiment.

Methods that are useful for *in vitro* mutagenesis for protein engineering purposes fall into two broad classes: those designed to change a specific codon in a specific way and those designed to produce random base-pair substitutions throughout the protein coding sequence. Clearly, the choice between these two classes of approach will depend on the amount of detailed information that is available about the structure–function relationships of the target protein and the questions being asked. Ten years ago, it would have been thought that the intrinsic complexity of protein structure would have resulted in a preference for the random mutagenesis approach. Surprisingly, the great majority of protein engineering experiments that have been carried out to date have been directed at changing particular amino acid residues in predetermined ways. This approach will continue to be popular because so many of the questions being asked about protein function refer to the roles of a limited number of amino acids, such as those involved in substrate or ligand binding, or in catalysis or those that are the targets of post-translational modification such as specific proteolysis, phosphorylation or glycosylation. It is possible that the use of random mutagenesis will increase as interest increases in the production of mutants that are compensatory to defects caused by specific mutagenesis and as less focused properties of proteins become the targets of research.

PRODUCTION OF MUTATIONS IN SPECIFIC CODONS

Protein engineering directed at specific codons usually requires the use of chemically synthesized DNA fragments as mutagens. One approach involves total or partial (cassette) synthesis of the double-stranded DNA segment containing the mutated codon. The attractions of this method lie in the fact that it should be, in principle, 100% efficient in producing the desired mutant; that a gene can be constructed with codons that are used in highly expressed genes in the proposed host organism; and that unique restriction endonuclease cleavage sites can be placed at convenient intervals. Two chemistries are used for the synthesis of the oligonucleotide fragments that are subsequently joined using DNA ligase to form the desired double-stranded DNA fragment. The first of these uses phosphodiester intermediates and synthesizes, simultaneously, a large number of relatively short oligonucleotides (about 15–20 nucleotides in length)[15,16]. These oligonucleotides, after purification, are enzymatically phosphorylated and then complementary subsets are ligated prior to assembly of the total gene. The second chemistry, using phosphoramidite intermediates, can be carried out as described above[16], but is more usually carried out in an automated synthesizer which assembles a small number of very long oligonucleotides (with 50 to 100 or more nucleotides)[17]. An example of gene synthesis and mutant production using long oligonucleotides produced in this manner is provided by the synthesis of a thymidylate synthetase gene[18]. This synthesis also illustrates a feature of the total synthesis approach to mutant genes; there is a low level (three base pairs in one thousand, in this case) of unwanted mutations in a synthetic gene isolated as a plasmic-propagated clone. These mutants, which are presumed to arise because of trace nucleotide contaminants in the synthetic building blocks, are subsequently corrected by replacement of appropriate cassettes or by oligonucleotide heteroduplex mutagenesis.

The most commonly used strategy for specific mutagenesis continues to involve the use of a single synthetic oligonucleotide containing the desired mutation which is annealed with a wild-type complementary DNA in a circular, single-stranded vector, after which the oligonucleotide is incorporated enzymatically into a double-stranded molecule using the single-stranded circle as template: oligonucleotide heteroduplex mutagenesis[19]. In principle, if there is no asymmetry in DNA synthesis or DNA repair *in vivo*, equal amounts of mutant and wild-type progeny should be produced. Although yields of mutants approaching 50% have been realized[19], often as little as 1% or less of the progeny are mutant. There are a variety of possible causes of low yields[14], including impurities in the target DNA[20]. One cause of low yield relates to the property of the Klenow fragment of *E. coli* DNA polymerase I where, as it moves along a DNA template during synthesis of a complementary strand, it displaces the strand of DNA downstream of the growing point. If the displaced region contains the mutagenic oligonucleotide primer, no mutant is produced[14]. The DNA polymerases produced by phage T4[21] and phage T7[22] do not strand displace and are now being used in oligonucleotide mutagenesis experiments.

When mutant yields are of the order of 1%, the mutant clones can readily be recognized using an oligonucleotide probe[19]. However, if yields are routinely 50% or higher, clones of mutant DNA can be identified directly by DNA sequence determination applied to only two or three samples. Thus, a number of strategies for enrichment of mutants have been developed[14]. These provide the basis for the oligonucleotide mutagenesis kits that are commercially available. Examples of these enrichment methods include: the use of a dU-containing template DNA strand for the construction of the heteroduplex double strand with a mutagenic oligonucleotide primer (*in vivo* repair during propagation of the DNA removes wild-type template sequence)[21]; the incorporation of thiophosphate-containing nucleotides in the enzymatically synthesized strand containing the mutagenic oligonucleotide (this renders the strand resistant to endonucleolytic cleavage by a restriction endonuclease and allows the *in vitro* removal of the wild-type template complementary to the mutagenic oligonucleotide)[23]; and procedures which, in addition to the mutagenic oligonucleotide, use a second primer or DNA fragment that contains a genetically selectable marker[24–26]. An alternate to the thiophosphate-containing nucleotide method for protecting the oligonucleotide-containing strand from endonuclease cleavage uses 5-methyl deoxycytidylic acid[27].

Because most recombinant DNA clones are propagated in double-stranded plasmids, a variety of different procedures have been developed for direct oligonucleotide mutagenesis of double-stranded DNA[14]. These methods all create a single-stranded target region in the plasmid (gapped duplex) to which the oligonucleotide is annealed prior to being incorporated into the double-stranded DNA using DNA polymerase and ligase. In general, these methods have not been as efficient as procedures using single-stranded phage vectors[14]. However, highly efficient procedures for producing oligonucleotide-directed mutants in double-stranded plasmids are now available. These are based on the dU-DNA or the thiophosphate-DNA procedures described above. The dU-DNA based method constructs a gapped duplex, where the continuous strand contains dU residues, by annealing dU-containing plasmid DNA that has been cleaved at one restriction endonuclease site with normal DNA from the same plasmid that has been cleaved twice[28]. Other procedures take advantage of the resistance of a DNA strand containing thiophosphate residues to cleavage by restriction endonucleases. The simplest of these uses alkali-denatured plasmid DNA as template for oligonucleotide mutagenesis where the oligonucleotide is used to act as primer for thiophosphate-containing DNA synthesis, followed by enzymatic degradation of the non-thiophosphate containing wild-type DNA[29]. A more complex procedure uses gapped duplex DNA where the mutagenic oligonucleotide primes gap-filling with thiophosphate DNA to provide resistance to restriction endonuclease cleavage[30]. While the availability of efficient methods for direct oligonucleotide mutagenesis of plasmids would appear to be appealing, it should be remembered that plasmid vectors containing an origin of single-stranded DNA synthesis (phagemid vectors) offer the convenience of propagating DNA as plasmids, and of being able to generate easily purified single-stranded DNA at will[31,32]. In addition, oligonucleotide mutagenesis applied to a single-stranded dU-DNA template involves very few enzymatic manipulations[21,22].

The advent of the polymerase chain reaction for *in vitro* amplification of DNA[33] has resulted in a number of procedures for incorporating mutagenic oligonucleotides into *in vitro* amplified DNA. The methods produce mutants with high efficiency and are applied to double-stranded DNA[34,35]. They offer an attractive new approach to oligonucleotide mutagenesis.

The strategies for mutagenesis discussed to this point are directed at the replacement of one specific amino acid codon with one other specific codon. Often, an analysis of the role of the amino acid residue at one position in a protein requires that several different replacements at that position be constructed. An example of this involves the replacement of the codon for a readily oxidizable methionine residue in subtilisin by codons for all 19 of the other common amino acids in order to determine which variant subtilisin has the best catalytic properties[36]. The mutants, in this case, were generated by insertion of double-stranded synthetic cassettes containing mixtures of nucleotides at the target codon. It is also possible to use an oligonucleotide with a mixture of nucleotides at one codon for heteroduplex mutagenesis to obtain a number of amino acid replacements at one codon[37].

An approach to replace one amino acid residue by a series of other residues that does not involve oligonucleotide mutagenesis uses the *in vivo* suppression of a nonsense mutation at the target codon by a specially constructed family of tRNA suppressors[38]. A set of amber suppressor genes have been constructed by chemical synthesis and each gene inserted into an *E. coli* strain. Transformation of these strains with a plasmid containing the gene of interest harbouring an appropriate amber mutation readily results in a family of specific mutants. At present 13 different amino acids can be inserted using this powerful system.

While this article is concerned with site-directed mutagenesis using genetic methods, it should be noted that major strides in the direct chemical synthesis of proteins make this an attractive route to systematic protein structure–function analysis[39]. A particular feature of this approach is the possibility of inserting unusual amino acids in the synthetic protein. As well as providing a route to totally synthetic proteins, chemical synthesis can also be used as a semisynthetic approach to structure–function analysis[40].

An alternate procedure to direct chemical synthesis of protein for introducing an unusual amino acid at a specific position uses an amber-suppressing tRNA to which the unusual amino acid is esterified *in vitro*[41]. The gene encoding the protein of interest with an amber mutation at the target codon is expressed in an efficient *in vitro* transcription–translation system in the presence of the synthetic amino-acyl tRNA. Sufficient protein (30–45 $\mu g\ ml^{-1}$ of reaction) can be obtained for purification and functional studies.

Direct mutagenesis with synthetic oligonucleotides

It has long been known that a single-stranded DNA fragment can be a specific mutagen if annealed to a complementary single-stranded genome target[42]. However, the process is relatively inefficient and requires a reasonably long fragment for success[14]. It was these factors that led to the development of a procedure for mutagenesis wherein a short synthetic mutagenic oligonucleotide is incorporated enzymatically into a longer DNA prior to transformation into a biological system[43]. However, the improvements in chemical synthesis that have occurred in the last decade have made long synthetic DNA fragments readily available. This has revived interest in procedures for direct mutagenesis using synthetic DNA fragments that are annealed to a single-stranded vector carrying the target gene. The procedure produces specific mutants with *E. coli*[44] or *Saccharomyces cerevisiae*[44,45] as the recipient host. A variant of the procedure uses a gapped duplex DNA as target for the mutagenic oligonucleotide[46]. A particularly striking direct approach to oligonucleotide mutagenesis involves the pseudoreversion of a defective chromosomal gene in *S. cerevisiae*[47].

Random point mutations

In general, the objective in producing random point mutations throughout the structural gene of a protein is to identify as many as possible of the amino acid residues in the protein that are required for function. Alternatively, it is possible to modify some general property of the protein, such as thermal stability, without substantially changing the specific function of the protein, such as its catalytic activity. In developing a method for random mutagenesis of a specific target, the problem is to achieve high efficiency in terms of the numbers of mutant progeny produced whilst ensuring that a large fraction of the mutants contain only one base-pair replacement per target gene. Both chemical and enzymatic methods have been developed for this type of mutagenesis.

Chemical methods for producing random point mutations can be divided into two classes: those which modify bases so that their hydrogen-bond pairing with a complementary base is changed and those which so damage a base that it can no longer form hydrogen bonds with a complementary base[6,14]. No one chemical reagent causes a mutation at every base pair in a DNA target and therefore several different reagents are used in a parallel set of experiments in order to obtain a complete set of mutations. In the case of chemical mutagenesis of a target DNA, single hits are achieved by limiting the amount of reaction. As a consequence, 90% or more of the molecules are not mutated. Two procedures have been developed for increasing the efficiency of mutant production under these circumstances. The first involves treatment of each of the separated DNA strands of the target with the reagent, after which the two mutagenized DNA strands are annealed and cloned[48]. The second procedure is elegant but more complex and involves electrophoretic separation, under partial denaturation conditions, of mutant DNA fragments[49].

A number of enzymatic procedures involving misincorporation of nucleotides during *in vitro* synthesis of DNA are available for random mutagenesis[6,14]. The enzymatic methods have two attractive features: experiments can be set up so that in four parallel reactions each of the four nucleotides is misincorporated, and protocols can be devised so that a single misincorporation occurs in each target molecule. A convenient procedure containing both these features has been described[50]. The polymerase chain reaction has been modified to favour nucleotide misincorporation. In this mode, it provides a convenient enzymatic route to *in vitro* random mutagenesis[51].

Synthetic oligonucleotides also can be used for random mutagenesis. The synthetic oligonucleotide is deliberately 'spiked' with a low level of the three 'wrong' nucleotides at each position. Such 'spiked' synthetic oligonucleotides have a number of attractive features as mutagens: the ends of the synthetic oligonucleotide define the target boundaries quite precisely; specific positions within the mutagenized window can be deliberately not mutated; and, at positions where mutations are induced, each possible replacement can be specified. Versions of the procedure where the mutated segment is introduced as a double-stranded cassette have been described[52–54], as have procedures where the oligonucleotide is annealed to the DNA target[55,56].

Conclusions

As will be seen from the above discussion, useful procedures continue to be developed for site-directed mutagenesis that can be applied to problems in protein engineering. Notable amongst these is the mutagenic oligonucleotide-primed polymerase chain reaction for making specific mutants[34,35]. This reaction, as well as being used for specific codon mutations, can be conveniently used to insert short synthetic regions[57] or long, naturally occurring sequences with synthetic ends[58]. The various 'spiked' oligonucleotide strategies are clearly going to be very useful tools in random or semi-random targeted mutagenesis.

As was noted earlier, site-directed replacement of specific codons will continue to be a dominant feature of protein engineering studies. However, it is clear that the era of specifically designed enzyme functions is to hand[12], as is the generation of new enzyme activities by random or semi-random mutagenesis of disabled genes[54,56]. The roles of different structural features of proteins can be analysed by deletion of single codons[59] or of segments of proteins[60]. Hybrid proteins can be constructed for functional analysis[58] or to attach useful features such as a cellulose-binding domain[61]. The general properties of proteins continue to be targets for investigation and these are most obviously illustrated by the diverse approaches to protein stability[13,62,63]. It is clear that protein engineering will continue to provide new insights into biology and molecular function and that it is beginning to offer useful novel tools to biochemistry and to industry[64].

Acknowledgements

Research in the author's laboratory is supported by grants from the Medical Research Council of Canada and the National Cancer Institute of Canada. The author is a Career Investigator of the Medical Research Council of Canada.

References

1 Fersht, A. (1984) *Enzyme Structure and Mechanism* 2nd edn, W.H. Freeman & Co.
2 Ferguson-Miller, S., Brautigan, D. L. and Margoliash, E. (1979) in *The Porphyrins* (Dolphin, D., ed.), Vol. 7, pp. 149–239, Academic Press
3 Kraut, J. (1977) *Annu. Rev. Biochem.* 46, 331–358
4 Phillips, D. C. (1966) *Sci. Am.* 215, 78–90
5 Joyce, C. M. and Steitz, T. A. (1987) *Trends Biochem. Sci.* 12, 288–292
6 Smith, M. (1986) *Phil. Trans. R. Soc. London, Ser. A* 317, 295–304
7 Lipscomb, W. N. (1983) *Annu. Rev. Biochem.* 51, 17–34
8 Gardell, S. J., Craik, C. S., Hilvert, D., Urdea, M. S. and Rutter, W. J. (1985) *Nature* 317, 551–555
9 Shortle, D. (1986) in *Protein Engineering. Applications in Science, Medicine and Industry* (Inouye, M. and Sathua, R., eds), pp. 233–241, Academic Press
10 Wetzel, R. (1986) in *Protein Engineering. Applications in Science, Medicine and Industry* (Inouye, M. and Sarma, R., eds), pp. 257–274, Academic Press
11 Alber, T. and Matthews, B. W. (1987) in *Protein Engineering* (Oxender, D. L. and Fox, C. F., eds), pp. 289–287, A. R. Liss, Inc.
12 Carter, P., Nilsson, B., Burnier, J. P., Burdick, D. and Wells, J. A. (1989) *Proteins* 6, 240–248
13 Matsumura, M., Signor, G. and Matthews, B. W. (1989) *Nature* 342, 291–293
14 Smith, M. (1985) *Annu. Rev. Genet.* 19, 423–463
15 Frank, R., Meyerhaus, A., Schwellnus, K. and Blocker, H. (1987) *Methods Enzymol.* 154, 221–249
16 Matthes, H. W. D., Staub, A. and Chambon, P. (1987) *Methods Enzymol.* 154, 250–287
17 Horvath, S., Firca, J. R., Hunkapiller, T., Hunkapiller, M. W. and Hood, L. (1987) *Methods Enzymol.* 154, 314–326
18 Climie, S. and Santi, D. V. (1990) *Proc. Natl Acad. Sci. USA* 87, 633–637
19 Zoller, M. J. and Smith, M. (1987) *Methods Enzymol.* 154, 329–350
20 Xu, S-Y. and Cue, D. (1990) *BioTechniques* 8, 168–169
21 Kunkel, T. A., Roberts, J. D. and Zakour, R. A. (1987) *Methods Enzymol.* 154, 367–382
22 Bebenek, K. and Kunkel, T. A. (1989) *Nucleic Acids Res.* 17, 5408
23 Taylor, J. W., Ott, J. and Eckstein, F. (1985) *Nucleic Acids Res.* 13, 8765–8785
24 Kramer, W. and Fritz, H-J. (1987) *Methods Enzymol.* 154, 350–367
25 Stanssens, P., Opsomer, C., McKeown, Y. M., Kramer, W., Zabeau, M. and Fritz, H-J. (1989) *Nucleic Acids Res.* 17, 4441–4454
26 Carter, P. (1987) *Methods Enzymol.* 154, 382–403
27 Vandeyar, M. A., Weiner, M. P., Hutton, C. J. and Batt, C. A. (1988) *Gene* 65, 129–134
28 Hofer, B. and Kuhlein, B. (1989) *Gene* 84, 153–157
29 Sugimoto, M., Esaki, N., Tanaka, H. and Soda, K. (1989) *Anal. Biochem.* 179, 309–311
30 Olsen, D. B. and Eckstein, F. (1990) *Proc. Natl Acad. Sci. USA* 87, 1451–1455
31 Dente, L. and Cortese, R. (1987) *Methods Enzymol.* 155, 111–119
32 Zagursky, R. and Baumeister, K. (1987) *Methods Enzymol.* 155, 139–155
33 Saiki, R. K., Scharf, S., Faloona, F., Mullis, K. B., Horn, G. T., Erlich, H. A. and Arnheim, N. (1985) *Science* 230, 1350–1354
34 Jones, D. H. and Howard, B. H. (1990) *BioTechniques* 8, 178–183
35 Higuchi, R. (1990) in *PCR Protocols: A Guide to Methods and Applications* (Innis, M. A., Gelfand, D. H. and Sninsky, J. J., eds), pp. 177–183, Academic Press
36 Wells, J. A., Vasser, M. and Powers, D. B. (1985) *Gene* 34, 315–323
37 Ner, S. S., Goodin, D. B. and Smith, M. (1988) *DNA* 7, 127–134
38 Kleina, L. G. and Miller, J. H. (1990) *J. Mol. Biol.* 212, 295–318
39 Kent, S. B. H. and Parker, K. F. (1988) in *Therapeutic Peptides and Proteins: Assessing the New Technologies* (Marshak, D. R. and Liu, D. T., eds), pp. 5–13, Cold Spring Harbor Laboratory
40 Raphael, A. L. and Gray, H. B. (1989) *Proteins* 6, 338–340

41 Anthony-Cahill, S. J., Griffith, M. C., Noren, C. J., Suich, D. J. and Schultz, P. G. (1989) *Trends Biochem. Sci.* 14, 400–403
42 Hutchison, C. A., III and Edgell, M. H. (1971) *J. Virol.* 8, 181–189
43 Hutchison, C. A., III, Phillips, S., Edgell, M. H., Gillam, S., Jahnke, P. and Smith, M. (1978) *J. Biol. Chem.* 253, 6551–6560
44 Burke, D. T. and Olson, M. V. (1986) *DNA* 5, 325–332
45 Walder, R. Y. and Walder, J. A. (1986) *Gene* 42, 133–140
46 Fritz, H-J., Hohlmaier, J., Kramer, W., Ohmayer, A. and Wippler, J. (1988) *Nucleic Acids Res.* 16, 6987–7000
47 Moerschell, R. P., Tsunasawa, S. and Sherman, F. (1988) *Proc. Natl Acad. Sci. USA* 85, 524–528
48 Kadonaga, J. Y. and Knowles, J. R. (1985) *Nucleic Acids Res.* 13, 1733–1745
49 Myers, R. M., Lerman, L. S. and Maniatis, T. (1985) *Science* 229, 242–247
50 Lehtovaara, P. M., Koivula, A. K., Bamford, J. and Knowles, J. K. C. (1988) *Protein Eng.* 2, 63–68
51 Leung, D. W., Chen, E. and Goeddel, D. V. (1989) *Technique* 1, 11–15
52 Porter, S. D. and Smith, M. (1986) *Nature* 320, 766–768
53 Hutchison, C. A., III, Nordeen, S. K., Vogt, K. and Edgell, M. H. (1986) *Proc. Natl Acad. Sci. USA* 83, 710–714
54 Oliphant, A. R. and Struhl, K. (1989) *Proc. Natl Acad. Sci. USA* 96, 9094–9098
55 Ner, S. S., Goodin, D. B. and Smith, M. (1988) *DNA* 7, 127–134
56 Hermes, J. D., Parekh, S. M., Blacklow, S. C., Koster, H. and Knowles, J. R. (1989) *Gene* 84, 143–151
57 Kammann, M., Laufs, J., Schell, J. and Gronenborn, B. (1989) *Nucleic Acids Res.* 17, 5404
58 Clackson, T. and Winter, G. (1989) *Nucleic Acids Res.* 17, 10163–10170
59 Luck, D. N., Gout, P. W., Beer, C. T. and Smith, M. (1989) *Mol. Endocrinol.* 3, 822–831
60 Fetrow, J. S., Cardillo, T. S. and Sherman, F. (1989) *Proteins* 6, 372–381
61 Ong, E., Greenwood, J. M., Gilkes, N. R., Kilburn, D. G., Miller, R. C., Jr and Warren, R. A. J. (1989) *Trends Biotechnol.* 7, 239–243
62 Burley, S. K. and Petsko, G. A. (1989) *Trends Biotechnol.* 7, 354–359
63 Nosoh, Y. and Sekiguchi, T. (1990) *Trends Biotechnol.* 8, 16–20
64 Arbige, M. V. and Pitcher, W. H. (1989) *Trends Biotechnol.* 7, 330–335

M. Smith is at the Biotechnology Laboratory and Department of Biochemistry, University of British Columbia, Vancouver, British Columbia, Canada V6T 1W5.

FROM ANALYSIS TO SYNTHESIS: NEW LIGAND BINDING SITES ON THE LACTATE DEHYDROGENASE FRAMEWORK. PART I

ANTHONY R. CLARKE, TONY ATKINSON AND J. JOHN HOLBROOK

In Part I of this article, the naturally evolved protein framework of lactate dehydrogenase is investigated by genetically introduced modifications which reveal the structural basis of its catalytic and substrate-binding properties. In Part II, this analytical information is exploited in the design of two modified forms of the enzyme; one which is specific for a new substrate and one which lacks allosteric regulation.

For many widely studied enzymes, there has existed for some time both a large body of kinetic and ligand-binding information to provide an understanding of overall behaviour, and X-ray crystallographic descriptions of the protein structure. The enzymologist, however, has lacked a versatile and routine way of linking these two areas of knowledge. Apart from a few instances in which a specific group in the protein could be modified by chemical means, the contribution of given residues of an enzyme to its catalytic mechanism, its control or its substrate preference was often a matter of intelligent guesswork.

The comparatively new-found ability of synthetic organic chemistry to make oligonucleotides for altering protein-coding sequences in DNA has released enzymology from its general inability to test the role of a chosen amino acid side chain[1]. Inevitably, early experiments which exploited this genetic technique were analytical in nature. The most incisive experiments used crystallographic structures to suggest potentially important residues in active-site regions. These were then substituted for residues of varying chemical make-up and the functional properties of these mutant forms were examined. The most useful mutations were those in which function was not completely destroyed, since these allowed comparative measurements of binding and catalysis[2–4]. From this range of results, the role of the original (natural) residue could usually be inferred.

This analytical application is now being superceded by attempts to design new properties into old enzymes[5–7], a challenge which demands a clear understanding of the principles connecting the structural features of an enzyme with the parts they contribute to activity. To illustrate this shift in emphasis from the analysis of naturally evolved catalysis (explained in Part I) to the synthesis of variant forms of an enzyme into which new properties are designed (Part II), we describe the outcome of a series of mutagenesis experiments on the glycolytic enzyme lactate dehydrogenase.

THE STARTING POINT

Lactate dehydrogenase (LDH) offers several advantages as a study

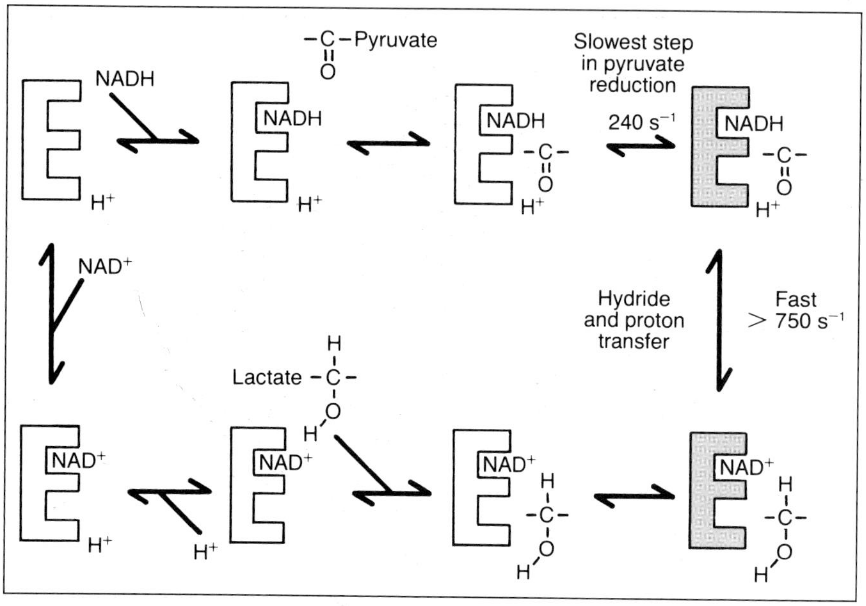

FIGURE 1

The catalytic cycle of lactate dehydrogenase. The eight kinetically identifiable intermediates through which the enzyme passes during one cycle of catalysis are shown schematically. The cycle represents the reaction at a pH below 7 (the pK_a of the His195–Asp168 couple), where the enzyme–coenzyme complexes and apo-enzyme are protonated. The enzyme proton donated to the carbonyl oxygen during the reduction is carried on His195. The shaded forms represent enzyme structures which have achieved the catalytically competent protein conformation mainly due to the closure of the active-site loop.

object for examining the principles of enzymatic catalysis. The reaction mechanism has been defined by classical steady-state methods and in a series of transient kinetic measurements (for review see Ref. 8); eight intermediate forms through which the enzyme passes during catalysis are summarized in Fig. 1. Identification of these steps is facilitated by the spectral characteristics of the intermediates; the free enzyme has high tryptophan fluorescence, the enzyme–NADH compound has low tryptophan fluorescence but enhanced coenzyme fluorescence and the enzyme–NADH–pyruvate complex has quenched coenzyme fluorescence. On conversion to enzyme–NAD^+–lactate, coenzyme absorbance at 340 nm is lost[8]. The reaction that converts pyruvate and NADH to lactate and NAD^+ consists of the concerted donation of a proton (H^+) from the active site His195 residue to the carbonyl oxygen atom and of a hydride ion (H^-) from the dihydronicotinamide ring of NADH to the carbonyl carbon atom.

There are crystal structures of three distinct enzyme intermediates in this cycle, the free enzyme[9], the enzyme–NADH (K. Piontek and M. G. Rossmann, pers. commun.) and the enzyme–NADH–substrate complex[10,11]. In the first two the active site is open to the solvent, in the last there are rearrangements of the protein structure induced by the binding of substrate. The most pronounced is the encasement of the reactive end of the coenzyme (the nicotinamide group) and the substrate molecule by a 13-residue 'loop' of polypeptide which closes over the catalytic site (see Fig. 2). This displaces water and 'solvates' the reactants with protein residues.

ANALYSIS OF THE CATALYTIC MECHANISM

The arrangement of the active site in the enzyme–NADH–pyruvate complex is shown as a stereographic pair in Fig. 3 and schematically in Fig. 4. Early pH-titration and chemical modification work showed His195 to be the proton donor/acceptor in the reaction, but the roles of the remaining groups were unknown. To elucidate these, we constructed mutant forms of the enzyme with substitutions of the active-site residues. The following results emerged.

Replacing the positively charged Arg109 by the neutral glutamine severely reduced the chemical reactivity of bound substrates; the rate of hydride transfer reduced to less than 0.07% of that in the wild type[12]. The binding affinity of pyruvate (and oxamate, an isosteric and isoelectronic analogue) was reduced much less sharply (to 5%) and coenzyme binding was unaffected. The simplest interpretation of these observations is that the arginine residue interacts most strongly with the transition-state structure in the reaction, reducing its free energy and accelerating the rate. It does this by supporting negative charge on the carbonyl oxygen to stabilize a polarized carbon–oxygen bond ($>C^{+}-O^{-}$). To support this conclusion, when substrate binds to the wild-type enzyme–NADH complex the coenzyme fluorescence is heavily quenched. This is attributed to a strongly polarized carbonyl group drawing electrons from the dihydronicotinamide ring to make it more like the unfluorescent NAD^{+}. In the mutant enzyme which lacks the charged arginine, the coenzyme fluorescence remains high, implying that the carbonyl group is less polarized.

Measurements were also performed on this mutant enzyme to establish how specific the effects of this mutation were. Might not such a substitution cause widespread distortion in the active site and thus nullify the above argu-

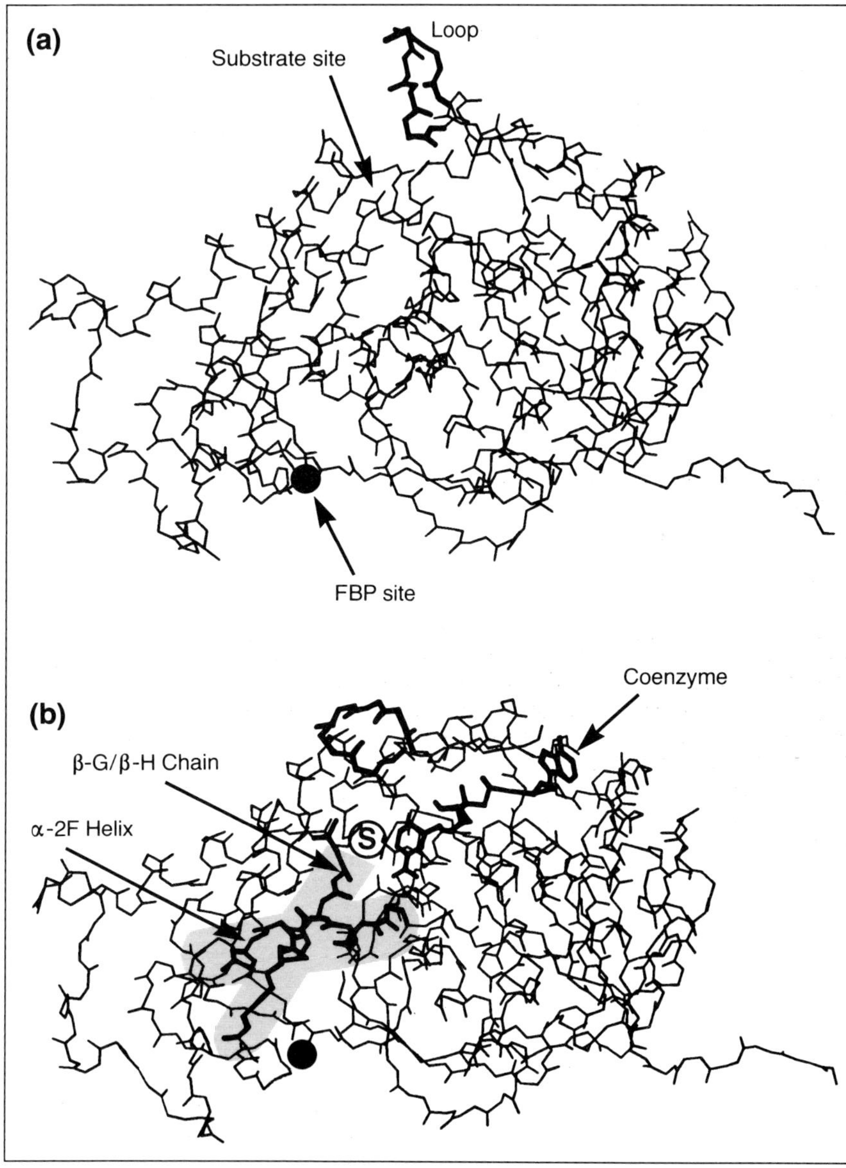

FIGURE 2

The enzyme subunit. Diagrams (**a**) and (**b**) show the polypeptide backbones of the apo-enzyme and the active enzyme–coenzyme–substrate complex, respectively. In (**a**) the active-site 'loop' is emboldened to emphasize its position lying away from the body of the protein. When coenzyme and substrate are bound (**b**) this 'loop' closes over the active site. Also marked in (**b**) are the binding positions of the coenzyme molecule (emboldened), the substrate (marked 'S') and the FBP effector (black circle, also shown in (**a**)). The structural connections (helix and extended chain) between the effector site and the active site are also highlighted. Structure (**a**) is based on dogfish apo-LDH[9] and (**b**) on the porcine H_4LDH:NAD–s–lac complex[11].

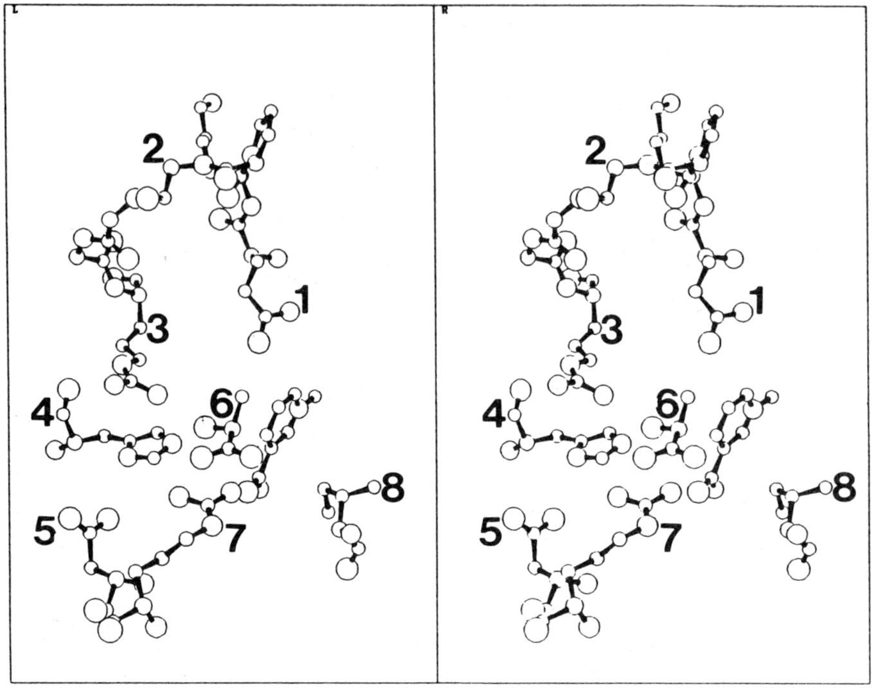

FIGURE 3

Stereographic views of the active site environment. In the catalytically active form of the enzyme the 'loop' isolates the active site from the solvent. The arrangement of protein residues around the substrate and coenzyme nicotinamide ring, shown here, is taken from the LDH:NAD–*s*–lac crystal structure. Key to structures: 1, Gln102; 2, loop residues 103–108; 3, Arg109; 4, His195; 5, Asp168; 6, lactate; 7, Arg171; 8, Ile250; 9, nicotinamide ring of NAD^+.

ments? To address this question, one of the alternative catalytic activities of LDH (for descriptions see Ref. 13) was examined; the addition of a sulphite ion to the nicotinamide group of NAD^+. This reaction proceeds by the nucleophilic attack of SO_3^{2-} on the C-4 position of the ring. It is promoted in the active site by the 'oily' environment of the nicotinamide ring, which 'forces' net positive charge towards C-4, and the binding of the sulphite ion in a suitable orientation to attack this locus. The Arg109→Gln mutation does not disturb this catalysis and so is unlikely to cause general and uninterpretable structural disturbance across the active site region.

One feature of the dehydrogenase active site that is analogous to the serine proteases is the presence of a carboxylic acid group, carried by Asp168, next to the catalytic histidine residue[14]. To test whether the role of aspartate lay in holding the histidine in the right orientation for catalysis or in providing a negative charge to make the imidazole group more basic, the residue was changed to an alanine (deleting the bond completely) and an asparagine (deleting the charge but still allowing a hydrogen-bonding interaction)[15]. Analysis of these proteins showed that both mutations caused a reduction in the binding affinity of pyruvate but not of lactate. This differential effect results from the requirement for the histidine to be protonated in forming the enzyme–NADH–pyruvate complex, but unprotonated in forming enzyme–NAD^+–lactate. Hence, the negatively charged aspartate will complement the buried positive charge on the imidazole group in the former complex, reducing its free energy and promoting pyruvate binding.

In addition to this effect, the mutations also lead to a lower rate of chemical reaction in the active site. This is expected from the explanation given of the role of Arg109 in stabilizing a transition state in which the carbonyl bond of pyruvate is polarized. Hence, like this arginine residue, a protonated and positively charged histidine would promote the formation of a negatively charged carbonyl oxygen in the substrate.

Two other aspects of these mutations deserve comment. Firstly, in the absence of substrate there is no measurable shift in the p*K* of the histidine residue on removing the neighbouring aspartate. The effect of the replacements is, therefore, only seen when substrate is bound; we interpret this to mean that the aspartate and histidine residues only form strong contacts in the closed, dehydrated active site. This view has recently been verified by X-ray structures of the enzyme–NADH and enzyme–NADH–oxamate forms of the pig muscle enzyme (oxamate is a close, but unreactive, analogue of pyruvate; C. R. Dunn, H. Muirhead and J. J. Holbrook, pers. commun.). Secondly, the properties of the alanine and asparagine mutants are very similar, despite the hydrogen-bonding ability of the former. It would appear that it is the negative charge and not the orienting effect of the wild-type aspartate residue which is important at this locus.

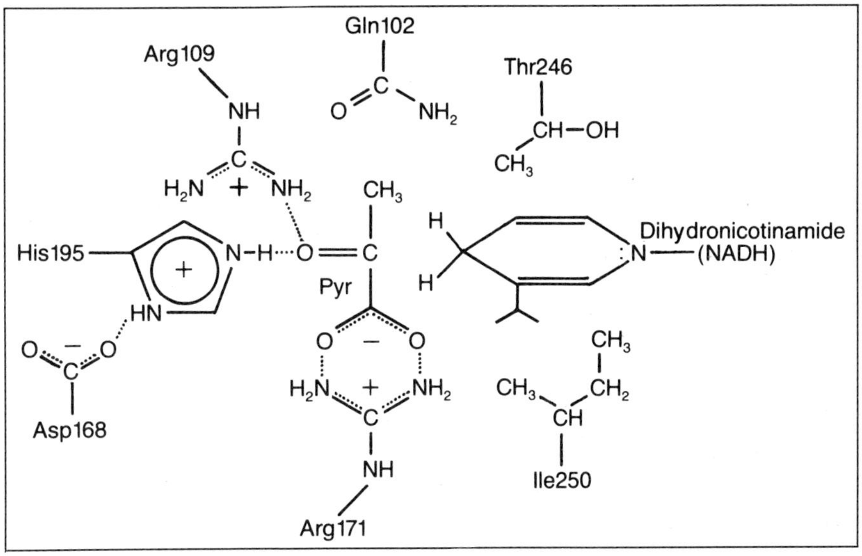

FIGURE 4

Summary of the functions of active-site residues. The roles of residues which envelope the catalytic centre and have been investigated by mutagenesis are shown schematically. To summarize the contributions of each: Arg171 provides a strong two-point interaction with the substrate carboxylate, ensuring efficient orientation of the substrate; His195 is the proton donor/acceptor in the redox step; when substrates are bound Asp168 interacts with His195 to stabilize its protonated form, thus selectively reducing the free energy of the enzyme–NADH–pyruvate complex; Ile250 provides a hydrophobic environment which strengthens the Arg171–substrate interaction and stabilizes NADH relative to NAD^+; Arg109 serves to polarize the carbonyl bond of the keto-acid and so promotes hydride transfer to the carbon and proton transfer to the oxygen; Gln102 and Thr246 form the environment of the substrate side chain and are, in part, responsible for substrate discrimination.

The three-dimensional structure of the enzyme shows the substrate carboxylate group to be bound to the guanidinium group of Arg171 in a coplanar arrangement. There is thus a two-point interaction between these 'forked' structures (see Figs 3 and 4). To maintain charge complementarity but reduce this to a one-point bond, the arginine residue was replaced by lysine. The effect on substrate binding is very large. The affinity of the enzyme for pyruvate is weakened to 0.05% of wild type or by 5 kcal (21 kJ) in the free energy of the interaction. In contrast, the affinity for NAD–SO_3^-, which only needs a point charge, is only slightly altered[16]. Two factors will contribute to this. The forked arginine–carboxylate interaction has two components (in net terms two charged hydrogen bonds), a lysine–carboxylate has only one. Also, the hydration potential of arginine (−20 kcal mol^{-1}; –84 kJ mol^{-1}) is much greater than that of lysine (−9.5 kcal mol^{-1}; –39.9 kJ mol^{-1}). Thus, as substrate binds, a greater number of bound water molecules may be displaced from the guanidinium group[17] than from the amine, leading to a greater entropic component in the arginine bond.

The great strength of the arginine–carboxylate bond is critical in substrate selection because of its ability to force the orientation of the substrate[18]. The wild-type enzyme is specific for pyruvate as against substrates with larger side chains on the grounds of binding affinity. For example, with one and two carbon side chains the K_m values for the wild type are 0.05 mM and 2.00 mM, respectively. The lysine mutant, however, is unable to discriminate on this basis; the K_m values with this enzyme are 120 mM and 60 mM. These results suggest that, by virtue of the tight interaction with Arg171, the wild-type enzyme can force the side chain of a keto-acid substrate into a recognition pocket formed by Gln102 and Thr246. With a weak lysine interaction this stringency is lost and with it binding specificity[18].

One of the prominent features of the coenzyme pocket in LDH is the hydrophobic environment of the nicotinamide ring. This has been invoked to explain the tighter binding of the neutral NADH over the charged NAD^+ and accords with the view that in an effective active site the free energy of the enzyme–reactant complex should be balanced with that of the enzyme–product[19]. In the case of the LDH reaction the unbound NADH–pyruvate pair has a much greater free energy than NAD^+–lactate. On the enzyme, however, the free energies of these pairs are almost equal, showing that the active site forms stronger contacts with NADH and pyruvate[15]. This is partly secured by Asp168 stabilizing the protonated, carbonyl-binding form of His195, but may also be due to this hydrophobic coenzyme pocket. To make this site more polar, Ile250 was replaced by asparagine, producing a protein in which NADH binding was, indeed, weaker (by around 1 kcal mol^{-1}; 4.2 kJ mol^{-1}) and which had a slightly reduced catalytic rate[20]. Perhaps more interestingly, this residue is a direct neighbour of Arg171, discussed above. Owing to this, the substitution of the hydrophobic isoleucine makes the environment of the arginine bond with the substrate carboxylate much more polar and reduces the enzyme's affinity for pyruvate to 0.1% of wild type. The magnitude of this change is surprisingly large and may reflect the unusual position of Arg171 in the enzyme–NADH complex, where it is bounded by the dihydronicotinamide ring of NADH, by Val233 and by Ile250. This environment is peculiarly non-polar for a residue of such high hydration potential and will lead to a high local free energy. On pairing with the oppositely charged and highly polar substrate carboxylate, the solvation requirements of the guanidinium group will be satisfied. This leads to a large reduction in free energy and a correspondingly high binding affinity.

Protein motion in catalysis

As stated earlier, X-ray structures of LDH show that the binding of the small substrate molecule induces marked rearrangements of the protein structure. This is reflected in the kinetics of the pyruvate reduction reaction. At room temperature the rate of catalysis is limited by a slow structural rearrangement of the enzyme–NADH–pyruvate complex[8]. Which, if any, of the several differences between the X-ray structures (see Fig. 2) accounts for this slow kinetic step? One way of answering this question is to use an optical probe which is sensitive to its environment, place this in the region of interest and observe whether the probe 'sees' its environment change at the same rate as the kinetic step in the reaction pathway. In practice, this strategy usually proves awkward; the region of interest more often than not contains no responsive and unambiguous natural chromophore and the likelihood of introducing an extrinsic probe at the right place in the protein by chemical means is low. To overcome these problems in LDH, all three natural tryptophan residues have been replaced by tyrosine, the residue of closest geometry and hydration potential[17]. This disrupts neither the catalytic properties of the enzyme nor its stability. The introduction of a single tryptophan residue at a chosen point in the protein then provides an unambiguous fluorescence signal[21,22] which is sensitive to structural rearrangements.

Of the structural rearrangements in the protein which may give rise to the slow kinetic step, the most obvious candidate is the closure of the active site 'loop' (see Fig. 2). To test this, tryptophan was introduced on the tip of the loop, at residue 106. The X-ray structures suggest that this residue is out in the solvent until substrate binds, whereupon the loop closes to bury it in the protein. When this variant of LDH reduces pyruvate, the rate at which tryptophan fluorescence is enhanced exactly matches the rate of catalysis[22]. The result shows that loop closure is a major element of the rate-determining step. This rearrangement of the protein structure induced by substrate binding is essential to catalysis on two counts. Firstly, it brings Arg109 into the active site to polarize the substrate carbonyl bond and, secondly, by surrounding the substrate with protein atoms it displaces water from the site of catalysis and allows the enzyme to discriminate between keto-acids with different side chains. In this case, 'induced fit' generates both catalytic power and substrate specificity.

The manipulation of this protein environment to select new substrate molecules will be the subject of the second part of this article.

ACKNOWLEDGEMENTS

The authors wish to thank the SERC (initiatives in Protein Engineering and Molecular Recognition), NATO (travel) and the Department of Trade and Industry for supporting this work.

REFERENCES

1 Winter, G., Fersht, A. R., Wilkinson, A. J., Zoller, M. and Smith, M. (1982) *Nature* 299, 756–758

2 Fersht, A. R., Shi, J-P., Knill-Jones, J., Lowe, D. M., Wilkinson, A. J., Blow, D. M., Brick, P., Carter, C., Waye, M. M. Y. and Winter, G. (1985) *Nature* 314, 235–238

3 Gardell, S. J., Craik, C. S., Hilvert, D., Urdea, M. S. and Rutter, W. J. (1985) *Nature* 317, 551–555

4 Sprang, S., Standing, T., Fletterick, R. J., Stroud, R. M., Finer-Moore, J., Xuong, N-H., Hamlin, R., Rutter, W. J. and Craik, C. S. (1987) *Science* 237, 905–913

5 Cronin, C. N., Malcolm, B. A. and Kirsch, J. F. (1987) *J. Am. Chem. Soc.* 109, 2222–2226

6 Craik, C. S., Largman, C., Fletcher, T., Roczniak, S., Barr, P. J., Fletterick, R. and Rutter, W. J. (1985) *Science* 228, 291–297

7 Winther, J. R., Kielland-Brandt, M. C. and Breddam, K. (1985) *Carlsberg Res. Commun.* 50, 273–284

8 Holbrook, J. J., Liljas, A., Steindel, S. J. and Rossmann, M. G. (1975) in *The Enzymes* (*Vol. II*) *3rd Edn* (Boyer, P. D., ed.), pp. 191–292, Academic Press
9 Adams, M. J., Haas, D. J., Jeffery, B. A., McPherson, A., Jr, Mermall, H. L., Rossmann, M. G., Schevitz, R. W. and Wonacott, A. J. (1969) *J. Mol. Biol.* 41, 159–188
10 Adams, M. J., Buehner, M., Chandrasekhar, K., Ford, G. C. Hackert, M. L., Liljas, A., Rossmann, M. G., Smiley, I. E., Allison, W. S., Everse, J., Kaplan, N. O. and Taylor, S. S. (1973) *Proc. Natl Acad. Sci. USA* 70, 1968–1972
11 Grau, U. M., Trommer, W. E. and Rossmann, M. G. (1981) *J. Mol. Biol.* 151, 289–307
12 Clarke, A. R., Wigley, D. B., Chia, W. N., Barstow, D. A., Atkinson, T. and Holbrook, J. J. (1986) *Nature* 324, 699–702
13 Burgner, J. W. and Ray, W. J., Jr (1984) *Biochemistry* 23, 3636–3648
14 Birktoft, J. J. and Banaszak, L. J. (1983) *J. Biol. Chem.* 258, 472–482
15 Clarke, A. R., Wilks, H. M., Barstow, D. A., Atkinson, T., Chia, W. N. and Holbrook, J. J. (1988) *Biochemistry* 27, 1617–1622
16 Hart, K. W., Clarke, A. R., Wigley, D. B., Chia, W. N., Barstow, D. A., Atkinson, T. and Holbrook, J. J. (1987) *Biochem. Biophys. Res. Commun.* 146, 346–353
17 Wolfenden, R., Andersson, P. M., Cullis, P. M. and Southgate, C. C. B. (1981) *Biochemistry* 20, 849–855
18 Hart, K. W., Clarke, A. R., Wigley, D. B., Waldman, A. D. B., Chia, W. N., Barstow, D. A., Atkinson, T., Jones, J. B. and Holbrook, J. J. (1987) *Biochim. Biophys. Acta* 914, 294–298
19 Albery, W. J. and Knowles, J. R. (1976) *Biochemistry* 15, 5631–5640
20 Wigley, D. B., Clarke, A. R., Dunn, C. R., Barstow, D. A., Atkinson, T., Chia, W. N., Muirhead, H. and Holbrook, J. J. (1987) *Biochim. Biophys. Acta* 916, 145–148
21 Waldman, A. D. B., Clarke, A. R., Wigley, D. B., Hart, K. W., Chia, W. N., Barstow, D. A., Atkinson, T., Munro, I. and Holbrook, J. J. (1987) *Biochim. Biophys. Acta* 913, 66–71
22 Waldman, A. D. B., Hart, K. W., Clarke, A. R., Wigley, D. B., Barstow, D. A., Atkinson, T., Chia, W. N. and Holbrook, J. J. (1988) *Biochem. Biophys. Res. Commun.* 150, 752–759

A. R. Clarke and J. J. Holbrook are at the Molecular Recognition Centre, School of Medical Sciences, University Walk, University of Bristol, Bristol BS8 1TD, UK. T. Atkinson is at the PHLS Division of Biotechnology, Centre for Applied Microbiology and Research, Porton, Salisbury, Wiltshire SP4 0JG, UK.

Lactate dehydrogenase catalyses the interconversion of an oxo-acid (pyruvate) and a hydroxy-acid (lactate) using the NADH/NAD^+ pair as a redox co-factor. The reaction, in the pyruvate-to-lactate direction, is achieved by reducing the carbonyl group of the oxo-acid substrate with a hydride ion (H^-) donated by NADH, and a proton (H^+) from the protein[2]. In Part I of this article[1], the amino acid residues which promote this chemistry were identified and their functions deduced. Presumably, these are necessary for the efficient turnover of any oxo-acid substrate in this enzyme and are summarized in the mechanism shown in Fig. 1. How lactate dehydrogenase has evolved its preference for the pyruvate/lactate pair, and where in the enzyme structure the specificity for the substrate side chain resides, are questions we address by using amino acid substitutions to adapt the protein structure to a new substrate. Such experiments provide a rigorous test of our understanding of the recognition process.

From Analysis to Synthesis: New Ligand Binding Sites on the Lactate Dehydrogenase Framework. Part II

Anthony R. Clarke, Tony Atkinson and J. John Holbrook

In Part I of this article, substrate-binding and catalysis in lactate dehydrogenase were examined by genetic modification of the protein structure and analysis of the functional consequences. In Part II, the conclusions are used in the design and synthesis of two modified forms of the enzyme; one in which the substrate specificity is shifted to produce a more effective malate dehydrogenase than that isolated from the host organism and one which no longer requires its allosteric activator (fructose 1,6-bisphosphate).

Adapting the enzyme to select a new substrate

The wild-type enzyme discriminates sharply between different keto-acids; extra methylene groups reducing the catalytic efficiency (measured by k_{cat}/K_m) by a factor of 10, a phenyl group by a factor of 50 and a carboxylate group by a factor of 1000 (Refs 2–4). Because of the magnitude of the loss in efficiency an attempt was made to redesign the active site to select a substrate of the last type, the oxaloacetate/malate pair[5]. Although the substrate bears an additional carboxylate group, the reaction has a similar redox potential to the pyruvate/lactate pair and so should preserve the energetics of the chemical conversion.

Inspection of the three-dimensional structure of the enzyme with an occupied active site (see Fig. 3 on p. 34) can only show which residues form the environment of the substrate. In this case the methyl side chain is directly enclosed by three residues: Thr246, Gln102 and Arg109 (see Fig. 1). Any one of these might account for the enzyme's recognition of its native substrate or this ability may be a result of more subtle and physically distant adaptations of the protein structure. Three possible

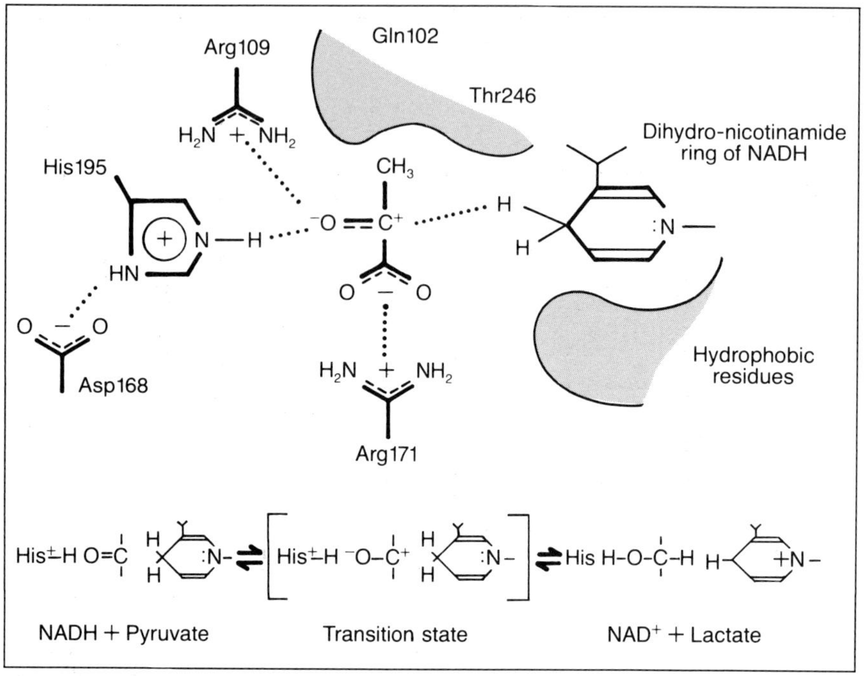

FIGURE 1

Schematic summary of the mechanism of lactate dehydrogenase (LDH) showing the LDH active site with bound substrate and coenzyme in the proposed transition state, and those residues in the protein which are important in this interaction. Evidence suggests[1] that the enzyme has evolved to stabilize the higher free-energy pair, NADH and pyruvate, more than NAD^+ and lactate. The result is that reactants and products, when bound to the enzyme, have a similar free energy. We propose that the transition state more resembles the NADH/pyruvate structure, but with a highly polarized, reactive carbonyl bond in the substrate. The substrate, transition state and product are represented in the lower portion of the figure.

factors responsible for the recognition of substrate structure by this dehydrogenase were examined: (1) the maintenance of overall charge balance in the active-site envelope, (2) the influence of substrate and active-site volumes, and (3) the effect of direct electrostatic complementarity between protein and ligand surfaces.

The idea that general charge balance is important arises from the observation that the enzyme can only form stable complexes when the combined charge on the active-site histidine, the substrate and the coenzyme nicotinamide ring is zero. For instance:

$$\text{His-H}^+/\text{ NADH / Oxamate}^-$$

is stable, but to form a complex with an alternative inhibitor, oxalate (a dicarboxylic acid), the coenzyme must carry a positive charge to form:

$$\text{His-H}^+/\text{NAD}^+/\text{Oxalate}^{2-}$$

Given this empirical observation, is it possible that the extra negative charge of the substrate, when forming the complex:

$$\text{His-H}^+/\text{NADH/Oxaloacetate}^{2-},$$

can be 'offset' by deleting a protein negative charge? Two mutations were evaluated: Asp197 →Asn, and Glu107 → Gln. Both residues are about 1 nm from the substrate side chain, but the former is just buried in the protein and the latter fully exposed to the solvent water. The result of these mutations is shown in Fig. 2. Removing the buried charge has a marked effect, shifting specificity towards oxaloacetate by a factor of 25, removing the exposed charge shifts it by a factor of only 2. Two conclusions can be drawn: (1) the overall protein charge in the environment of the substrate site can contribute to the selection of differently charged substrate molecules; and (2) this effect is more pronounced with buried charges because of the low dielectric constant of the protein interior compared to solvent water. These mutations, however, do not have a sufficiently potent effect to reverse the specificity.

Removing bulk from the active site has a more dramatic effect in shifting the relative efficiency of the enzyme from pyruvate towards the larger oxaloacetate. Replacing Thr246 by Gly allows a greater space to accommodate the carboxylate group and achieves a shift in specificity of more than a factor of 3000. This produces an enzyme which is slightly more effective against oxaloacetate than against the original substrate but which, despite binding the new substrate tightly, can only turn it over slowly (1 s^{-1} compared to 250 s^{-1} for the wild-type enzyme with pyruvate). In fact the change in specificity is largely caused by a reduction in the efficiency of the enzyme with pyruvate, attributable to the exposure of the active site to solvent water.

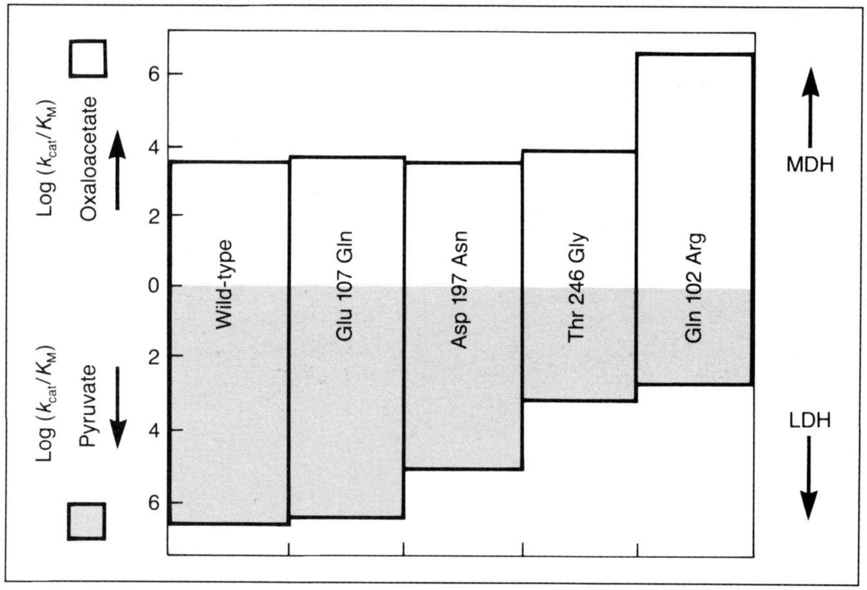

FIGURE 2

Substrate selection in the redesigned enzymes. The efficiencies of the wild-type and mutant enzymes against the natural substrate, pyruvate, and the new substrate, oxaloacetate, are as the ratio of k_{cat}/K_m. The chart emphasizes the effectiveness of each mutation in converting the enzyme to a malate dehydrogenase (MDH) (i.e. a malate/oxaloacetate oxido-reductase).

The provision of a direct protein counter ion for the substrate side chain proves to be the most effective way of switching the preference of the enzyme. This is achieved by replacing Gln102 with Arg to form the interaction modelled in Fig. 3 and results in an enzyme which is 8400 times more efficient with oxaloacetate than with pyruvate. The shift in specificity is about 7 orders of magnitude, giving a k_{cat}/K_m for oxaloacetate equal to that of the wild-type enzyme against pyruvate.

To break down this 'switch' into components: the loss of efficiency against pyruvate is approximately 10^4-fold and the gain against oxaloacetate is 10^3-fold. In energetic terms the enzyme–transition-state complex formed with pyruvate is less stable by 5.5 kcal (23.1 kJ), an observation which can be explained by the necessity to close the loop and accommodate the positive charge of Arg102 in an uncharged protein interior. The environment of the highly basic and polar guanidinium group:

$$-NH-C(NH_2)=NH_2^+$$

would be formed by hydrophobic elements (substrate methyl group, coenzyme dihydronicotinamide ring and Thr246) and bases (His195 and Arg109). However, when the catalytically active complex forms with oxaloacetate in the active site, this guanidinium group is 'solvated' by the substrate carboxylate group and the requirement for ion-charge complementarity is fulfilled.

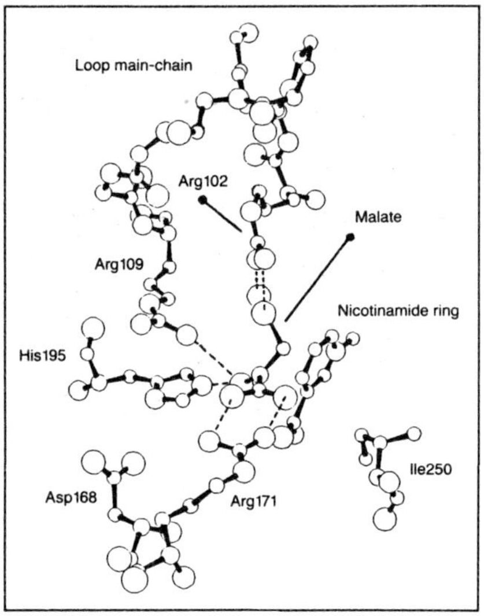

FIGURE 3

A model of the active site in the Gln102 → Arg mutant. The most effective of the redesigned enzymes is shown with malate occupying the active site. The view is equivalent to that of the wild-type enzyme shown in Fig. 3 on p. 34 and emphasizes the proposed interaction between Arg102 and the side chain carboxylate group of the substrate.

It is significant that this mutation increases protein bulk in the active site but, paradoxically, the resulting enzyme is effective against a larger substrate. This suggests that steric or packing considerations with a mobile surface loop are less important in substrate recognition in this dehydrogenase than are those of polarity. In this respect, the recognition between substrate and enzyme is more akin to that between solute and solvent, rather than lock and key.

MANIPULATING THE MECHANISM OF EFFECTOR CONTROL

The experiments described in this article were performed on the cloned enzyme from the thermophilic bacterium *Bacillus stearothermophilus*. In addition to binding coenzyme and substrate, this lactate dehydrogenase has a third ligand site which can be examined and modified by protein engineering techniques. This site binds fructose 1,6-bisphosphate (FBP), and one of the few functional differences between mammalian and bacterial lactate dehydrogenases is that some of the latter bind substrates tightly only in the presence of this allosteric activator. This prokaryotic regulation makes the enzyme responsive to the levels of available glucose in the cell, there being a build up of FBP when glucose is plentiful. This intermediate stimulates LDH and directs the production of lactate. In conditions of glucose starvation LDH is switched off and the products of glycolysis are further oxidized.

A molecular mechanism for this metabolic switch was proposed from the results of a series of experiments in which the state of subunit assembly was correlated with the binding of effector and of substrate[6–8]. At cellular protein concentrations the enzyme is a dimer with poor substrate affinity. These dimers do not easily associate to form tetramers because this brings together four, positively charged pockets (one on each subunit to form two symmetry-related pairs) within repulsive range (see Fig. 4). In the presence of FBP, however, two of these bisphosphates can satisfy all four basic sites and 'staple' the tetramer together across the dimer–dimer interface in a manner similar to the 2,3-diphosphoglycerate molecule in haemoglobin. The FBP-tetramer complex has a pyruvate affinity 50-fold higher than the dimer. The activation is split into two components, half is due to the dimer–dimer interaction and half to the FBP–protein interaction (i.e. each contributes a sevenfold increase in pyruvate affinity).

The basic pocket which binds a phosphate group of the regulatory FBP molecule is formed at the end of two structural 'rods', an α-helix (α) and an extended chain (β) (see Fig. 2 on p. 33, and Fig. 4). The other ends of the rods are 20 Å away and form the substrate-binding site. The extended chain contributes His188 to the regulatory site and His195 to the substrate site. The α-helical rod has Arg173 in the regulatory site and Arg171 in the substrate site. It is proposed that the occupation of the regulatory site with an FBP phosphate group will rearrange the relative positions of these rods and optimize the positions of residues 171 and 195 for substrate binding.

As a test of this mechanism a modified enzyme was synthesized in which Arg173 at the effector site was replaced with Gln[9]. This mutation does not prevent FBP binding to the protein and has no effect on the catalytic properties of unactivated dimeric form. It does, however, completely block the FBP-activation signal by breaking the strong guanidinium–phosphate contact at the regulator end of the α-helix. The new enzyme does not rearrange the helix position in response to the binding of FBP and manifests no site-to-site communication; a result confirmed by Matsuzawa and co-workers on an FBP-sensitive LDH from another bacterial source[10].

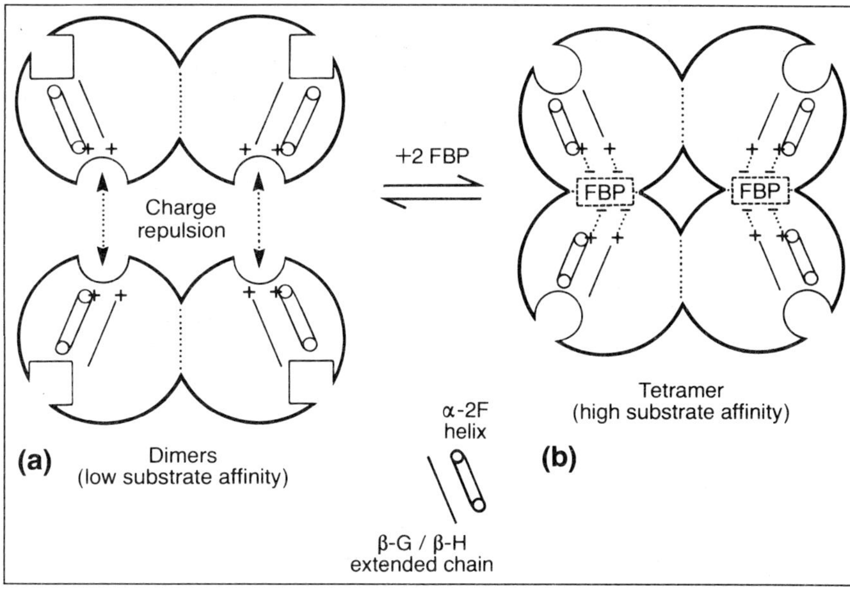

FIGURE 4

Activation of the enzyme by fructose 1,6-bisphosphate (FBP). Depicted in (**a**) is the dimeric form of the enzyme in the absence of FBP, a species with poor substrate affinity ($K_{m(pyruvate)}$ = 2.5 mM). The tetramer is unstable under these conditions owing to charge repulsion between the positively charged effector sites. When the dimers combine with two molecules of FBP the charges in these sites are neutralized and the tetrameric state of the protein shown in (**b**) is five orders of magnitude more stable. This FBP–tetramer complex has high substrate affinity ($K_{m(pyruvate)}$ = 0.05 mM).

As predicted from the proposed activation mechanism, the mutation also has a constructive element. The removal of positive charge from the, originally very basic, regulator sites reduces their mutual repulsion. This allows dimers to come together (in the absence of the neutralizing FBP) to form the more active and environmentally more stable tetramer, partially removing the effector requirement. At the level of applied enzymology, the manipulation of this regulatory response in LDH offers the opportunity of controlling the fermentative output of certain bacteria and is used in making unregulated enzymes for industrial catalysis.

THE PROSPECTS FOR REDESIGNING PROTEINS

The preceding sections describe how, given a firm understanding of the structural properties of an enzyme which make it an efficient catalyst, it is possible to manipulate it to accept a new substrate or to modify its regulatory properties. This can be done with a large measure of predictability and with the maintenance of fast enzymatic reaction rates. None of the modified proteins have lost their environmental robustness and, like the wild type, all are expressed as 25–30% of the total soluble protein of the *E. coli* host cells.

Although the aim of this series of manipulations was modest, the results show sufficient promise to encourage further attempts to build new properties into pre-existing enzyme frameworks. These attempts will rely on guidance from the active-site arrangements and catalytic mechanisms of well-studied 'model' enzymes[11–14], from the structure of antibodies raised to transition-state analogues (so-called 'abzymes'[15]) and, increasingly, from computational chemistry as the ability to predict molecular behaviour from structural properties improves.

Success in rational protein design will have many consequences. One will be an increased understanding of the link between active-site structure and substrate specificity, so furthering our knowledge of molecular recognition in biology. Another will be the possibility of exploiting the stereo- and regio-specific qualities of enzyme catalysis in industry. The use of readily redesigned and thermostable enzymes, encoded by cloned and over-expressing genes, will provide versatile catalysts for the synthesis of high-value organic compounds.

Acknowledgements

The authors wish to thank Smith Klein and French (Research) UK Ltd, Porton International plc, Genzyme (UK) Ltd, the SERC (initiatives in Protein Engineering and Molecular Recognition), NATO (travel) and The Department of Trade and Industry for supporting this work.

References

1 Clarke, A. R., Atkinson, T. and Holbrook, J. J. (1989) *Trends Biochem. Sci.* 14, 101–105; see also p. 31 of this book
2 Holbrook, J. J., Liljas, A., Steindel, S. J. and Rossmann, M. G. (1975) in *The Enzymes, 3rd Edn, Vol. 11* (Boyer, P. D., ed.), pp. 191–292, Academic Press
3 Hart, K. W., Clarke, A. R., Wigley, D. B., Waldman, A. D. B., Chia, W. N., Barstow, D. A., Atkinson, T., Jones, J. B. and Holbrook, J. J. (1987) *Biochim. Biophys. Acta* 914, 294–298
4 Clarke, A. R., Smith, C. J., Hart, K. W., Wilks, H. M., Chia, W. N., Lee, T. V., Birktoft, J. J., Banaszak, L. J., Barstow, D. A., Atkinson, T. and Holbrook, J. J. (1987) *Biochem. Biophys. Res. Commun.* 148, 15–23
5 Wilks, H. M., Hart, K. W., Feeney, R., Dunn, C. R., Muirhead, H., Chia, W. N., Barstow, D. A., Atkinson, T., Clarke, A. R. and Holbrook, J. J. (1988) *Science* 242, 1541–1544
6 Clarke, A. R., Waldman, A. D. B., Munro, I. and Holbrook, J. J. (1985) *Biochim. Biophys. Acta* 828, 375–379
7 Clarke, A. R., Atkinson, T., Campbell, J. W. and Holbrook, J. J. (1985) *Biochim. Biophys. Acta* 829, 387–396
8 Clarke, A. R., Evington, J. R. N., Dunn, C.R., Atkinson, T. and Holbrook, J. J. (1986) *Biochim. Biophys. Acta* 870, 112–126
9 Clarke, A. R., Wigley, D. B., Barstow, D. A., Chia, W. N., Atkinson, T. and Holbrook, J. J. (1987) *Biochim. Biophys. Acta* 913, 72–80
10 Matsuzawa, H., Michida, M., Kuwai, K., Ito, Y. and Ohta, T. (1988) *FEBS Lett.* 233, 375–378
11 Estell, D. A., Graycar, T. P., Miller, J. V., Powers, D. B., Burnier, J. P., Ng, P. G. and Wells, J. A. (1986) *Science* 233, 659–663
12 Fersht, A. R., Shi, J-P., Knill-Jones, J., Lowe, D. M., Wilkinson, A. J., Blow, D. M., Brick, P., Carter, C., Waye, M. M. Y. and Winter, G. (1985) *Nature* 314, 235–238
13 Gardell, S. J., Craik, C. S., Hilvert, D., Urdea, M. S. and Rutter, W. J. (1985) *Nature* 317, 551–555
14 Sprang, S., Standing, T., Fletterick, R. J., Stroud, R. M., Finer-Moore, J., Xuong, N-H., Hamlin, R., Rutter, W. J. and Craik, C. S. (1987) *Science* 237, 905–913
15 Massey, R. J. (1987) *Nature* 328, 457–458

A. R. Clarke and J. J. Holbrook are at the Molecular Recognition Centre, School of Medical Sciences, University Walk, University of Bristol, Bristol BS8 1TD, UK. T. Atkinson is at the Division of Biotechnology, Centre for Applied Microbiology and Research, Porton, Salisbury, Wiltshire SP4 0JG, UK.

SUBTILISIN: AN ENZYME DESIGNED TO BE ENGINEERED

JAMES A. WELLS AND DAVID A. ESTELL

Almost every property of subtilisin, a simple bacterial serine protease, has been altered by protein engineering including its catalysis, substrate specificity, pH/rate profile, and stability to oxidative, thermal and alkaline inactivation.

It was nearly a century ago that Emil Fischer first demonstrated enzyme specificity, yet we are only beginning to understand the chemical forces and energetics that govern it. The reason for this is clear – enzymes are large and complicated catalysts. Often present in regulated multisubunit complexes and sometimes in association with membranes or other cellular structures, enzymes frequently require cofactors or metal ions to carry out multistep reactions on elaborate substrates. Most enzymes are expressed in minute amounts, and no generic solution is available for the expression of large amounts of an *active* enzyme from its cloned gene. Enzyme assays can be cumbersome and the three-dimensional structures of fewer than 100 different enzymes have been determined to date (out of a possible $>10^5$). Thus, it is rare when nature builds an enzyme that is as easy to study as subtilisin.

This single domain serine protease (M_r 27 500) exhibits typical Michaelis–Menten kinetics and its mechanism is well understood (Refs 1, 2, and references therein). It is easily assayed (usually requiring only a spectrophotometer) and the active site may be readily probed using different synthetic peptide substrates. Expression of the subtilisin *BPN′* gene isolated from *B. amyloliquefaciens*[3], in a protease-deficient strain of *B. subtilis* produces more than gram per liter amounts of secreted subtilisin BPN′ (this represents ~30% of the total secreted protein). Collectively, over 50 high resolution X-ray structures of subtilisin have been determined in seven different laboratories (Refs 2, 4–13, and references therein). These delineate the binding of substrate, transition state[4], products[5], three different protease inhibitors[6–8], and define the structural consequences for natural variation[2,6–9] and site-specific mutations[11–13] (R. Bott, M. Ultsch, B. Katz and T. Kossiakoff, unpublished). All this presents the enzymologist with a molecular framework model that is tailored for protein engineering and structure–function analyses. In addition, it should be possible to make subtilisin more useful commercially because nature has not selected it to function optimally in cleaning solutions where it is used today.

CATALYSIS

The catalytic apparatus in subtilisin consists of the well-known catalytic triad (Ser221, His64 and Asp32), an oxyanion binding site (Asn155 and the main chain amide of Ser221) and other binding determinants further away from the scissile peptide bond of the substrate (Figs 1 and 2).

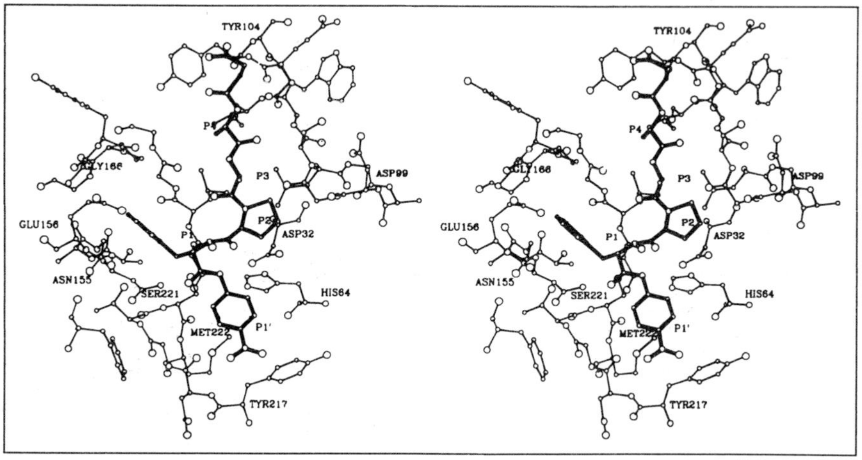

FIGURE 1

Stereoview showing a model substrate, *N*-succinyl-L-Ala-L-Ala-L-Pro-L-Phe-*p*-nitroanilide (shown with thicker bonds), bound to the active site of subtilisin BPN′. Residues in the catalytic triad (Ser221, His64 and Asp32) and other pertinent residues are labeled. Protease substrate nomenclature may be represented as:

$$\begin{array}{c} \quad\quad O\ H \\ \quad\quad \|\ \ | \\ NH_2\text{-}Pn\ldots P2\text{-}P1\text{-}C\text{-}N\text{-}P1'\text{-}P2'\ldots Pn'\text{-}COOH, \end{array}$$

where the scissile peptide bond is between the P1 and the P1′ substrate residues[33]. Accordingly, substrate residues are labeled P4 through P1′ starting with the most amino-terminal Ala residue and ending with the *p*-nitroanilide moiety. The model is based upon a 2.0 Å X-ray structure containing bound products (R. Bott and M. Ultsch, unpublished) and the graphic was produced by Dr R. Bott.

Together these provide subtilisin with a catalytic turnover rate constant (k_{cat}) that is at least 10^9–10^{10} greater than the first order spontaneous hydrolytic rate for amide substrates[14]. In the rate-determining acylation step for hydrolysis of peptide bonds by subtilisin (Fig. 2), the hydroxyl oxygen of the catalytic Ser221 transfers its proton (boxed in Fig. 2) to the catalytic His64 coincident with attack on the carbonyl group of scissile peptide bond (shaded in Fig. 2). This step produces a tetrahedral oxyanion intermediate (E·S‡) which is stabilized by at least two hydrogen bonds (dotted lines): one to Asn155 and another to His64. The tetrahedral intermediate collapses to the acyl-enzyme intermediate (E-Ac) as the proton on His64 is transferred to the leaving group amino. Hydrolysis of the acyl-enzyme to generate the free enzyme occurs by attack of a water molecule from the position of the amine using the reverse of the steps above. In the overall mechanism Ser221 serves as a catalytic nucleophile; His64 functions as a general acid–base; and Asp32 acts to position the correct tautomeric form of His64 in the E·S‡ complex and to provide electrostatic stabilization.

To compare the contribution to the enzymatic rate enhancement for residues in the catalytic triad relative to solvent, in our laboratory we have replaced them with alanine residues by site-directed mutagenesis and analysed the kinetic consequences[14,15]. Single alanine replacements for the Ser221, His64, or Asp32 reduced the values of k_{cat} by factors of 2×10^6, 2×10^6 and 3×10^4, respectively, with only small changes in K_m (increased up to twofold). When all three residues are simultaneously converted to alanine, the k_{cat} is still reduced by a factor of 2×10^6 from the wild-type enzyme (again there is essentially no change in K_m), thus demonstrating that once any one of the three residues is mutated, the catalytic function of the others is severely impaired. Although further studies are necessary to clarify the catalytic mechanisms for the catalytic triad mutants, these data underscore the highly cooperative role they play in the enzyme mechanism. Despite strong kinetic synergy within the catalytic triad, the kinetic analyses of

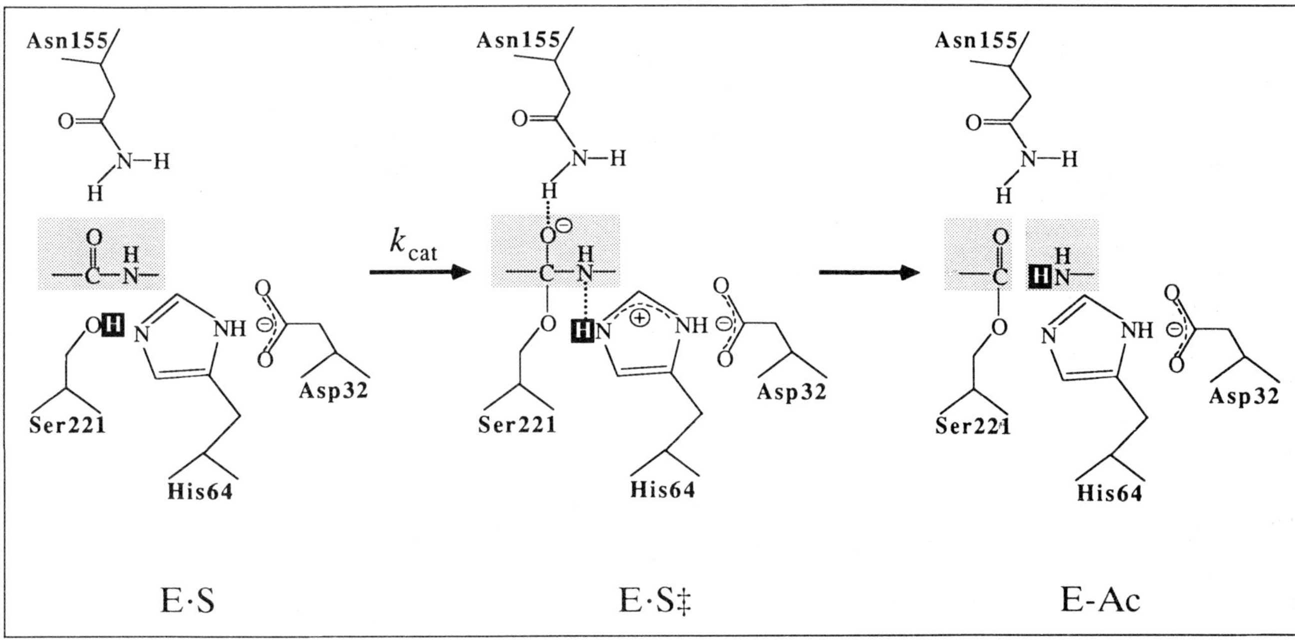

FIGURE 2

Diagram showing the rate-limiting acylation step (which defines k_{cat}) for hydrolysis of peptide bonds by subtilisin BPN′. For subtilisin BPN′, deacylation is >33 times faster than acylation[15] (see text for description). *Reproduced, with permission, from Ref. 14.*

active site mutants also suggest two evolutionary pathways (out of six possible) in which stepwise installation of the catalytic residues give progressive increases in k_{cat}[14,15]. One pathway proceeds by installing Ser221, then His64 and finally Asp32 and the second pathway by introducing Ser221, then Asp32 and finally His64.

The residual activity of the triple mutant (k_{cat} ~10^3 above the non-enzymatic rate) demonstrates that contacts outside of the catalytic triad are important in catalysis. Indeed, when the hydrogen bond between the tetrahedral oxyanion and Asn155 is eliminated by site-directed mutagenesis[16,17], k_{cat} is reduced by a factor of 10^2–10^3 concomitant with only small changes in K_m. X-ray crystallographic studies[4] suggest two additional hydrogen bonds that form in the transition state complex: one between the carbonyl of Ser125 and the amide NH of the P1 substrate residue, and a second between the amide NH from Ser221 and the oxyanion. Based upon kinetics of hydrolysis of substrates modified in the P1 position[18] or substrates extended from the P2 to P4 positions[1,2,4], it is estimated that these substrate subsites contribute separately to k_{cat} factors that range from 10 to 100. It is not clear that these transition-state binding contacts function independently from each other and the catalytic triad. However, it is evident that in addition to nucleophilic catalysis from the triad, a network of binding interactions contribute significantly to stabilize the transition state of the reaction.

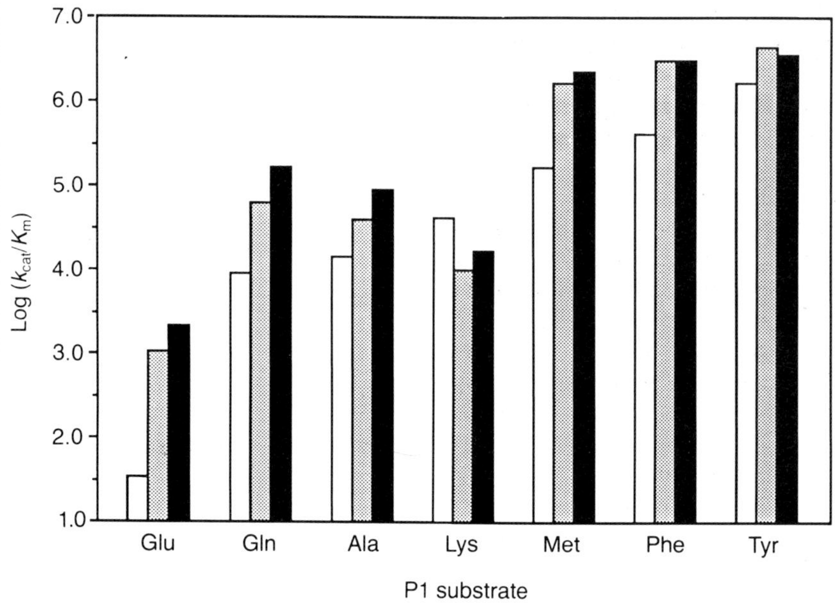

Figure 3

Specificity profiles for the *B. amyloliquefaciens* (□), *B. licheniformis* (■), and *B. amyloliquefaciens* triple mutant, Gln156→Ser:Gly169→Ala:Tyr217→Leu, (▒) subtilisins. The log (k_{cat}/K_m) are given for these three enzymes towards P1 substrates having the form *N*-succinyl-L-Ala-L-Ala-L-Pro-L-[X]-*p*-nitroanilide, where X is the P1 residue and is varied as shown on the abscissa. *Data is taken from Ref. 19.*

Engineering Substrate Specificity

One of the most striking aspects of the serine proteases is their wide diversity of substrate specificities that are coupled to a single catalytic mechanism. Subtilisins are generally characterized as being broadly specific but differ markedly against selected substrates. For example, the *B. amyloliquefaciens* and *B. licheniformis* enzymes have very similar structures[7,8] but differ by factors of 10–50 in catalytic efficiency (k_{cat}/K_m) toward various P1 substrates (Fig. 3)[19]. These enzymes have 86 amino acid differences out of 275 (including a deletion), yet only three substitutions can obviously affect the direct contacts with tetra-amino acid *p*-nitroanilide substrates. The amino acid side chain from 156 or 217 can make direct van der Waals contact with either the P1 side chain or *p*-nitroanilide moieties, respectively; Ala169 contacts Ala152 which in turn contacts the P1 substrate side chain (Fig. 1). When the three *B. licheniformis* substitutions (Glu156→Ser,Gly169→Ala, and Tyr217→Leu) were introduced simultaneously into the *B. amyloliquefaciens* subtilisin, the substrate specificity of the triple mutant approached that of the *B. licheniformis* enzyme towards small, large and charged P1 substrates (Fig. 3)[19]. Furthermore, even the individual kinetic parameters (k_{cat} and K_m) of the triple mutant closely matched those of the *B. licheniformis* enzyme.

Dissecting the triple mutant into component single mutants allowed us to identify the sources of the kinetic differences between the two natural variant subtilisins[19]. The higher k_{cat} values in the *B. licheniformis* enzyme could be attributed primarily to the Tyr217→Leu substitution, and altered K_m values primarily to the Glu156→Ser and Gly169→Ala mutations. The differences in charged substrate specificity (see below) were caused predominantly by the Glu156→Ser mutation. The effects of single mutants were roughly additive on log (k_{cat}/K_m), in contrast to the non-additivity observed within the catalytic triad. Although these data show that changes in direct contact residues largely account for the specificity differences, other substitutions further removed from the substrate binding site must account for the remaining discrepancy between the triple mutant and the *B. licheniformis* subtilisin (Fig. 3). These studies demonstrate the feasibility for recruiting into a single subtilisin the most desirable specificity properties contained within its homologous family. For example, it may be possible to install the most advantageous specificity properties from one variant into the most stable variant in the family to produce a hybrid enzyme with the optimal stability and specificity.

Hydrophobic effects account primarily for binding of uncharged P1 substrate side chains to subtilisin[1]. For example, analysis of the catalytic efficiency (k_{cat}/K_m) of subtilisin BPN′ toward a homologous set of ten substrates in which only the P1 residue was varied showed a strong correlation between transition state stabilization energy (expressed as a function of log $[k_{cat}/K_m]$) and the hydrophobic free energy of the P1 substrate side chain[18]. It seemed possible that by filling the P1 binding cleft with more hydrophobic side chains from the enzyme, large P1 substrates may be sterically excluded but k_{cat}/K_m for smaller hydrophobic substrates may be enhanced. Model building showed that substitutions for Gly166 (conserved in almost all *Bacillus* subtilisins) located at the back of the P1 binding cleft could extend their side chains to interact with the P1 substrate side chain. Indeed, the catalytic efficiency toward a TyrP1 substrate (which just fills the wild-type P1 binding cavity) was progressively reduced by as much as a factor of 5×10^3 by increasing the volume of the uncharged side chain introduced at 166 (Fig. 4)[17]. Substrates containing progressively smaller uncharged P1 side chains were more efficiently hydrolysed by mutant enzymes with increasingly larger side chain substitutions at 166 (Fig. 4). From the combined packing volumes of the optimal enzyme–substrate pairs, it was possible to estimate a total productive binding volume available for side chains within the P1 cleft of 160 ± 30 Å^3 (a leucyl side chain is about 100 Å^3). Furthermore, the effects upon transition-state stabilization energy for exceeding this productive binding volume was estimated to be about 10.92 kJ mol^{-1} (+2.6 kcal mol^{-1}) per 100 Å^3 in excess volume. This value is probably an underestimate for what to expect in other enzymes because the P1 binding cleft in subtilisin is rather open and so steric hindrance may be more easily accommodated. Although specific structural features for side chains at 166, such as β-branching, deviated occasionally from the above trends (Fig. 4), X-ray structures for five different substitutions at position 166 show that the enzyme undergoes only small local changes in conformation[10] (R. Bott and M. Ultsch, unpublished).

In addition to engineering complementary steric and hydrophobic interactions between enzyme and substrate, larger enhancements in k_{cat}/K_m toward specific substrates have resulted from designing electrostatic complementarity[20]. Molecular modeling[20] and X-ray crystallography[5] showed that charged substitutions at positions 156 or 166 could form salt bridges with oppositely charged P1 substrate side chains. Sixteen mutant (and wild type) subtilisins containing various charged and neutral substitutions at 156 and 166 were analysed kinetically for

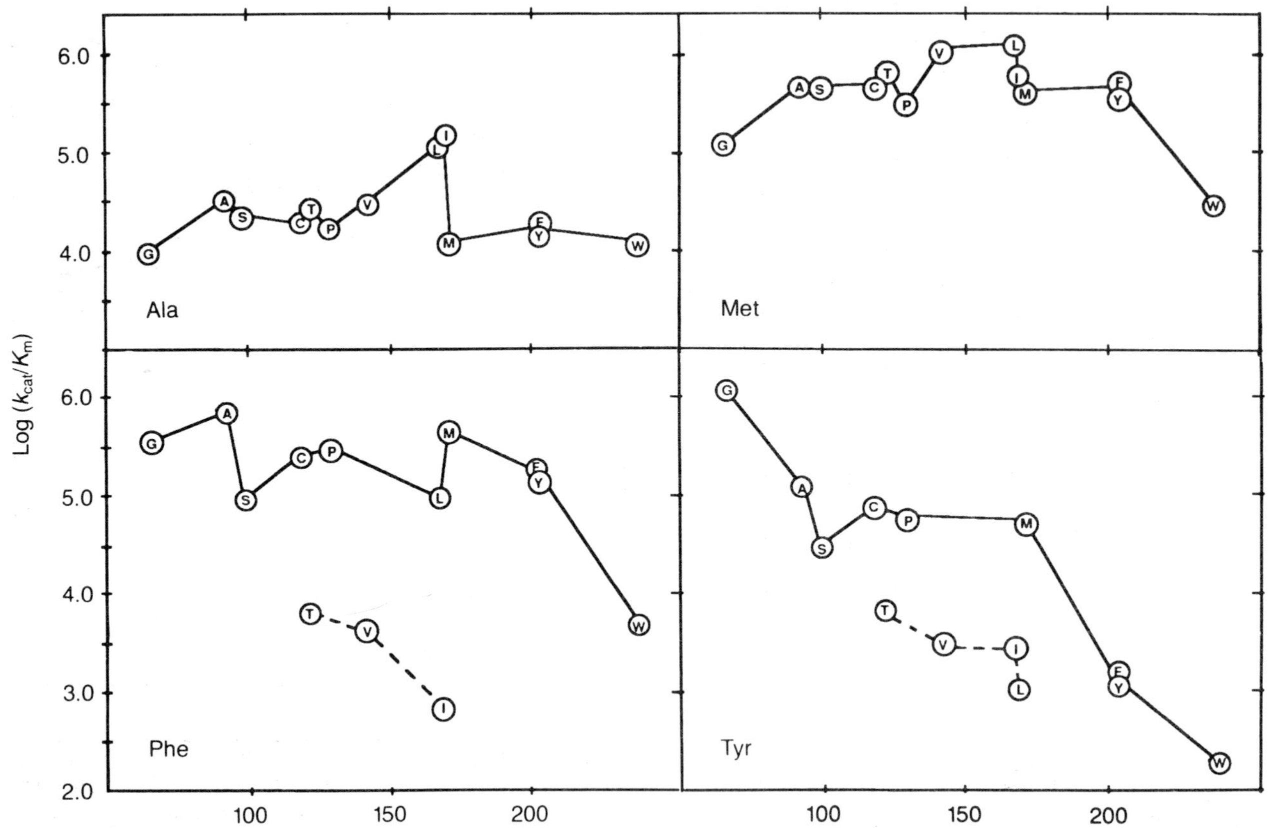

FIGURE 4

Effect of molecular side chain volume (Å^3) at position 166 on log (k_{cat}/K_m) toward various P1 substrates. The single-letter amino acid code is used to identify the residue at position 166 and substrates are described in Fig. 3. *Reproduced, with permission, from Ref. 18.*

hydrolysis of four homologous P1 substrates (Glu, Gln, Met, and Lys)[20]. As the charge state of the P1 binding site was varied from −2 to +1 (by site-directed mutagenesis), the average k_{cat}/K_m for the GluP1 substrate increased about tenfold per unit charge relative to its neutral homolog Gln, whereas the LysP1 substrate decreased, by a factor of ten, relative to the MetP1 substrate (Fig. 5). In particular, the average free energies for complementary enzyme–substrate electrostatic interactions at positions 156 and 166 were estimated to be -7.56 ± 2.1 and 9.66 ± 2.52 kJ mol^{-1} (-1.8 ± 0.5 and -2.3 ± 0.6 kcal mol^{-1}) (at 25°C in 0.1 M Tris·HCl, pH 8.6), respectively. The combined effects of mutations at these sites were roughly additive. In one of the most striking examples, k_{cat}/K_m toward the GluP1 substrate was increased 1900-fold from wild type by combining Glu156→Gln and Gly166→Lys mutations. These studies demonstrate the feasibility for engineering subtilisin's specificity toward preferential cleavage at acidic or basic P1 side chain residues, and exemplify the possibilities for engineering short-range electrostatic interactions in other enzyme or receptor binding sites.

An alternative strategy to engineering substrate specificity is one that we call *substrate-assisted catalysis*[15]. In this approach substrates are distinguished primarily by their ability to participate directly in the catalytic mechanism. Molecular modelling studies showed that the imidazole nitrogens from a bound HisP2 substrate side chain could be virtually superimposed upon the imidazole nitrogens of the catalytic His64. As discussed briefly above, when His64 is mutated to Ala there is a drop of $\sim 10^6$ in k_{cat} and only a fourfold increase in K_m for hydrolysis of Ala or GlnP2 substrates. In striking contrast, the k_{cat} value for the homologous HisP2 substrate is reduced only by a factor of 2×10^2 relative to wild type, and K_m is increased 15-fold. Thus the relative k_{cat}/K_m value for hydrolysing the HisP2 versus the AlaP2 substrate is over 200 for the His64→Ala mutant, compared to 1/4 for wild-type subtilisin, representing a change in substrate preference of over 800-fold in favor of hydrolysing the HisP2 substrate. The His64→Ala enzyme is the most substrate-selective subtilisin mutant yet produced and derivatives of it may be useful in peptide mapping or site-specific proteolysis of heterologous fusion proteins.

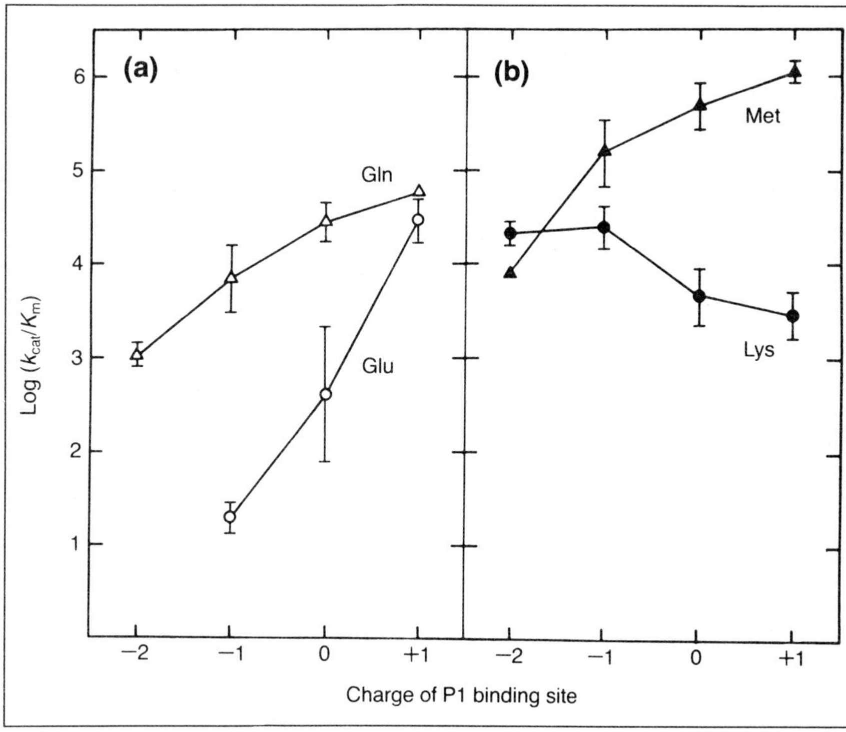

FIGURE 5

Change in the log (k_{cat}/K_m) ratio as a function of the net charge in the P1 binding site towards P1 substrates (see Fig. 3) containing Glu or Gln **(a)** and Met or Lys **(b)**. Each data point represents the average log (k_{cat}/K_m) value (and error bars one standard deviation) for each charge class. Each data point is composed of kinetic determinations from two to five mutant enzymes (or wild type) where the charge at position 156, 166 (or both) are varied. *Data is taken from Ref. 20.*

ENGINEERING THE pH PROFILE OF SUBTILISIN

The activity of subtilisin increases between pH 6 and 8 ($pK_a \sim 7.2$) as His64 is deprotonated. Chemical modification studies in chymotrypsin[21] showed that alteration of the surface charge of the enzyme could substantially shift the acidic leg of the pH profile. Introduction of negatively charged functional groups at lysines (which would stabilize the protonated form of the catalytic His57) raised the pK_a of His57 while derivatization with positive functional groups had the opposite effect.

This approach has been expanded by changing the surface charge of subtilisin using site-directed mutagenesis (Ref. 22, and references therein). Substitutions for two surface acidic residues (Asp99 and Glu156) located more than 10 Å from the catalytic His64 with Ser and then Lys resulted in progressive downward shifts in the activity dependent pK_a. As predicted, these effects were most dramatic at low ionic strength (0.001 mM imidazole buffer, 25°C) and were almost fully screened at high ionic strength (1 M KCl, although this depended to some extent on the salt used). The combined effects of the single mutants were cumulative in the double mutants, and in the most extreme case the double mutant containing Asp99→Lys and Glu156→Lys had its pK_a shifted down one pH unit (at low ionic strength). These effects also paralleled a progressive reduction in k_{cat}/K_m down to 1% for the double mutant containing Asp99→Lys and Glu156→Lys in the ability to hydrolyse the substrate benzoyl-Val-Gly-Arg-*p*-nitroanilide relative to succinyl-Ala-Ala-Pro-Phe-*p*-nitroanilide. These long-range electrostatic effects have had somewhat predictable effects on the acidic leg of the pH profile. However, the substitution Met222→Lys (within ~5 Å of the His64) gave results that were opposite to those predicted strictly on the basis of electrostatic effects on His64 but probably reflect steric and/or electrostatic effects on the oxyanion intermediate (D. Estell and J. Wells, unpublished).

The activity of subtilisin BPN′ decreases between pH 9.5 and 11 with an apparent pK_a of ~10.5. The alkaline effects on activity are reversible over short times, and spectral analysis showed that tyrosine ionization correlates with the reversible loss of activity[1]. Although it was not possible to identify a particular tyrosine residue, or exclude the possible effects of lysine deprotonation, two tyrosine residues are located in the substrate binding site: Tyr104 in the P4 binding site and Tyr217 in the P1′ binding site (Fig. 1). Indeed, mutagenesis of Tyr104→Phe extended the alkaline leg (D. Estell, T. Graycar, D. Powers and J. Wells, unpublished). For example, at pH 10.6 the k_{cat}/K_m ratio for the Tyr104→Phe enzyme was actually increased 1.2-fold relative to its k_{cat}/K_m at pH 8.6, whereas the k_{cat}/K_m for wild-type subtilisin BPN′ was halved. Thus, both the acidic and basic limbs of the pH/rate profile can be engineered. When coupled with appropriate stabilizing mutations, it should be possible for these engineered enzymes to function stably over greater extremes of pH.

A TEST CASE FOR THEORETICAL ANALYSIS

Subtilisin's high resolution X-ray structure, well understood enzyme mechanism and the availability of kinetic data from a large number of site-specific mutations (collectively over 400) make it highly amenable to theoretical analysis. Theories that predict changes in free energies of binding and catalysis closely simulated the results from mutagenesis of Asn155 (Refs 23 and 24). A free-energy perturbation method[23] predicted the kinetic effects of an Asn155→Ala mutation by a thermodynamic cycle in which the free-energy differences between wild-type and mutant subtilisins were calculated from energies of the free state (E + S), bound state (E·S) and transition state (E·S‡). The calculated free-energy differences ($\Delta\Delta G_{binding}$ = 0.46±3.36 kJ mol^{-1} [0.11±0.8 kcal mol^{-1}] and $\Delta\Delta G_{cat}$ = 14.28±4.62 kJ mol^{-1} [3.4±1.1 kcal mol^{-1}]) agreed closely with those determined experimentally from ratios of k_{cat} and K_m ($\Delta\Delta G^{expt}_{binding}$ = 1.68 kJ mol^{-1} [0.4 kcal mol^{-1}] and $\Delta\Delta G^{expt}_{cat}$ = 15.54 kJ mol^{-1} [3.7 kcal mol^{-1}]; J. Wells and D. Estell, unpublished). The same mutation has been studied by a combination of the empirical valence bond and a free-energy perturbation method[24] that allows one to locate the actual transition state in the enzyme active site and to obtain a consistent thermodynamic cycle. This approach also closely predicted the empirical values ($\Delta\Delta G_{binding}$ = 0.84±2.52 kJ mol^{-1} [0.2±0.6 kcal mol^{-1}] and $\Delta\Delta G_{cat}$ = 17.22±2.1 kJ mol^{-1} [4.1±0.5 kcal mol^{-1}]). Both theoretical studies provided 'true' predictions because the kinetic data were made available only after the computer simulations were finished. The latter approach has also rationalized the kinetic effects of the His64→Ala mutation (A. Warshel and J-K. Hwang, pers. commun.).

The results of long-range electrostatic effects upon the pK_a of His64 have been simulated by two theoretical treatments based upon Coulomb's law. Using a linearized Poisson–Boltzmann model[25], it was possible to account for the magnitude and ionic strength dependences resulting from the surface charge mutations at positions 99 and 156. In another approach[26], using an algorithm of Warwicker and Watson that considers both molecular charge and shape, the magnitude of the electrostatic effects at these and at two other positions in subtilisin were reasonably predicted. Calculations of the dielectric constants (ε) at the protein surface either by long-range effects on the pK_a of His64 (Refs 22 and 26) or short-range effects on substrate specificity[20] both give values of ε ranging from 40 to 80 (roughly a polarity between ethanol and water).

It is premature to conclude that we understand the theoretical basis for binding, catalysis and electrostatics based upon these early successes. However, the subtilisin system provides an extensive data base from which such theories may be tested.

OXIDATIVE STABILITY

Oxidative inactivation of subtilisin poses a severe limit to its industrial utility. Nearly 20 years ago it was shown that rapid inactivation of subtilisin by millimolar concentrations of H_2O_2 occurs coincident with the oxidation of Met222 to the sulfoxide derivative[27]. Although not directly involved in the enzyme mechanism, this methionine is conserved in all known *Bacillus* subtilisins (seven sequences) and occupies a partially buried position next to the catalytic Ser221 (Fig. 1).

It was not possible to predict which oxidatively resistant amino acid could replace Met222 to improve oxidative stability yet maintain high catalytic function. To find the optimal replacement, we decided to substitute this position with all nineteen possibilities which was facilitated by a cassette mutagenesis strategy[28]. Specific activities of the corresponding mutant enzymes ranged from 138 to 0.3% of wild type[29]

The Met222→Cys mutant had the highest specific activity resulting from a nearly twofold increase in k_{cat}. The most active mutant subtilisins had small amino acid substitutions (Cys, Ala, Ser, Gly) and were not the ones usually considered homologous to Met such as Ile, Val, or Leu. Bulky, β-branched or charged amino acids at position 222 yielded weakly active enzymes. X-ray crystallographic studies showed that oxidation of Met-222 occludes the oxyanion binding site and also sterically impairs substrate binding[11]. These observations help to explain the observed reduction in k_{cat} and increase in K_m upon chemical oxidation. Finally, mutants containing Ala or Ser at position 222 were resistant to 1 M H_2O_2 over extended periods of time[29].

It is reasonable to ask why nature apparently compromises the oxidative stability of subtilisin by maintaining Met222 for only a small increase in catalytic efficiency. *Bacillus* cells may protect subtilisin from oxidative inactivation by expressing oxygen free-radical scavenging enzymes because H_2O_2 added to crude culture supernatants is rapidly disproportioned without inactivating subtilisin.

Stability to Irreversible Inactivation

Factors known to disrupt the conformational integrity of subtilisin (high temperature, denaturants, high pH or calcium chelating agents) cause irreversible inactivation leading to autolytic digestion which limits the industrial use of subtilisin. Accordingly, disulfide bonds have been introduced into subtilisin (normally cysteine free) in attempts to stabilize the enzyme[30,31] (C. Mitchinson and J. Wells, unpublished). However, to date, engineered disulfides have not substantially increased the autolytic stability of subtilisin. The basis for this is not yet understood because it is unclear how much the molecule needs to unfold before becoming a substrate for irreversible inactivation and autolysis. As a further complication, the mere introduction of cysteine residues can be destabilizing when they disrupt pre-existing stabilizing interactions[30]. In most (but not all) cases, the disulfide mutants are more stable than their reduced counterparts. However, autolytic stability of the disulfide mutants are generally the same or below that of wild type, except in a rare case and under artificial conditions where subtilisin is destabilized by removing enzyme-bound calcium by chelation[31] (C. Mitchinson and J. Wells, unpublished).

Despite these results much has been (and will be) learned by engineering disulfides into subtilisin. First, the disulfide mutants of subtilisin are secreted and their disulfide bonds are formed spontaneously. Thus, subtilisin should be a useful model system for studying factors which influence disulfide bond formation *in vivo*. Next, it is possible to evaluate the redox potential of the engineered disulfides[30] (C. Mitchinson and J. Wells, unpublished) and relate these values to disulfide bond geometries determined crystallographically[12]. For example, redox potential measurements show that the disulfide bond between positions 24 and 87 is stronger than that between 22 and 87 by about -3.36 kJ mol^{-1} (-0.8 kcal mol^{-1}) (Ref. 29). Based on the known disulfide-bond geometries, the difference predicted solely from dihedral bond strain is −9.66 kJ mol^{-1} (−2.3 kcal mol^{-1}) (Ref. 12). Such data should help to refine theories that predict disulfide-bond strengths in proteins. The structures of engineered disulfides between positions 22 and 87 or 24 and 87 (at 2.0 Å resolution) show that they are accommodated by small local structural perturbations; these include a slight movement in the main chain which allows for less strain than that predicted from the modeling. Finally, scanning calorimetry has been applied to try to evaluate the change in thermodynamic stability of the molecule that results from one of these disulfide bonds (between Cys22 and Cys87)[31]. However, this data does not give rigorously the change in thermodynamic stability because the subtilisin has yet to be evaluated under reversible folding conditions. Mutants have also been produced to attempt to engineer stabilizing salt bridges into subtilisin (C. Erwin, B. Barnett and J. Sullivan, unpublished). The preliminary results for ten different salt-bridge mutants indicate that, although most of these have not substantially increased autolytic stability, two may have increased stability.

A practical alternative to the site-specific approach to improving the stability of proteins is the use of random mutagenesis coupled to a phenotypic screen. *Bacillus* colonies expressing randomly mutated subtilisin genes can be grown on filters that will bind the subtilisin that is secreted from each colony. The filters can be transferred to assay plates under high temperature or high pH (or other abiotic conditions) and more stable subtilisin mutants can be identified by larger proteolysis 'halos'. Using this approach, a more stable mutant was isolated under high temperature conditions (Asn218→Ser)[13]. This mutation occurred at one end of a β-hairpin structure, and X-ray crystallographic analysis showed this mutation causes the shortening of hydrogen bonds across the chains of the β-hairpin. A similar approach[32] was taken to isolate three mutant subtilisins (Met50→Phe, Ile107→Val, and Lys213→Arg) that were slightly more stable under alkaline conditions. The triple mutant was considerably more alkaline stable than any of its parent single mutants showing that the effects of the mutations are cumulative. It is noteworthy that although the triple mutant undergoes autolysis at half the rate of wild type at pH 12, the mutant is only marginally more stable under neutral and high temperature conditions (B. Caldwell and J. Wells, unpublished). These mutations have caused virtually no change in kinetic parameters, except for Asn218→Ser which is located in the P1′ binding site and caused a twofold increase in k_{cat}/K_m toward a synthetic peptide substrate.

Such mutations show that more stable variants of subtilisin BPN′ *can* be produced, and thus the enzyme has not evolved to its upper limit of autolytic stability. In fact, nature has used some of these very same substitutions in other subtilisin variants (Asn218→Ser is present in protease K [a fungal subtilisin][8], and Met50→Phe is present in most other *Bacillus* subtilisins). All of these sites, each located on the surface of subtilisin, have been substituted in at least one or more natural subtilisin variants. Although it is possible to rationalize the structural basis for some of the stabilizing effects of these mutations, it is fair to say that one would not have predicted these results. A large collection of such random mutants should expand our understanding of the relationship between autolytic stability and subtilisin's structure.

Summary

Large, additive and qualitatively predictable changes in substrate specificity can be produced by alteration of steric, hydrophobic and electrostatic complementarity in the binding site. Alternatively, the strong kinetic synergy observed within the catalytic triad can be harnessed to engineer specificity toward HisP2 substrates by substrate-assisted catalysis. It is feasible to engineer both the acidic and basic limbs of pH profile of the enzyme. The electrostatic effects upon the pH profile, and upon catalysis, have been accurately simulated by various theoretical analyses. Mutant enzymes that have been studied crystallographically show only small structural changes in the immediate vicinity of the substitution. Finally, the stability of subtilisin to oxidants, heat and alkali has been improved by mutagenesis. Thus, subtilisin is both a useful enzyme for unlocking the basic principles that link structure with function, and a simple molecular scaffold that can be engineered for practical applications.

Acknowledgements

We are grateful to our colleagues at Genentech and Genencor who made available their unpublished results, contributed their support and encouragement, and to Drs Tony Kossiakoff, Ron Wetzel, Rick Bott and Paul Carter for critical review of this manuscript.

References

1 Markland, F. S. and Smith, E. L. (1971) in *The Enzymes, Vol. 3* (P. D. Boyer, ed.), pp. 516–608, Academic Press
2 Kraut, J. (1977) *Annu. Rev. Biochem.* 46, 331–358
3 Wells, J. A., Ferrari, E., Henner, D. J., Estell, D. A. and Chen, E. Y. (1983) *Nucleic Acids Res.* 11, 7911–7925
4 Robertus, J. D., Kraut, J., Alden, R. A. and Birktoft, J. J. (1972) *Biochemistry* 11, 4293–4303
5 Poulos, T. L., Alden, R. A., Birktoft, J. J., Freer, S. T. and Kraut, J. (1976) *J. Biol. Chem.* 251, 1097–1103
6 Hirono, S., Akagawa, H., Mitsui, Y. and Iitaka, Y. (1984) *J. Mol. Biol.* 178, 389–413
7 Bode, W., Papamokos, E., Musil, D., Seemueller, U. and Fritz, M. (1986) *EMBO J.* 5, 813–818
8 McPhalen, C. A., Svendsen, I., Jonassen, I. and James, M. N. G. (1985) *Proc. Natl Acad. Sci. USA* 82, 7242–7246
9 Pahler, A., Banarjee, A., Dattagupta, J. K., Fujiwara, T., Linder, K., Pal, G. P., Snck, D., Weber, G. and Saenger, W. (1984) *EMBO J.* 3, 1311–1314
10 Bott, R., Ultsch, M., Wells, J., Powers, D., Burdick, D., Struble, M., Burnier, J., Estell, D., Miller, J., Graycar, T., Adams, R. and Power, S. (1987) in *Biotechnology in Agricultural Chemistry*, pp. 137–147, ACS Symposium Series 334
11 Bott, R. R., Ultsch, M., Kossiakoff, A., Graycar, T. P., Katz, B. and Power, S. (1988) *J. Biol. Chem.* 263, 7895–7906
12 Katz, B. A. and Kossiakoff, A. (1986) *J. Biol. Chem.* 261, 15480–15485
13 Bryan, P. N., Rollence, M. L., Pantoliano, N. W., Wood, J., Finzel, B. C., Gilliland, G. L., Howard, A. J. and Poulos, T. L. (1986) *Proteins: Structure, Function and Genetics* 1, 326–334

14 Carter, P. and Wells, J. A. (1988) *Nature* 332, 564–568
15 Carter, P. and Wells, J. A. (1987) *Science* 237, 394–399
16 Wells, J. A., Cunningham, B. C., Graycar, T. P. and Estell, D. A. (1986) *Philos. Trans. R. Soc. Lond. A* 317, 415–423
17 Bryan, P., Pantoliano, M. W., Quill, S. G., Hsiao, H. Y. and Poulos, T. (1986) *Proc. Natl Acad. Sci. USA* 83, 3743–3745
18 Estell, D. A., Graycar, T. P., Miller, J. V., Powers, D. B., Burnier, J. P., Ng, P. G. and Wells, J. A. (1986) *Science* 233, 659–663
19 Wells, J. A., Cunningham, B. C., Graycar, T. P. and Estell, D. A. (1987) *Proc. Natl Acad. Sci. USA* 84, 5167–5171
20 Wells, J. A., Powers, D. B., Bott, R. R., Graycar, T. P. and Estell, D. A. (1987) *Proc. Natl Acad. Sci. USA* 84, 1219–1223
21 Valenzuela, P. and Bender, M. L. (1971) *Biochem. Biophys. Acta* 250, 538–548
22 Russell, A. J. and Fersht, A. R. (1987) *Nature* 328, 496–500
23 Rao, S. N., Singh, U. C., Bash, P. A. and Kollman, P. A. (1987) *Nature* 328, 551–554
24 Hwang, J. K. and Warshel, A. (1987) *Biochemistry* 26, 2669–2673
25 Gilson, M. K. and Honig, B. H. (1987) *Nature* 330, 84–86
26 Sternberg, M. J. E., Hayes, F. R. F., Russell, A. J., Thomas, P. G. and Fersht, A. R. (1987) *Nature* 330, 86–88
27 Stauffer, C. E. and Etson, D. (1969) *J. Biol. Chem.* 244, 5333–5338
28 Wells, J. A., Vasser, M. and Powers, D. B. (1985) *Gene* 34, 315–323
29 Estell, D. A., Graycar, T. P. and Wells, J. A. (1985) *J. Biol. Chem.* 260, 6518–6521
30 Wells, J. A. and Powers, D. B. (1986) *J. Biol. Chem.* 261, 6564–6570
31 Pantoliano, M. W., Ladner, R. C., Bryan, P. N., Rollence, M. L., Wood, J. F. and Poulos, T. L. (1987) *Biochemistry* 26, 2077–2082
32 Cunningham, B. C. and Wells, J. A. (1987) *Protein Eng.* 1, 319–325
33 Schechter, I. and Berger, A. (1967) *Biochem. Biophys. Res. Commun.* 27, 157–162

J. A. Wells is at the Department of Biomolecular Chemistry Genentech, Inc., 460 Point San Bruno Boulevard, South San Francisco, CA 94080, USA. D. A. Estell is at the Research Department, Genencor, Inc., 180 Kimball Way, South San Francisco, CA 94080, USA.

β-Galactoside Transport in *E. coli*: A Functional Dissection of *Lac* Permease

H. Ronald Kaback, Eitan Bibi and Paul D. Roepe

The polytopic membrane protein *lac* permease harnesses energy from the electrochemical H^+ gradient to transport β-galactosidases against a concentration gradient. Although high-resolution structural information is still lacking, the permease is thought to possess 12 membrane-spanning α-helical segments. Various experimental strategies, including site-directed mutagenesis, have been employed to probe the function of this membrane protein at the molecular level.

Accumulation of β-galactosides against a concentration gradient in *Escherichia coli* is dependent on the lactose (*lac*) permease, a hydrophobic protein found in the cytoplasmic membrane that catalyses the coupled translocation of a single β-galactoside with a single H^+ (i.e. β-galactoside/H^+ symport or cotransport) (see Refs 1–3 for reviews). Thus, under physiological conditions, where the H^+ electrochemical gradient across the cytoplasmic membrane ($\Delta\bar{\mu}_{H^+}$) is interior-negative and alkaline, *lac* permease utilizes free energy released from downhill translocation of H^+ to drive accumulation of β-galactosides against a concentration gradient. In the absence of $\Delta\bar{\mu}_{H^+}$, the permease catalyses the converse reaction, utilizing free energy from downhill translocation of β-galactosides to drive uphill translocation of H^+ and generating $\Delta\bar{\mu}_{H^+}$, the polarity of which depends upon the direction of the substrate concentration gradient. As such, *lac* permease is a prototypic membrane protein representative of a wide variety of biological machines in both prokaryotic and eukaryotic membranes that transduce free energy stored in an electrochemical gradient into work in the form of a concentration gradient.

The *lac* permease is encoded by the *lacY* gene, the second structural gene in the *lac* operon, which has been cloned and sequenced. Moreover, the permease has been solubilized from the membrane, purified to homogeneity, reconstituted into phospholipid vesicles (i.e. proteoliposomes) and shown to catalyse all of the translocation reactions typical of the β-galactoside transport system in intact cells and right-side-out membrane vesicles with comparable turnover numbers. Therefore, the product of *lacY* is solely responsible for β-galactoside transport *in vivo*.

This article will discuss recent experiments conducted to further our understanding of the structure and function of *lac* permease at the molecular level. It should be emphasized, however, that structural information at the atomic level is difficult to obtain with hydrophobic membrane proteins[4], the great majority of which, and *lac* permease in particular, have yet to be crystallized. Since this level of structural resolution is a prerequisite for definitive mechanistic considerations, conclusions derived from some of the studies to be described must be regarded as speculative.

Structure of *lac* permease

Circular dichroic measurements on purified *lac* permease indicate that the protein is over 80% helical in conformation, an estimate consistent with the hydropathy profile of the permease which suggests that approximately 70% of its 417 amino acid residues are found in hydrophobic domains with a mean length of 24±4 residues[5]. These considerations led to a secondary structure model in which the permease is composed of a hydrophilic N-terminus on the inner surface of the membrane followed by 12 hydrophobic segments in α-helical conformation that traverse the membrane in zig-zag fashion connected by hydrophilic domains (loops) with a 17-residue C-terminal hydrophilic tail on the inner surface (Fig. 1). Support for the general features of the model was obtained subsequently from laser Raman spectroscopy[6] and Fourier transform infrared spectroscopy (P. D. Roepe, H. R. Kaback and K. J. Rothschild, unpublished), immunological studies[7–13], limited proteolysis[14,15] and chemical modification[16], but none of these approaches is able to differentiate between the 12-helix structure and another model[6] containing 14 transmembrane helices.

Recently, Calamia and Manoil[17] have elegantly provided exclusive support for the topological predictions of the 12-helix model by analys-

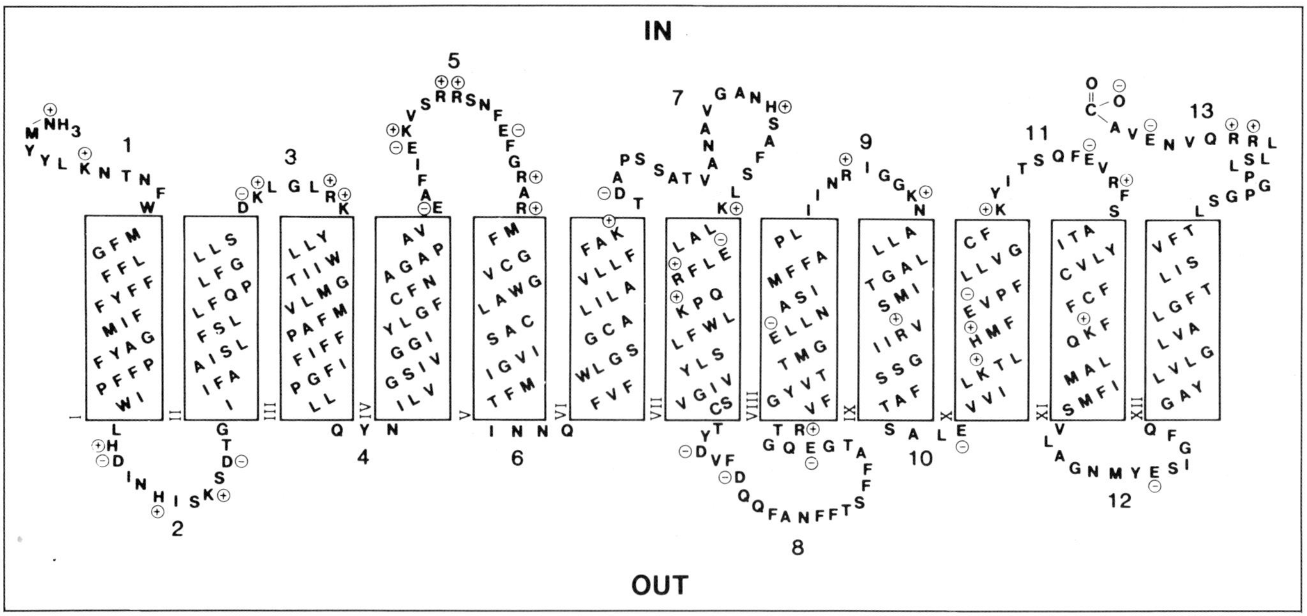

Figure 1

Secondary-structure model of *lac* permease based on the hydropathy profile of the protein. The single amino acid code is used, and hydrophobic segments are shown in boxes as transmembrane α-helical domains connected by hydrophilic loops.

ing an extensive series of *lac* permease–alkaline phosphatase (*lacY–phoA*) gene fusions. The activity of alkaline phosphatase fused to a cytoplasmic membrane protein reflects the disposition of the fusion junction sites with respect to the cytoplasmic or periplasmic surfaces of the membrane. Alkaline phosphatase activity in cells independently expressing each of 36 *lacY–phoA* fusions distinguishes between the two models proposed for *lac* permease and strongly favors the model containing 12 transmembrane domains.

Although early evidence (see Ref. 18) suggested that oligomerization might be important for *lac* permease activity, the permease is probably fully functional as a monomer[19]. Purified *lac* permease or purified cytochrome *o* was reconstituted into proteoliposomes under conditions in which both proteins are fully functional, and the preparations were examined by freeze-fracture electron microscopy. In nonenergized proteoliposomes, both proteins appear to reconstitute as monomers based on: (1) the variation of intramembrane particle density with protein concentration; (2) the ratio of particles corresponding to each protein in proteoliposomes reconstituted with a known ratio of permease to oxidase; and (3) the dimensions of the particles observed in tantalum replicas. Importantly, the initial rate of $\Delta\tilde{\mu}_{H}{}^{+}$-driven lactose transport in proteoliposomes varies linearly with the ratio of *lac* permease to phospholipid, particularly over the range at which there is statistically between 0 and 3 molecules of permease per proteoliposome. Furthermore, no change is observed in either the size or distribution of *lac* permease molecules when the proteoliposomes are energized.

Lac permease exhibits a notch or cleft under high magnification[19], an observation independently documented by Li and Tooth[20] using completely different techniques. The presence of a solvent-filled notch in the permease may have important implications with regard to the molecular mechanism of β-galactoside/H^+ symport, as the barrier within the permease may be thinner than the full thickness of the membrane. Therefore, the number of amino acid residues in the protein directly involved in translocation may be fewer than is generally thought to be the case.

IN VIVO EXPRESSION OF THE *LAC Y* GENE IN TWO SEGMENTS LEADS TO FUNCTIONAL *LAC* PERMEASE

Recently[21], the *lacY* gene was restricted into two approximately equal-size fragments which were subcloned individually or together under separate *lac* operator/promoters in plasmid pT7-5. Under these conditions, *lac* permease is expressed in two portions: the N-terminus, the first six putative transmembrane helices and most of putative loop 7; and the last six putative transmembrane helices and the C-terminus (Fig. 2). Cells harboring pT7-5 encoding both fragments transport lactose at about 30% the rate of cells expressing intact permease to a comparable steady-state level of accumulation. In contrast, cells expressing either half of the permease independently do not transport lactose. Radio-

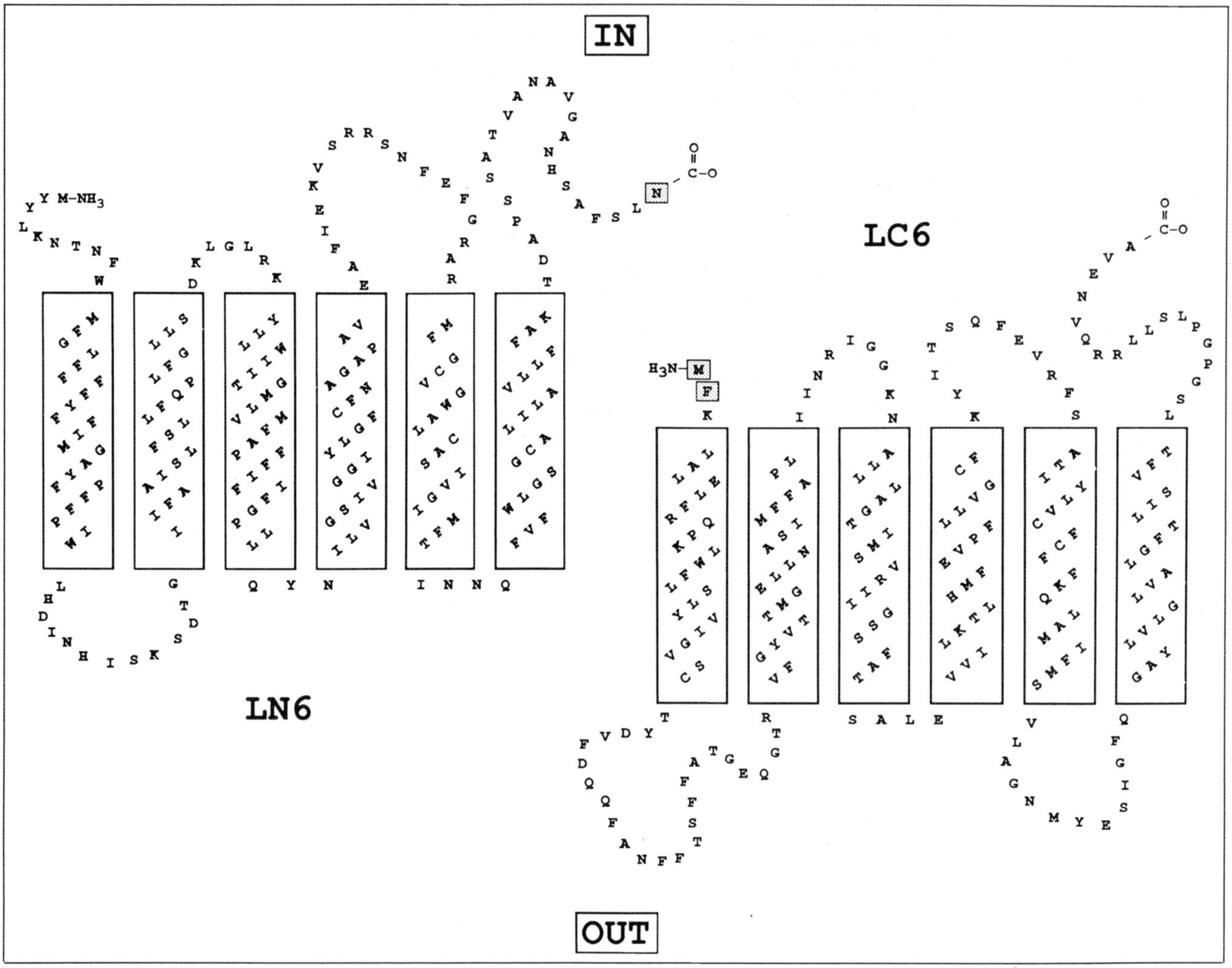

FIGURE 2

Secondary structure of *lac* permease showing portions encoded by N6 and C6 *lacY* gene fragments[21]. New amino acid residues inserted are shown in shaded boxes.

active labeling and immunoblotting experiments demonstrate that intact permease is completely absent from the membrane of cells expressing *lacY* fragments either individually or together. Thus, transport activity must result from an association between independently synthesized pieces of *lac* permease. When the gene fragments are expressed individually, the N-terminal portion of the permease is observed sporadically and the C-terminal portion cannot be detected. When the gene fragments are expressed together, polypeptides identified as the N- and C-terminal moieties of the permease are found in the membrane. The results are consistent with the conclusion that the N- or C-terminal halves of *lac* permease are proteolysed when synthesized independently and that association between the two complementing polypeptides leads to a more stable, catalytically active complex.

SITE-DIRECTED MUTAGENESIS

Notwithstanding the importance of high-resolution structure, it has become apparent that oligonucleotide-directed site-specific mutagenesis can be used to delineate amino acid residues that are important for active transport (see Refs 18 and 22 for reviews). Out of the more than 100 site-directed mutants in *lac* permease, about 70% exhibit little change in activity. Therefore, it is unlikely that individual amino acid replacements indiscrimately cause deleterious conformational changes in the protein.

Given the limited scope of this discussion, the reader is referred to other more extensive reviews[18,22] for descriptions of site-directed mutants in Cys, Tyr, Pro and Trp residues of the permease, as well as truncation mutants in the C-terminus and mutants that exhibit altered ligand-binding properties.

The most provocative findings from site-directed mutagenesis studies on *lac* permease began when the four His residues in the protein were mutagenized[23–25]. Replacement of His35 and His39 (putative loop 2; Fig. 1) with Arg or replacement of His205 (putative loop 7) with Arg*, Asn or Gln has no effect on active lactose transport, whereas replacement of His322 (putative helix X) with Arg, Asn, Gln or Lys causes dramatic loss of activity. Consistently, permease constructed with His322 as the only His residue exhibits properties identical to wild-type permease[26]. Strikingly, permease mutated at His322 catalyses downhill lactose influx at high substrate concentrations without concomitant H^+ translocation.

Efflux, exchange and counterflow are useful for studying permease turnover because specific steps in the overall kinetic cycle can be delineated. Permease with Arg, Asn, Gln or Lys in place of His322 is markedly defective in all translocation reactions that involve protonation or deprotonation (see Fig. 3). Furthermore, the primary kinetic effect of $\Delta\bar{\mu}_{H^+}$ (i.e. a decrease in apparent K_m for lactose) is not observed. Interestingly, permeases with Asn, Gln or Lys in place of His322 catalyse downhill efflux, as well as influx, but both processes occur without concomitant H^+ translocation[25].

Since His322 may be directly involved in lactose-coupled H^+ translocation and this residue is located in putative helix X (Fig. 1), attention focused on Glu325, which should be on the same face of helix X as His322 and may be ion-paired with this residue. In addition, structure–function studies on chymotrypsin[27] and other serine proteases have led to the notion that acidic amino acid residues may function with His as

*Although initial experiments[23] indicated that permease with Arg205 is defective in transport, the mutant construct was found to have two additional mutations in the 5′ end of the gene. When the secondary mutations are removed, Arg205 permease exhibits normal activity[25].

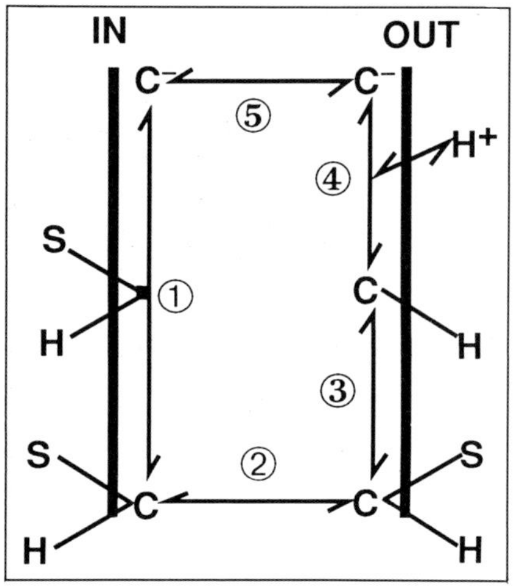

FIGURE 3

Schematic representation of reactions involved in lactose efflux, exchange and counterflow. C represents *lac* permease; S is substrate (lactose). The order of substrate and H^+ binding at the inner surface of the membrane is not implied.

components of a charge-relay system, a mechanism that might be adapted in part to H^+ translocation. For these reasons, Glu325 was subjected to site-directed mutagenesis[28,29]. Permease with Ala, Gln, Val, His, Cys or Trp in place of Glu325 catalyses downhill influx of lactose without H^+ translocation, but does not catalyse either active transport or efflux. Remarkably, the rate of equilibrium exchange with the altered permeases is at least as great as that observed with wild-type permease. Moreover, permease mutated at position 325 catalyses counterflow at the same rate and to the same extent as wild-type permease, but the internal concentration of [^{14}C]lactose is maintained for a prolonged period due to the defect in efflux. It is also noteworthy that permease mutated at position 325 catalyses counterflow three or four times better than wild-type permease when the external lactose concentration is below the apparent K_m.

The results can be rationalized by the simple kinetic scheme shown in Fig. 3. Efflux down a concentration gradient is thought to consist of a minimum of five steps: (1) binding of substrate and H^+ on the inner surface of the membrane (order unspecified); (2) translocation of the ternary complex to the outer surface; (3) release of substrate; (4) release of H^+; (5) return of the unloaded permease to the inner surface. Alternatively, exchange and counterflow with external lactose at saturating concentrations involve steps 1–3 only. Furthermore, release of H^+ appears to be rate-limiting for the overall cycle (see Ref. 19).

All steps in the mechanism that involve protonation or deprotonation appear to be blocked in the His322 mutants. Therefore, it seems reasonable to suggest that protonation of His322 may be involved in step 1. In contrast, replacement of Glu325 results in a permease that is defective in all steps involving net H^+ translocation but catalyses exchange and counterflow normally. Thus, permease mutated at position 325 is probably blocked in step 4 (i.e. it is unable to lose H^+).

Experiments in which Glu325 was replaced with Asp have yielded unexpected results (P. D. Roepe, D. Mechling, L. Patel and H. R. Kaback, unpublished). Permease with Asp325 is partly uncoupled and catalyses symport about 30% as well as wild-type permease. The observation is not unexpected, as the side-chain containing the carboxylate is about 1.5 Å shorter in Asp relative to Glu. Surprisingly, however, Asp325 permease catalyses equilibrium exchange below pH 7.7 normally, but as ambient pH is increased, exchange activity is progressively and reversibly inhibited with a mid-point at about pH 8.5. The findings indicate that translocation of the fully loaded permease does not tolerate a negative charge at position 325 and suggest that the carboxylate at position 325 undergoes protonation and deprotonation during lactose/H^+ symport. The contrasting observation, that equilibrium exchange with wild-type permease is insensitive to pH over the same range, is consistent with the notion that Glu325 is strongly hydrogen-bonded to His322.

Replacement of Arg302 with Leu, His or Lys (putative helix IX; Fig. 1) yields permease with properties similar to those of permease mutated at His322[30]. In marked contrast, replacement of Ser300 with Ala (helix IX),

Ser306 with Ala (helix IX) or Cys333 with Ser (helix X) has no significant effect on permease activity, thereby highlighting the specificity of Arg302, His322 and Glu325 and providing further support for the contention that single amino acid changes do not indiscriminately cause conformational alterations. Furthermore, by molecular modeling of putative helices IX and X, it can be shown that the guanidino group in Arg302 may be sufficiently close to His322 to participate in hydrogen-bonding with the imidazole ring that, in turn, may be hydrogen-bonded to the carboxylate of Glu325 (Fig. 4). Minimally, therefore, the putative charge-relay in the permease would involve interactions between Arg302, His322 and Glu325.

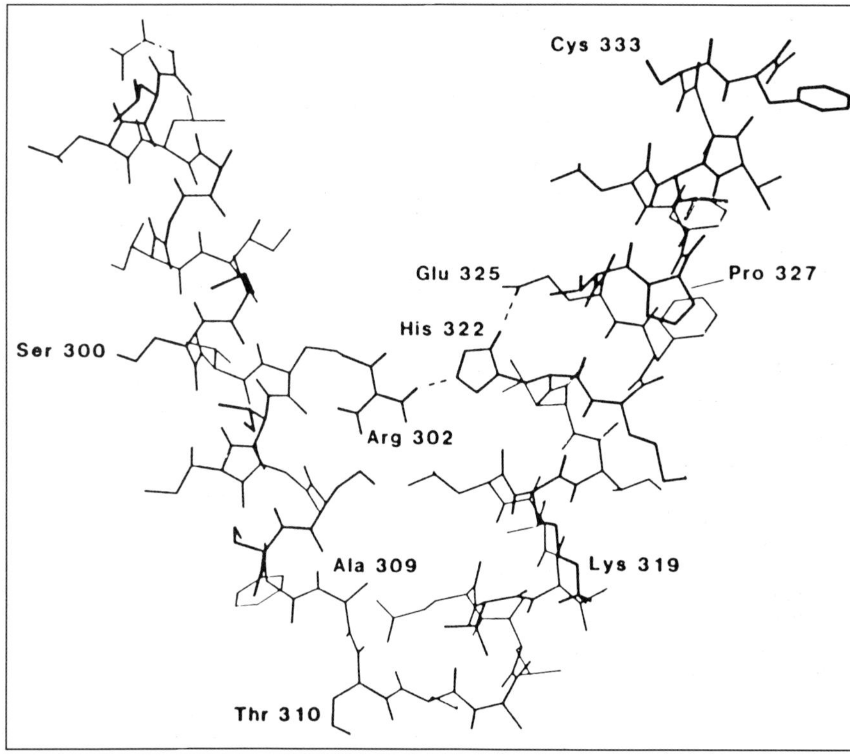

FIGURE 4

Modified molecular model of putative helices IX and X in *lac* permease. The study was performed by Dr Vincent Madison on an Evans–Sutherland computer on the basis of the hydropathy profile of Foster *et al.*[5], except that Ala309 and Thr310 were transferred from helix IX to hydrophilic segment 10, which connects helices IX and X (cf. Fig. 1). See Ref. 30 for details.

As evidenced by binding studies with the high-affinity ligand *p*-nitrophenyl-α,D-galactopyranoside (NPG), permease mutated at position 325 binds with a K_d approximating that of wild-type permease[29]. This is consistent with the observation that counterflow, a process that exhibits an apparent K_m similar to that observed for active transport, is intact in the mutants, but is in marked contrast to findings with permeases mutated at Arg302 or His322, which exhibit significantly decreased affinities[25]. It is therefore tempting to speculate that the pathways for H^+ and lactose may overlap (i.e. that Arg302 and His322 may also be components of the substrate binding site in addition to being involved in H^+ translocation; see also Refs 31, 32) and that protonation of His322 may be required for high-affinity binding.

If Arg302, His322 and Glu325 are sufficiently close to hydrogen-bond and function as components of a charge-relay system, the polarity, distance and orientation between the residues should be critical[33]. The importance of polarity between His322 and Glu325 was studied by interchanging the residues; the modified permease is inactive in all modes of translocation. The effect of distance and/or orientation between His322 and Glu325 was investigated by interchanging Glu325 with Val326, thereby moving the carboxylate one residue around putative helix X. The resulting permease molecule is also completely inactive, and control mutations indicate that a Glu residue at position 326 inactivates the permease. The wild-type orientation between His and Glu was then restored by further mutation to introduce a His residue into position 323 or by interchanging Met323 with His322. The resulting permease molecules contain the wild-type His/Glu orientation, but the putative His/Glu ion-pair is rotated about the helical axis by 100° relative to Arg302 in putative helix IX. Both mutants are inactive for all modes of translocation. The results provide support for the contention that the polarity between His322 and Glu325 and the geometric relation between Arg302, His322 and Glu325 are critical for permease activity. In addition, they suggest that perturbation of the putative His322–Glu325 ion pair alone is insufficient to account for inactivation (i.e. Glu322–His325 should remain ion-paired) and are consistent with the proposed role of His322 and Glu325 as components of a H^+ relay.

Previous studies[30] suggested that replacement of Lys319, which should be on the same face of putative helix X as His322 and Glu325, with Leu has no effect on permease activity; however, this conclusion has been found recently to be incorrect (P. D. Roepe, B. Persson, L. Patel, J. Lee and H. R. Kaback, unpublished). Rather, Leu319 permease is defective in lactose/H^+ symport. The mutant is unable to catalyse lactose accumulation or efflux, but it exhibits downhill lactose influx and catalyses equilibrium exchange at about half the rate of wild-type permease. Unlike the Glu325 mutants, which catalyse exchange at least as well as wild-type permease, Leu319 permease binds NPG poorly (see also Ref. 31) and is defective in counterflow activity. The results suggest that Lys319 is important for substrate recognition, as well as lactose-coupled H^+ translocation. It is also noteworthy that the double mutant Leu319–Ala325 catalyses equilibrium exchange.

Although the notion that a type of charge-relay mechanism may be involved in lactose/H^+ symport is consistent with the results described above, recent experiments question the notion that His322 is obligatory for lactose-coupled H^+ translocation. From studies on permease mutants with Tyr or Phe in place of His322, King and Wilson have concluded: that sugar-dependent H^+ transport is observed, albeit with low efficiency[34,35]; that melibiose efflux, in particular, remains coupled to H^+ translocation in mutants; and that reactions involving exchange are limiting for lactose but not melibiose efflux suggesting that slow exchange is substrate-specific in the His322 mutants[36] (*lac* permease will transport the α-galactosidase melibiose in addition to lactose). In addition, Brooker[37] has shown that permease with Val in place of Ala177 *and* Asn in place of His322 catalyses lactose-dependent H^+ influx with a stoichiometry close to unity. Taken at face value, the results are difficult to reconcile with the contention that His322 is obligatory for lactose-driven H^+ translocation. However, it should be emphasized that all of the His322 mutants isolated thus far are completely defective with regard to sugar accumulation. Thus, whatever its precise role in the mechanism, a His residue at position 322 appears to be absolutely required for accumulation against a concentration gradient.

On a broader level, H^+- and Na^+-coupled symport are conceptually and thermodynamically analogous, but it is unclear whether the two types of transport occur by the same general mechanism. Another cotransporter, melibiose (*mel*) permease uses H^+, Na^+ or Li^+ as the coupling cation depending on the sugar transported, and like *lac* permease, is inactivated by diethylpyrocarbonate. Pourcher *et al.*[38] replaced each of the seven His residues in *mel* permease with Arg and demonstrated that His94 alone is important for transport and binding. Thus, *mel* permease and *lac* permease require a single specific His residue for activity, although the residues are located at opposite ends of the respective molecules. Importantly, however, *mel* permease with His94 replaced with Arg does not exhibit an uncoupled phenotype, and the defect observed may be due specifically to a loss in ability to bind substrate. As outlined above, much of the data with *lac* permease is consistent with the idea that a charge-relay mechanism may be operative. On the other hand, it is not obvious how the same mechanism can be directly involved in Na^+ symport unless N or O atoms that can coordinate with Na^+ are present in the pathway in addition to the minimal structures necessary for a charge-relay or a 'H^+-wire' of the type suggested by Onsager[39]. Despite certain marked kinetic differences between H^+- and Na^+-coupled sugar translocation, a few lines of evidence suggest that substrate-coupled Na^+ and H^+ translocation may occur by similar mechanisms: (1) NPG binding studies suggest that H^+ and Na^+ compete for a common binding site on *mel* permease; (2) Na^+ is a competitive inhibitor of melibiose-coupled H^+ translocation via *mel* permease; and (3) certain mutations in *melB* have been isolated and characterized that alter the cation specificity of *mel* permease. Boyer[40] suggested recently that H_3O^+, rather than H^+, may be the symported species. Appropriately placed N or O atoms in symporters like *lac* or *mel* permease could provide cation-binding domains akin to those in the crown ethers or cryptates, both of which form coordination complexes with Na^+ and H_3O^+. In this context, however, some of the translocation reactions catalysed by *lac* permease exhibit a significant D_2O effect (see Ref. 19) which is not expected if coordination with H_3O^+ is the rate-limiting step in translocation. Thus, although the contrast between H^+- and Na^+-coupled symport is of singular importance, the mechanistic relationship between the two is presently unclear.

Summary

In summary, the *lac* permease, a prototypic membrane transport protein from the cytoplasmic membrane of *E. coli*, has been solubilized, purified to homogeneity, reconstituted into proteoliposomes and shown to be fully functional as a monomer. A secondary-structure model for the protein containing 12 hydrophobic transmembrane domains in α-helical conformation has been presented, and strong exclusive support for the topological predictions of the model has been obtained by analysis of a series of *lac* permease–alkaline phosphatase fusion proteins. When fragments encoding the N- and C-terminal halves of the permease are subcloned and expressed from the same plasmid, functional permease is obtained and polypeptides corresponding to the N- and C-terminal halves of the protein are observed in the membrane. Site-directed mutagenesis is being used to delineate amino acid residues that are important for substrate recognition and/or coupling between lactose and H^+ translocation. Moreover, studies using the technique have led to the suggestion that Arg302 (and possibly Lys319), His322 and Glu325 are sufficiently close to hydrogen-bond and may be involved in a type of charge-relay mechanism. However, it is clear that in order to define the mechanism of lactose/H^+ symport, high-resolution structural information is essential.

References

1 Kaback, H. R. (1983) *J. Membr. Biol.* 76, 95–112
2 Kaback, H. R. (1986) in *Physiology of Membrane Disorders* (Andreoli, T. E., Hoffman, J. F., Fanestil, D. D. and Schultz, S. G., eds), pp. 387–408, Plenum Press
3 Wright, J. K., Seckler, R. and Overath, P. (1986) *Annu. Rev. Biochem.* 55, 225–248
4 Deisenhofer, J. and Michel, H. (1989) *Science* 245, 1463–1473
5 Foster, D. L., Boublik, M. and Kaback, H. R. (1983) *J. Biol. Chem.* 258, 31–34
6 Vogel, H., Wright, J. K. and Jähnig, F. (1985) *EMBO J.* 4, 3625–3631
7 Carrasco, N., Tahara, S. M., Patel, L., Goldkorn, T. and Kaback, H. R. (1982) *Proc. Natl Acad. Sci. USA* 79, 6894–6898
8 Carrasco, N., Viitanen, P., Herzlinger, D. and Kaback, H. R. (1984) *Biochemistry* 23, 3681–3687
9 Carrasco, N., Herzlinger, D., Mitchell, R., DeChiara, S., Danho, W., Gabriel, T. F. and Kaback, H. R. (1984) *Proc. Natl Acad. Sci. USA* 81, 4672–4676
10 Seckler, R., Wright, J. K. and Overath, P. (1983) *J. Biol. Chem.* 258, 10817–10820
11 Seckler, R. and Wright, J. K. (1984) *Eur. J. Biochem.* 142, 269–279
12 Seckler, R., Möröy, T., Wright, J. K. and Overpath, P. (1986) *Biochemistry* 25, 2403–2409
13 Herzlinger, D., Viitanen, P., Carrasco, N. and Kaback, H. R. (1984) *Biochemistry* 23, 3688–3693
14 Goldkorn, T., Rimon, G. and Kaback, H. R. (1983) *Proc. Natl Acad. Sci. USA* 80, 3322–3326
15 Stochaj, V., Bieseler, B. and Ehring, R. (1986) *Eur. J. Biochem.* 158, 423–428
16 Page, M. G. P. and Rosenbusch, J. P. (1988) *J. Biol. Chem.* 263, 15906–15914
17 Calamia, J. and Manoil, C. *Proc. Natl Acad. Sci. USA* (in press)
18 Kaback, H. R. (1989) *Harvey Lectures* 83, 77–103
19 Costello, M. J., Escaig, J., Matsushita, K., Viitanen, P. V., Menick, D. R. and Kaback, H. R. (1987) *J. Biol. Chem.* 262, 17072–17082
20 Li, J. and Tooth, P. (1987) *Biochemistry* 26, 4816–4823
21 Bibi, E. and Kaback, H. R. *Proc. Natl Acad. Sci. USA* (in press)
22 Roepe, P. D. and Kaback, H. R. *CRC Crit. Rev. Biochem.* (in press)
23 Padan, E., Sarkar, H. K., Viitanen, P. V., Poonian, M. S. and Kaback, H. R. (1985) *Proc. Natl Acad. Sci. USA* 82, 6765–6768
24 Püttner, I. B., Sarkar, H. K., Poonian, M. S. and Kaback, H. R. (1986) *Biochemistry* 25, 4483–4485
25 Püttner, I. B., Sarkar, H. K., Padan, E., Lolkema, J. S. and Kaback, H. R. (1989) *Biochemistry* 28, 2525–2533
26 Püttner, I. B. and Kaback, H. R. (1988) *Proc. Natl Acad. Sci. USA* 85, 1467–1471
27 Blow, D. M., Birktoft, J. J. and Hartley, B. S. (1969) *Nature* 221, 337–340
28 Carrasco, N., Antes, L. M., Poonian, M. S. and Kaback, H. R. (1986) *Biochemistry* 25, 4486–4488
29 Carrasco, N., Püttner, I. B., Antes, L. M., Lee, J. A., Larigan, J. D., Lolkema, J. S. and Kaback, H. R. (1989) *Biochemistry* 28, 2533–2539
30 Menick, D. R., Carrasco, N., Antes, L., Patel, L. and Kaback, H. R. (1987) *Biochemistry* 26, 6638–6644
31 Collins, J. C., Permuth, S. F. and Brooker, R. J. (1989) *J. Biol. Chem.* 264, 14698–14703
32 Franco, P. J., Eelkema, J. A. and Brooker, R. J. (1989) *J. Biol. Chem.* 264, 15988–15992

33 Lee, J. A., Püttner, I. B., Carrasco, N., Antes, L. M. and Kaback, H. R. (1989) *Biochemistry* 28, 2540–2544
34 King, S. C. and Wilson, T. H. (1989) *J. Biol. Chem.* 264, 7390–7394
35 King, S. C. and Wilson, T. H. (1989) *Biochim. Biophys. Acta* 982, 253–264
36 King, S. C. and Wilson, T. H. (1990) *J. Biol. Chem.* 265, 3153–3160
37 Brooker, R. J. (1990) *J. Biol. Chem.* 265, 4155–4160
38 Pourcher, T., Sarkar, H. K., Bassilana, M., Kaback, H. R. and Leblanc, G. (1990) *Proc. Natl Acad. Sci. USA* 87, 468–472
39 Onsager, L. (1969) *Science* 166, 1359–1364
40 Boyer, P. D. (1988) *Trends Biochem. Sci.* 13, 5–7

H. R. Kaback, E. Bibi and P. D. Roepe are at the Howard Hughes Medical Institute, Molecular Biology Institute and Department of Physiology, University of California Los Angeles, Los Angeles, CA 90024-1570, USA.

SIMILAR AMINO ACID SEQUENCES REVISITED

RUSSELL F. DOOLITTLE

The rapid accumulation of protein sequences, many bearing unexpected resemblances to each other, is providing a new perspective on evolution.

Some time ago, I wrote an article[1] with the alliterative title 'Similar Amino Acid Sequences: Chance, Convergence or Common Ancestry?' (Unhappily, *Science* magazine headers could not accommodate so many characters, and 'Convergence' was dropped from the title.) The main message of the paper was that, although one must always be cautious about low-level resemblances, a surprisingly high fraction of existing protein sequences seemed similar enough to other protein sequences that common ancestry could be inferred. Convergence, where similar sequences are the result of selection for particular structures, was thought to be rare, mostly because there are so many different combinations that can yield equivalent structures. Further, the point was made that, as more sequences were collected, the fraction belonging to definable families would rise, and eventually it should prove possible to count the root stocks of most contemporary proteins.

In the interval since that article was written the number of known protein sequences has grown almost tenfold, and we might inquire whether the trend has continued and whether any of the root stocks have been identified. Certainly the number of unexpected homologies uncovered by data base searching has reached the point where the sequence community takes 'unexpected match-ups' in stride; indeed, many investigators are truly disappointed if a search of their newly determined sequence fails to uncover some exotic relative.

In retrospect, we probably shouldn't have been so surprised that so many similarities were being found. All living organisms must trace back to a common ancestor, and it is reasonable to think that some very early ancestor had a relatively small genome coding for a relatively small inventory of prototypic proteins. Most contemporary gene products are the result of past gene duplications and subsequent divergence resulting from gradual amino acid replacement. As a result, many proteins have already been grouped into families. There are sequences for scores of each of four major protease families and correspondingly large numbers of protease inhibitors. There are vast numbers of protein kinases, all apparently descended from a common ancestor[2], and we can anticipate a similar multitude of protein phosphatases.

Hindsight not withstanding, some of the unanticipated sequence resemblances are nothing less than awe-inspiring. Consider, for example, the remarkable family of G protein-linked receptor proteins. When the β_2-adrenergic receptor was cloned and sequenced, it was found to be approximately 25% identical to bovine rhodopsin[3]. This is a degree of resemblance that leaves little doubt of common ancestry, even ignoring the fact that both proteins are transmembrane signaling devices that interact intracellularly with G proteins (transducins).

Since that initial finding, more than a dozen other receptor proteins have been found to have related sequences, including a wide assortment of receptors for biological amines[4–8], biologically active peptides[9–11] and, in the slime mold, cyclic AMP[12]! Moreover, there is good reason to think that in vertebrates the primary odor detection system involves receptors belonging to this family[13], and family membership may eventually be found to number in the thousands. Already its members encompass animals and fungi, and it will be very interesting to find whether they exist in plants.

And what about bacteriorhodopsin from halobacter, which is well known to bind the same ligand (retinal) and to respond to the same isomerization event as animal rhodopsins? Is this a functional convergence, or should we presume common ancestry? Halobacter are members of the kingdom archaebacter, and in that setting the role of rhodopsin is the light-pumped extrusion of protons. Coincidentally, bacteriorhodopsin resides in the cytoplasmic membrane and possesses seven transmembrane segments[14]. Early attempts at sequence alignment with animal rhodopsins were equivocal in that the bacteriorhodopsin sequence at 248 amino acids is about a hundred residues shorter than the animal equivalents. Even without the complication of differences in length, the average rate of change of these proteins is such that statistical measures of significance for the other seven-spanner proteins are strained by fungal–animal comparisons, and it may be that the eukaryote–archaebacter distance is beyond our provable measure for this particular protein. Nonetheless, simultaneous alignment of the various visual pigments and pharmacologic receptors with bacteriorhodopsin by a progressive alignment procedure[15] has convinced me that bacteriorhodopsin does indeed share common ancestry with this group. Ancillary considerations[14] like the binding of retinal and the presence of seven membrane-spanning segments reinforce the point. The shorter length is consistent with bacteriorhodopsin not interacting with an intracellular transducin.

A number of other protein families have emerged as a result of computer searching in the last several years that are deserving of special comment. The serpin family is a group of proteins sharing common ancestry, many of which have roles other than the inhibition of serine proteases, the function for which the group was named originally[16]. Thus, it was found[17] that ovalbumin, a common eggwhite protein of no known function, is homologous with α_1 antitrypsin and antithrombin 3, both known to be protease inhibitors. Subsequently, it was found[18] that the hormone precursor angiotensinogen is also a member of this family, and after that an assortment of other family members emerged, protease inhibitors and non-inhibitors alike. Representatives have been found in many animals and at least one plant (Table I).

Table I

Some heterofunctional protein families uncovered by data base searching

Serpins[a]		Lipocalins[b]	β-2 Glycoprotein-type[c]	EGF-containing[d] proteins	
α_1-antitrypsin	leuserpin-2	retinol-binder	cartilage proteoglycan	thyroid peroxidase	LDL receptor
α_1-antichymotrypsin	ovalbumin	purpurin	complement B reg	complement C9	sea urchin reg protein
α_2-antiplasmin	angiotensinogen	complement C8, gamma	complement C2	factor VII	factor IX
antithrombin 3	thyroxine-binder	α_1-microglobulin	complement C4-binder	protein S	urokinase
protein C inhibitor	corticosteroid-binder	β-lactoglobulin	factor XIIIb	thrombomodulin	*Drosophila* notch
monocyte PA inhibitor	endosperm protein Z	orosomucoid	complement factor H	cartil matrix protein	vaccinia virus
C1 inhibitor			decay accelerating factor	factor X	lin 12, *C. elegans*

[a]Serpins are named thus because several are known to be serine protease inhibitors.
[b]Lipocalins bind lipophilic agents.
[c]The β-2 glycoprotein unit is about 60 residues long and contains a disulfide bond.
[d]The EGF-unit, first observed as epidermal growth factor, consists of 40–45 residues with three disulfide bonds.

Then there are the lipocalins, a family of proteins with an affinity for nonpolar lipophilic substances. Several of these sequences have been known for some time, but the resemblances among them were not so obvious that anyone grouped them until Pervaiz and Brew[19], searching the sequence of a retinol-binding protein, pulled the far-flung family together (Table I). A similar situation unfolded for the immunoglobulin family, membership in which was originally restricted to antibodies, but which is now known to comprise a host of extracellular and cell surface proteins[20–21].

My earlier article[1] also considered the problem of identifying hybrid or mosaic proteins of mixed ancestry, the concern being that shuffled segments would be difficult to pin down because of the limiting statistics of small numbers. As it happens, this is an area that has exploded in the last few years, and examples abound, certain sequence motifs appearing over and over again in diverse proteins. Many have been recognized because they contain disulfide bridge-involved cysteines that are conserved even when most of the surrounding residues have been replaced[22].

The epidermal growth factor unit is a good example (Table I). More often than not these 40–45 residue segments are set off in the genome by introns, their diaspora in so many proteins being referred to as 'exon shuffling'[23]. Several other definable units are used over and again in various settings, including the 'kringle' structures of blood coagulation factors and zinc fingers in many DNA-binding proteins.

The elucidation of 'new' families from already known sequences has not been limited to eukaryotes. In this regard, a systematic consensus search of a data base resulted in the uncovering of a large family of bacterial activator proteins[24], and a similar study gathered in an unsuspected clutch of bacterial signaling proteins[25]. As in the cases of the other heterofunctional families mentioned above, the data are most readily interpreted as divergence from common ancestors rather than convergence to similar sequences.

The buzz-word that is emerging more and more in the sequence comparison field is 'redundancy'. This is a happy situation in that redundancy is what allows the reconstruction of past events. Common ancestry is the essence of evolution, and nowhere is Darwin's notion of 'descent with modification' more apparent than in the amino acid sequences unravelling before us.

References

1 Doolittle, R. F. (1981) *Science* 214, 149–159
2 Hanks, S. K., Quinn, A. M. and Hunter, T. (1988) *Science* 241, 42–52
3 Dixon, R. A. F., Kobilka, B. K., Strader, D. J., Benovic, J. L., Dohlman, H. G., Frielle, T., Bolanowski, M. A., Bennett, C. D., Rands, E., Diehl, R. E., Mumford, R. A., Slater, E. E., Sigal, I. S., Caron, M. G., Lefkowitz, R. J. and Strader, C. D. (1986) *Nature* 321, 75–79
4 Bunzow, J. R., Van Tol, H. H. M., Grandy, D. K., Salon, J., Christie, M., Machida, C. A., Neve, K. A. and Civelli, O. (1988) *Nature* 336, 783–787
5 Julius, D., MacDermott, A. B., Axel, R. and Jessell, T. M. (1988) *Science* 241, 558–564
6 Kobilka, B. K., Matsui, H., Kobilka, T. S., Yang-Feng, T. L., Francke, U., Caron, M. G., Lefkowitz, R. J. and Regan, J. W. (1987) *Science* 238, 650–656
7 Kubo, T., Fukuda, K., Mikami, A., Maeda, A., Takahashi, H., Mishina, M., Haga, T., Haga, K., Ichiyama, A., Kangawa, K., Kojima, M., Matsuo, H., Hirose, T. and Numa, S. (1986) *Nature* 323, 411–416
8 Gocayne, J., Robinson, D. A., Fitzgerald, M. G., Chung, F-Z., Kervalage, A. R., Lentes, K-U., Lai, J., Wang, C-D., Fraser, C. M. and Venter, J. C. (1987) *Proc. Natl Acad. Sci. USA* 84, 8296–8300
9 Masu, Y., Nakayama, K., Tamaki, H., Harada, Y., Kuno, M. and Nakanishi, S. (1987) *Nature* 329, 836–838
10 March, L. and Herskowitz, I. (1988) *Proc. Natl Acad. Sci. USA* 85, 3855–3859
11 Burkholder, A. C. and Hartwell, L. H. (1985) *Nucleic Acids Res.* 13, 8463–8475
12 Klein, P. S., Sun, T. J., Saxe, C. L., III, Kimmel, A. R., Johnson, R. L. and Devreotes, P. N. (1988) *Science* 241, 1467–1472
13 Lancet, D. and Pace, U. (1987) *Trends Biochem. Sci.* 12, 63–66
14 Ovchinnikov, Y. A. (1982) *FEBS Lett.* 148, 179–191
15 Feng, D-F. and Doolittle, R. F. (1987) *J. Mol. Evol.* 25, 351–360
16 Carrell, R. and Travis, T. (1985) *Trends Biochem. Sci.* 10, 20–24
17 Hunt, L. T. and Dayhoff, M. O. (1980) *Biochem. Biophys. Res. Commun.* 95, 864–871
18 Doolittle, R. F. (1983) *Science* 222, 417–419
19 Pervaiz, S. and Brew, K. (1987) *FASEB J.* 1, 209–214
20 Williams, A. F. and Barclay, A. N. (1988) *Annu. Rev. Immunol.* 6, 381–405
21 Hayashida, H., Kuma, K-I. and Miyata, T. (1988) *Proc. Japan Acad. Ser. B* 64, 113–118
22 Doolittle, R. F. (1985) *Trends Biochem. Sci.* 10, 233–237
23 Gilbert, W. A. (1985) *Science* 228, 823–824
24 Henikoff, S., Haughn, G. W., Calvo, J. M. and Wallace, J. C. (1988) *Proc. Natl Acad. Sci. USA* 85, 6601–6606
25 Kofoid, E. C. and Parkinson, J. S. (1988) *Proc. Natl Acad. Sci. USA* 85, 4981–4985

R. F. Doolittle is at the Center for Molecular Genetics M-034, University of California, San Diego, La Jolla, CA 92093, USA.

The surprise of protein crystallography in the 1960s was that protein structures could be determined at all. Each new structure was received with amazement for its complexity, and with fascination for its unexpected new features. In the 1970s the surprise was the variety of structures found, although the majority seemed to cluster into a few general types. In the 1980s the surprise is that protein crystallography has become an indispensible everyday tool of protein chemistry, because of the increased speed of determining structures. This mini-review attempts to summarize the state of protein crystallography as the field moves into the 1990s.

PROTEIN CRYSTALLOGRAPHY: MORE SURPRISES AHEAD

DAVID EISENBERG AND CHRISTOPHER P. HILL

As of 1989, over 400 protein structures have been determined, with some 100 solved during the last year. The advances in protein crystallography that have led to this explosion of information are surveyed, and some frontiers of the science are briefly noted.

NUMBER OF STRUCTURES

The January 1989 issue of the *Protein Data Bank Newsletter* lists over 400 sets of protein coordinates, about 100 of which were deposited in 1988. Not all of these are distinctly different structures. For example the holdings include some 16 coordinate sets of mutant T4 lysozymes from Brian Matthews' laboratory and 17 coordinate sets for rhinovirus complexed with different antiviral agents from Rossmann's lab. While it is not all that meaningful to define *distinctly different* structures, there seem to be about 120 different protein folds among the 400 deposited structures. There are also a significant number of structures in industrial labs and elsewhere that have been determined but not yet deposited. We hope these will join the other coordinates in the data bank, because the value of the database increases as it grows.

STAGES OF PROTEIN STRUCTURE DETERMINATION

What accounts for the accelerating pace of structure determination? To get some idea of the causes of this explosion of information, we survey the steps in the determination of a protein structure, with emphasis on recent developments.

To determine a protein structure, you must pass through five basic stages (see Fig. 1). Detailed descriptions of these steps are given by dozens of authors in two recent volumes of *Methods of Enzymology*[1], which are strongly recommended. Older reviews are still useful for a broad overview[2,3].

GROWING CRYSTALS

As in other aspects of life, the first step may be the hardest: in this

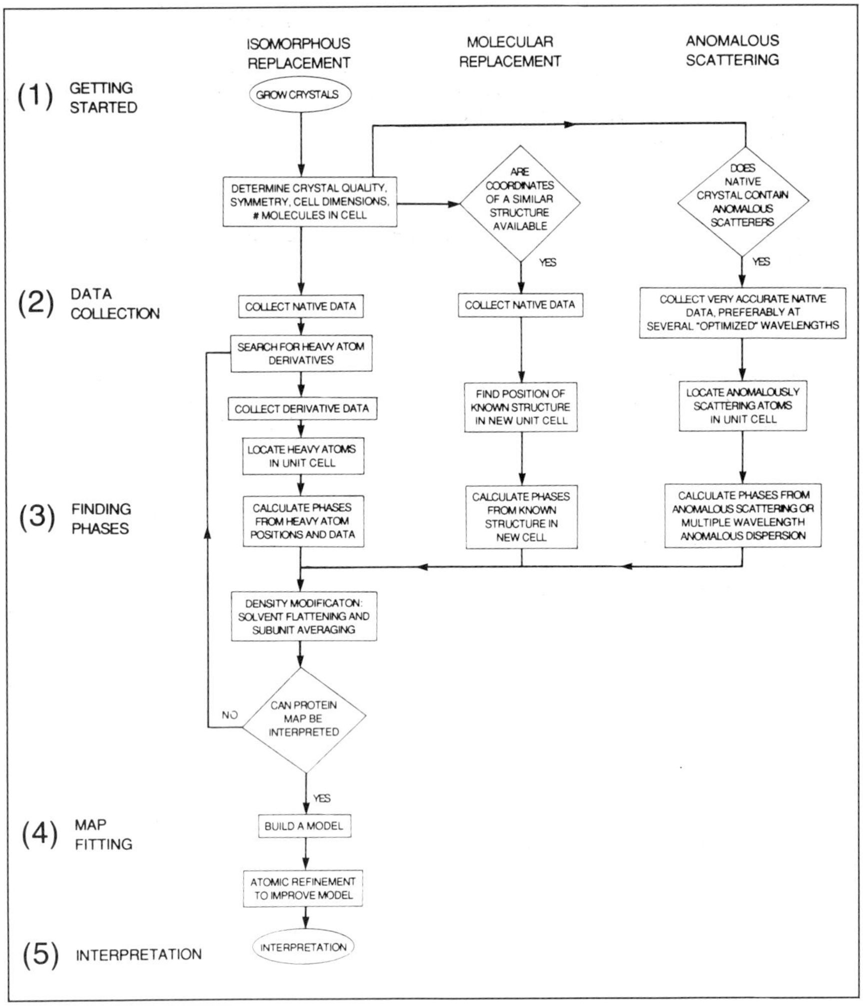

FIGURE 1

The five basic stages in the determination of a protein structure. (1) Growing and characterizing the crystals. (2) Collecting the X-ray data, that is, the intensities of the diffraction spots. (3) Getting over the phase problem, the annoyance that intensities are only half the quantities needed to calculate the electron density, the picture of the protein. Missing are the phases, one for each diffraction spot, which cannot yet be measured directly. (4) Building a model of the protein into the electron density using computer graphics. Then, improving the model by atomic refinement, in which atomic positions are adjusted to minimize the discrepancy between observed intensities and those calculated from the model. (5) Interpreting the final model.

case, you must grow X-ray grade crystals. These must be several tenths of a millimetre in each dimension, and need to be well enough ordered to give good X-ray diffraction spots. The basic difficulty is that good crystals grow from protein solutions under a narrow range of conditions. Many variables must be sampled to find these conditions, among them: protein concentration, pH, concentration of precipitant (usually ammonium sulfate or some other salt, polyethylene glycol, or an alcohol), added ions or effectors and their concentrations, temperature, and the method and kinetics of variation of concentrations or temperature. In short, you must sample a multidimensional space starting with an amount of protein that is often too small for a comprehensive search. Suggestions for approaching this problem have been put forward[4,5].

Some crystallization methods are depicted in Fig. 2. The hanging drop is the most widely used, because many conditions can be sampled with relatively little protein (some 50–200 crystallizing conditions can be sampled with 5 mg of protein). Both hanging and sitting drop methods can be automated[6,7], and it seems likely that such automated crystal searches will account for the next increase in speed of protein structure determination.

Whatever the method chosen, your chances of growing good crystals depends more on the purity and homogeneity of the protein than any other factor. If another protein is present, it may precipitate first, forming thousands of heterogeneous nuclei for your protein to grow on when its solubility is reduced, thus producing microscopic crystals. The same result will be obtained if your protein is partially denatured or in multiple forms,

METHOD		COMMENT	ADVANTAGES	DRAWBACKS
Hanging Drop	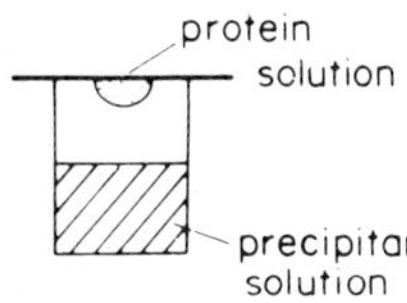	As drop and reservoir equilibrate by vapor diffusion, concentration of precipitant and protein increase in drop.	Small quantities of protein are required. (5 x mg for 50-200 drops). Rapid survey of possible conditions.	Crystals often small. Not as repeatable as other methods.
Sitting Drop	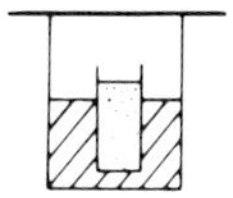	Used for drops of greater than 20μl volume.	Larger crystals possible.	Requires more protein.
Dialysis	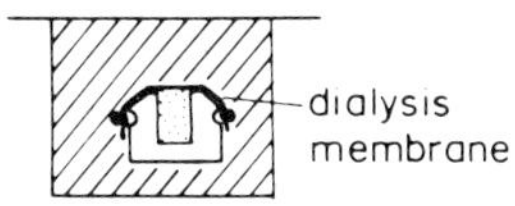	Precipitant diffuses across membrane into protein solution.	Large crystals possible.	Requires more protein and dexterity than hanging drops.
Seeding	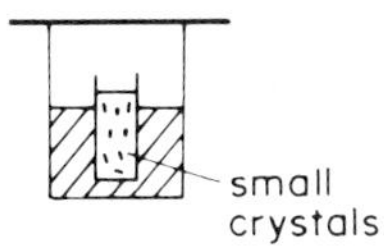	Small or microscopic crystal introduced into protein solution. Can be used with any of the above techniques, figure shows a sitting drop.	Used for enlarging crystals.	Need a small crystal to start with.
Batch	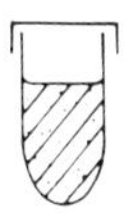	Precipitant is mixed with protein solution and left to stand.	Numerous large crystals possible.	Requires lots of protein.
Free Capillary Diffusion		Protein solution is layered on the precipitant solution. Capillary sealed with wax or by melting glass.	Excellent repeatability. Can be the best method if nucleation requires much higher protein concentration than growth.	Curvature of capillary makes it difficult to recognize microcrystals.
Temperature Gradient	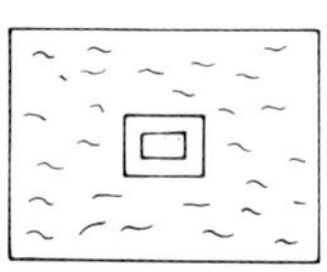	Protein solution slowly cools in a dewar flask packed in a box full of insulation material.	Can work if protein solubility varies with temperature.	Requires lots of protein. Takes time to set up.

FIGURE 2

Some of the methods used for the crystallization of a protein.

because it can itself provide many nuclei. Since the process of crystallization involves placing small protein samples in contact with surfaces, there are many opportunities for denaturation of the protein as you are searching for crystals. Addition of a mild detergent[8] may help to avoid some surface denaturation. If, on the other hand, your protein does not form nuclei easily, larger crystals may be obtained by introducing very small crystals into fresh crystallizing solutions as seeds for further growth[9].

TABLE I
Data collection

Method	Comment	Advantages	Drawbacks
Film:			
Precession	Crystal and film move. A screen allows only certain reflections to be recorded.	Gives an undistorted view of the reciprocal lattice. Good for finding space group, unit cell dimensions and screening for heavy atom derivatives.	No longer used for full data collection because it takes too many crystals and too much time.
Rotation	With the film stationary the crystal is rotated in small (0.5–3.0°) increments.	Efficient because many spots are recorded on each film. Film has high spatial resolution.	Film has high intrinsic noise; it must be developed and densitometered.
Laue	Polychromatic X rays with stationary crystal.	Most of the data collected from one crystal very quickly.	In developmental stage; requires a synchrotron.
Counter methods:			
Diffractometer	Crystal and detector move; one diffraction spot is recorded at a time.	Very accurate data possible, good for measuring anomalous differences.	Inefficient for large unit cells, including all but the smallest proteins.
Area detector	Commercial models use either multiwire proportional counter or television technology.	Combines efficiency of rotation geometry with accuracy of an electronic detector. Usually the method of choice.	Limited spatial resolution is a problem for very large unit cells.

CHARACTERIZATION AND DATA COLLECTION

Assuming that you have been skilful and lucky enough to grow large crystals, your next step is to characterize the diffraction pattern of the crystal. This will tell you the resolution of the electron density that you can hope to determine, as well as the size and symmetry of the unit cell of the crystal. This information is necessary to select the method of data collection (Table I). Small unit cells (i.e. < 5 nm on each edge) are often best suited for a diffractometer. Unit cells with lengths 5–20 nm are often best suited for collection on an area detector, because there are many more diffraction spots than for the small unit cell, and many can be collected at once on the area detector. For even larger unit cells, such as some viruses, the high flux and collimation of synchrotron radiation may be desirable, using film as the detector because of its greater ability to separate the closely spaced diffraction spots. The availability of synchrotron X-ray sources and of commercial area detectors have greatly speeded X-ray data collection. Developments in hardware and software are continuing and within the next decade we can hope for a new generation of detectors which will combine high speed with increased spatial resolution and lower background.

PHASE DETERMINATION

Having collected the X-ray intensities, you still have to defeat the phase problem (Table II). Max Perutz' discovery in the 1950s that heavy atoms bound within protein crystals can supply the phases[10] remains the main route to solving new structures. But this *isomorphous replace-*

ment method is increasingly supplemented by other techniques. The development of new methods is another of the main reasons that structure determination is faster than it used to be, because they avoid the tedious search for the heavy atom adducts needed for isomorphous replacement.

When a similar protein structure is already known, *molecular replacement* techniques can often locate the molecule in the new unit cell and thus lead to the phases for the new structure[13]. Or, if the structure contains a heavier atom, phases may be estimated from *anomalous scattering*. This method requires especially accurate measurement of the small differences in intensity (typically 1–8%) between diffracted X rays which have been reflected from the front and the back of the same set of Bragg planes. There are two variations: Hendrickson's method[16] of resolved anomalous scattering requires very accurate intensity measurements at one wavelength; the MAD method[17] requires intensity measurements at several wavelengths. Synchrotrons yield high intensity X rays of variable wavelength. Thus as synchrotron X-ray diffraction becomes increasingly available, the phase problem may be effectively defeated by anomalous scattering measurements.

Table II
Methods of determining phases

Method	Requirements	Comment	Refs
Multiple isomorphous replacement (MIR)	At least two heavy atom derivatives.	The most general method.	10
Single isomorphous replacement (SIR)	One very good heavy atom derivative.	Requires additional phase information, e.g. from solvent flattening, subunit averaging or anomalous scattering.	11, 12
Molecular replacement	A known similar structure, whose coordinates are used to position molecule in unit cell and to calculate phases.	Eliminates need for heavy atom derivatives.	13
Non-crystallographic symmetry averaging	Several copies of protein molecule in asymmetric unit.	Yields excellent phases when many subunits can be averaged, as in some viruses.	14, 15
Resolved anomalous scattering	Very accurate data. An anomalously scattering group in native crystal.	Requires additional phase information, e.g. from model-building or density modification.	16
Multiple-wavelength anomalous dispersion (MAD)	An anomalously scattering group in native crystal. A tunable X-ray source (i.e. a synchrotron).	Could become the standard method for metalloproteins.	17

Initial phases from isomorphous replacement and the other methods are often poor. These give degraded electron density that is sometimes too blurred even to trace the polypeptide chain. To improve the phases, a variety of methods have been devised. Some, involving general chemical assumptions, are collected under the term *density modification*. *Solvent flattening*[12] means applying the reasonable assumption that the large solvent regions around the protein have a uniform electron density. *Subunit averaging*[14,15] means applying the assumption that chemically equivalent subunits have the same structure. Both of these assumptions can improve the initial phases to the point that you can trace the polypeptide chain in the electron density and can identify the side chains. Subunit averaging increases in power as the number of identical protein chains increases.

ATOMIC REFINEMENT

Once a polypeptide model has been built with the aid of computer graphics[18,19], the phases can again be improved by refining the model to fit the measured intensities of diffraction, or to fit some combination of the measured intensities and a potential energy function. It is during

TABLE III

Some frontiers of protein crystallography

	Requirements	Potential	Comments	Refs
Time-resolved crystallography	Sub-second data collection, possible only with Laue method using polychromatic synchrotron X rays.	Direct observation of intermediates on a reaction pathway or conformational change.	All molecules in the crystal must change together. Triggering of such concerted changes is a difficult problem.	23, 24
Membrane proteins	Crystal growth remains the major limitation.	Structures are needed to illuminate such events as transport, signaling and energy transduction.	Importance of work in this area was recognized by the 1988 Nobel prize for chemistry, awarded to Deisenhofer, Huber and Michel for the light reaction center structure.	25
Large assemblies	Heterogeneity can make crystals hard to grow. Synchrotron data collection is almost essential. Phase determination may be difficult, except in cases of high non-crystallographic symmetry.	Learning new types of biological organization.	Results to date include the spectacular successes with viruses and heroic efforts with ribosomes.	26
Drug design	Structure of target enzyme should be determined at high resolution with as many inhibitors as possible.	A major practical pay-off of crystallography.	Offers possibility for collaboration of crystallographers with chemists and theorists.	27
Designed peptides	Crystals can be difficult to grow. Tight packing can make isomorphous replacement difficult.	Tests of rules for protein folding and elucidation of small folding motifs.	The peptide may be a fragment of a naturally occurring protein or the result of *de novo* design.	28, 29
Directed mutagenesis, biochemistry and structure	Good collaboration between crystallographers and biochemists. Rapid data collection is useful, because in general many crystals will be studied.	Assessing the role of specific side-chains in structure, catalysis, recognition and stability.	This approach is widely used already and is certain to grow.	
Combined methods	NMR or computational models to aid X-ray structure, and vice versa.	A deeper interplay of methods, yielding energetic and dynamic information as well as structural.	As distinctions among X-ray, NMR and computational models are blurred, care must be taken to catalog what is model and what is known structure.	30

this process of *atomic refinement* that much of the valuable biochemical information is established. During the 1980s refinement methods grew far more powerful, with the introduction of restrained least squares methods[20] and combined X-ray and molecular dynamics techniques[21]. These methods permit more detailed information to be extracted from electron density maps. For example, Smith *et al.*[22] found that in four high resolution structures, 6–13% of the side chains are in multiple, discrete conformations.

FRONTIERS OF PROTEIN CRYSTALLOGRAPHY

Each new protein structure allows us to correlate kinetic, spectroscopic and other observations on that system. Moreover, every new structure also contributes to our understanding of two of the central problems of current biochemistry. One is the protein folding problem; by careful examination of an accurate structure, new information is gained on the relationship of amino acid sequence to three-dimensional structure. Similarly, most protein structures contribute to our understanding of molecular recognition, and the stability of biological complexes, another of the important, general questions of current biochemistry.

These problems are at the frontier of protein crystallography, but there are many other problems being addressed by this technique. A few are listed in Table III. Some involve efforts to use crystallography to get time-dependent pictures of proteins[23,24]. Another area is the study of protein complexes with other cellular components, particularly lipids and nucleic acids. Still other problems of much current research are protein-targeted drug design, designed peptides, directed mutagenesis, and combinations of X-ray crystallography with NMR, computational, and spectroscopic analysis. The next decade is certain to see advances in all of these areas, with protein crystallography playing a central role, increasingly as an everyday tool of the protein laboratory.

Has the excitement gone out of protein crystallography? The first three months of this year (1989) have seen the publication of four new protein folds[31–34] and the first crystal structure of a protein from the AIDS (HIV) virus[35]. These structures suggest there are many surprises ahead.

ACKNOWLEDGEMENTS

The authors gratefully acknowledge support from NIH research grants GM31299 and GM 39558.

REFERENCES

1 Wyckoff, H. W., Hirs, C. H. W. and Timasheff, S. N. (eds) (1985) *Methods Enzymol.* 114–115
2 Blundell, T. L. and Johnson, L. N. (1976) *Protein Crystallography*, Academic Press
3 McPherson, A. (1982) *The Preparation and Analysis of Protein Crystals*, John Wiley & Sons
4 Carter, C. W., Jr and Carter, C. W. (1979) *J. Biol. Chem.* 254, 12219–12223
5 Cox, M. J. and Weber, P. C. (1988) *J. Cryst. Growth* 90, 318–324
6 Kelders, H. A., Kalk, K. H., Gros, P. and Hol, W. G. J. (1987) *Protein Engineering* 1, 301–303
7 Cox, M. J. and Weber, P. C. (1987) *J. Appl. Crystallogr.* 20, 366–373
8 McPherson, A., Koszelak, S., Axelrod, H., Day, J., Williams, R., Robinson, L., McGrath, M. and Cascio, D. (1986) *J. Biol. Chem.* 261, 1969–1975

9 Thaller, C., Eichele, G., Weaver, L. H., Wilson, E., Karlsson, R. and Jansonius, J. N. (1985) *Methods Enzymol.* 114, 132–135
10 Green, D. W., Ingram, V. M. and Perutz, M. F. (1954) *Proc. R. Soc. London Ser. A* 255, 287–307
11 Blow, D. M. and Rossmann, M. G. (1961) *Acta Crystallogr.* 14, 1195–1202
12 Wang, B-C. (1985) *Methods Enzymol.* 115, 90–112
13 Rossmann, M. G. (ed.) (1972) *The Molecular Replacement Method* Gordon & Breach
14 Bricogne, G. (1976) *Acta Crystallogr.* A32, 832–847
15 Rossmann, M. G., Arnold, E., Erickson, J. W., Frankenberger, E. A., Griffith, J. P., Hecht, H-J., Johnson, J. E., Kamer, G., Luo, M., Mosser, A. G., Rueckert, R. R., Sherry, B. and Vriend, G. (1985) *Nature* 317, 145–153
16 Hendrickson, W. A., Smith, J. L. and Sheriff, S. (1985) *Methods Enzymol.* 115, 41–55
17 Guss, J. M., Merritt, E. A., Phizackerley, R. P., Hedman, B., Murata, M., Hodgson, K. O. and Freeman, H. C. (1988) *Science* 241, 806–811
18 Jones, T. A. (1985) *Methods Enzymol.* 115, 157–171
19 Richardson, J. S. and Richardson, D. C. (1985) *Methods Enzymol.* 115, 189–206
20 Hendrickson, W. A. (1985) *Methods Enzymol.* 115, 252–270
21 Brünger, A. T., Kuriyan, J. and Karplus, M. (1987) *Science* 235, 458–460
22 Smith, J. L., Hendrickson, W. A., Honzatko, R. B. and Sheriff, S. (1986) *Biochemistry* 25, 5018–5027
23 Moffat, K., Szebenyi, D. and Bilderback, D. (1984) *Science* 223, 1423–1425
24 Hajdu, J., Machin, P. A., Campbell, J. W., Greenhough, T. J., Clifton, I. J., Zurek, S., Gover, S., Johnson, L. N. and Elder, M. (1987) *Nature* 329, 178–181
25 Deisenhofer, J., Epp, O., Miki, K., Huber, R. and Michel, H. (1985) *Nature* 318, 618–624
26 Harrison, S. C. (1984) *Trends Biochem. Sci.* 9, 345–351
27 Hol, W. G. J. (1986) *Angewandte Chemie* 25, 767–778
28 Landschulz, W. H., Johnson, P. F. and McKnight, S. L. (1988) *Science* 240, 1759–1764
29 Eisenberg, D., Wilcox, W., Eshita, S. M., Pryciak, P. M., Ho, S. P. and DeGrado, W. F. (1986) *Proteins: Structure, Function, Genetics* 1, 16–22
30 Braun, W., Epp, O., Wüthrick, K. and Huber, R. (1989) *J. Mol. Biol.* 206, 669–676
31 Parker, M. W., Pattus, F., Tucker, A. D. and Tsernoglou, D. (1989) *Nature* 337, 93–96
32 Weber, P. C., Ohlendorf, D. H., Wendoloski, J. J. and Salemme, F. R. (1989) *Science* 243, 85–88
33 Ke, H., Thorpe, C. M., Seaton, B. A., Marcus, F. and Lipscomb, W. N. (1989) *Proc. Natl Acad. Sci. USA* 86, 1475–1479
34 Hough, E., Hansen, L. K., Birknes, B., Jynge, K., Hansen, S., Hordvik, A., Little, C., Dodson, E. and Derewenda, Z. (1989) *Nature* 338, 357–360
35 Navia, M. A., Fitzgerald, P. M. D., McKeever, V. M., Leu, C-T., Heimbach, J.C., Herber, W. K., Sigal, I. S., Darke, P. L. and Springer, J. P. (1989) *Nature* 337, 615–620

D. Eisenberg and C. P. Hill are at the Molecular Biology Institute and the Department of Chemistry and Biochemistry, University of California, Los Angeles, CA 90024, USA.

Catalysis in Enzyme Crystals

J. Hajdu, K. R. Acharya, D. I. Stuart, D. Barford and L. N. Johnson

New developments in X-ray crystallography may produce three-dimensional movies of catalysis and structural changes.

Twenty years ago studies on lysozyme[1] (the first enzyme structure to be solved) and its inhibitor complexes led immediately to a proposal for the catalytic mechanism of the enzyme. Today more than 50 different enzyme structures are known. Crystallographic studies, however, have been confined either to unliganded enzymes or to stable enzyme–ligand complexes. Direct observation of conversion of substrate to product (or other transient events) in the crystal has, until recently, remained an unfulfilled dream, largely because data collection times with conventional X-ray techniques are long (days to months) while molecular events relevant to biological processes happen on the picosecond to kilosecond time scale. This problem may be circumvented to some degree by working at very low temperatures in order to freeze out intermediates and match the timescale of the experiment with the timescale of the data collection[2–4]. However, suitable cryoprotectants are not always available and cooling may change the mechanism of the reaction, limiting the application of this approach. The recent development of new techniques at synchrotron radiation sources, particularly the white beam Laue method, allows a dramatic reduction in the time needed for data collection. Transient phenomena (e.g. structural changes, catalysis or phase transitions) down to the millisecond time scale and below may now be investigated by X-ray crystallography in certain cases.

An enzyme crystal is a highly ordered three-dimensional arrangement of protein molecules with a number of well defined and stable contacts. The lattice contacts are similar to, but less extensive than, contacts between subunits in oligomeric proteins and may have a similar effect on protein function in the crystal as subunit–subunit contacts have on oligomeric proteins in solution. The lattice allows for some flexibility and protein molecules may undergo limited conformational changes. Protein crystals contain 20–80% water representing a somewhat more dense environment than the cytosol (70–80% w/w water). Many enzymes retain their activity in the crystal[5]. A three-dimensional network of channels (Fig. 1) provides access for diffusing substrates. The restriction imposed by the lattice on the motion of the protein (or of the substrate in the crystal) may result in reduction or even absence of activity in some cases. A few enzymes are believed to be more active in the crystal than in solution (e.g. papain[5]).

In an X-ray experiment, the resulting image is both a spatial average of all the molecules in the crystal volume irradiated, and a time average of the structures existing during the period of data collection. Consider an

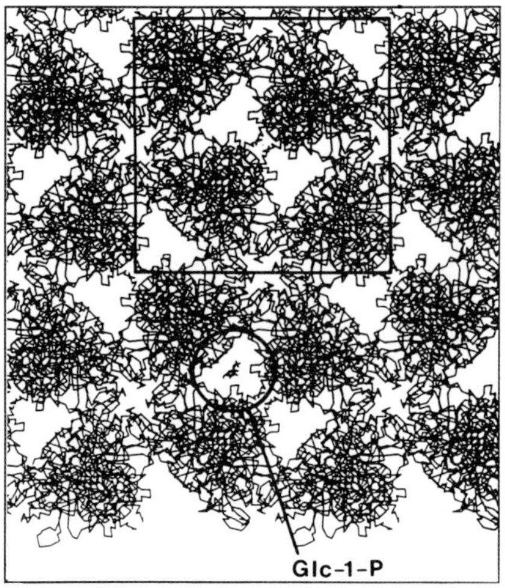

FIGURE 1

The crystal lattice of tetragonal phosphorylase b crystals viewed down the c axis. The protein is shown as an α-carbon trace. A unit cell is marked in the centre of the figure. Glucose 1-phosphate (Glc-1-P), a substrate of the enzyme, is drawn in one of the solvent filled channels which run the whole length of the crystal. *Adapted from a drawing by P. J. McLaughlin.*

enzyme which catalyses a three-step reaction in the crystal. For the sake of simplicity, it is assumed that the reaction is irreversible and that all components bind with the same binding constant to the enzyme in the crystal. The distribution of the various components in the crystal as a function of time is shown in Fig. 2. The ratio of the rate constants k_1, k_2 and k_3 determines the shape of the curves whereas the absolute value of these constants determines the time scale of the reaction. If, as in Fig. 2b, reaction 2 is much slower than the others, then the intermediate produced in the preceding step (compound B) will accumulate transiently and dominate the crystal for some time. With fast data collection, it may be possible to dissect this reaction and obtain a set of structures representing various stages along the reaction coordinate. In favourable cases (e.g. points marked with cross-hatching in Fig. 2) a single dominant structure may emerge. In other regions mixtures of structural states will be observed and it will be necessary to develop techniques that allow deconvolution of the average structure into its components. In certain cases it may be possible to achieve favourable volume averaging by inducing a transient population inversion in the crystal so that almost all molecules adopt the same structural state. Magnetic reorientation (depending on the symmetry of the crystal), lasing, optical pumping, pH jump, temperature jump and photodissociation are some of the possibilities. With enzymatic reactions, varying the temperature is often the easiest way of creating a population inversion since the steps governing the formation and breakdown of intermediates have different activation energies and enthalpies.

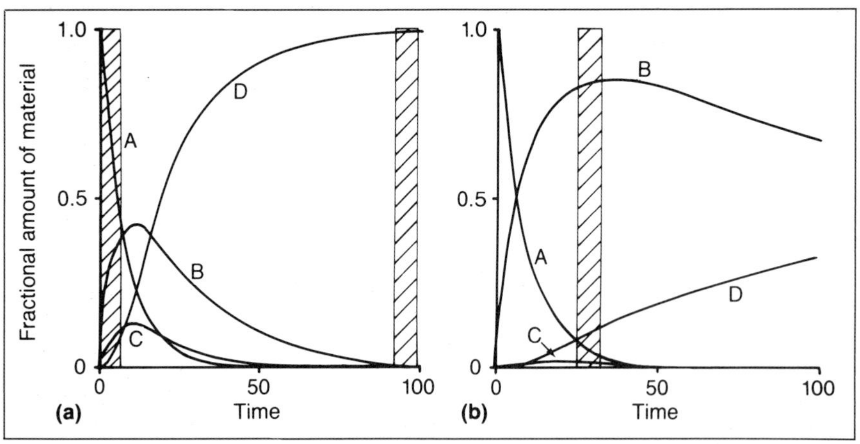

FIGURE 2

Transient accumulation of intermediates in a set of three consecutive reactions:

$$\mathbf{A} \xrightarrow{k_1} \mathbf{B} \xrightarrow{k_2} \mathbf{C} \xrightarrow{k_3} \mathbf{D}$$

(**a**) when $k_1 = k_2 = k_3$, and (***b***) when $k_2 < k_1 = k_3$ (specifically, $k_1 = k_3 = 20\ k_2$)

Hatched areas show times when volume averaging is more or less favourable in the crystal (see text).

THE SYNCHROTRON RADIATION SOURCE

Particle accelerators were originally built for high-energy research into the sub-atomic structure of matter. Around 1970 it was recognized that the electromagnetic radiation produced during the acceleration of charged particles (the so called synchrotron radiation) could be used for biological and material science research. In a given accelerating field, the effect is greater with particles of low mass/charge ratio. The first dedicated synchrotron radiation source (the SRS at Daresbury, UK, 1981) uses electrons. The radiation is confined to a narrow cone around the instantaneous direction of the flight (it is tangential to the curved electron orbit). At the SRS the main storage ring (96 m in circumference) accelerates electrons to 2 GeV energy with a circulating electron current of several hundred milliamperes. Electrons are held in a closed orbit by 16 bending magnets providing an intense and continuous radiation from X rays to radio frequencies. Below about 1 Å wavelength, radiation emission from the main bending magnets falls rapidly. It is possible to introduce much tighter bending on short sections of high field. These more complicated particle orbits are produced by special magnets known as wigglers[6] and undulators[7]. The result is an altered spectral profile shifted into the region of interest for crystallography (0.2–2.5 Å). The radiation is intense, pulsed, well collimated, polarized and tuneable over a wide energy range. It has had a significant impact on research in protein crystallography (see for example, Ref. 8). The advent of high-intensity synchrotron radiation sources had made the real time study of structural changes possible[9]. Early results include the kinetic study of muscle[10,11], of microtubule polymerization[12] and real time studies of phase transitions in phosphatidylethanolamine[13]. Application of time resolved studies to three-dimensional structures is now underway. Data may be recorded in two different ways: one method uses a single wavelength (monochromatic techniques) and the other a much broader spectral range (Laue or white X-radiation techniques).

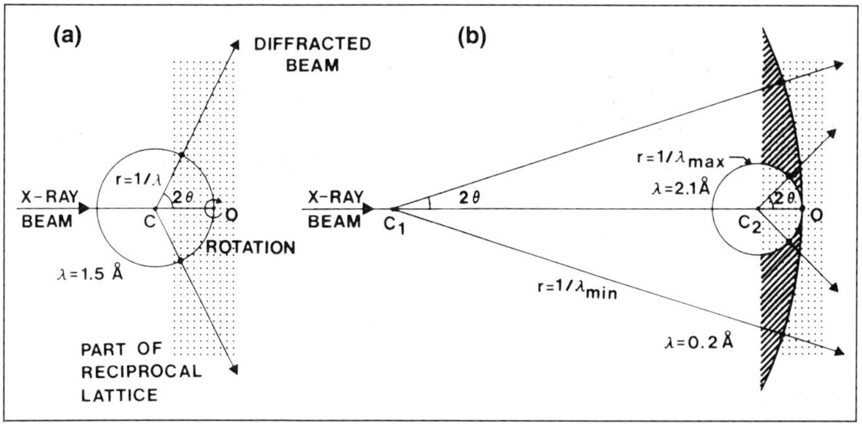

Figure 3

Diffraction geometry for monochromatic and Laue methods. The diagram shows the Ewald construction, a simple geometric construction which encompasses Bragg's law and which is a useful aid in the description of diffraction (see Ref. 31). A sphere is drawn with centre C and a radius equivalent to the reciprocal of the wavelength. The reciprocal lattice is placed with origin O at the point where the straight through the beam cuts the sphere. Diffraction occurs whenever a reciprocal lattice point cuts the sphere. (**a**) **Monochromatic method:** With a single wavelength (e.g. 1.5 Å) only a few reciprocal lattice points cut the sphere and there are only a few diffracted rays recorded on the film. The reciprocal lattice must be rotated (achieved by rotating the crystal) in order to bring other spots into the diffracting position. (**b**) **Laue method:** The reciprocal lattice is illuminated with a spectrum of X rays with wavelengths varying from 0.2 Å to 2.1 Å. All reciprocal lattice points between these limits (shaded region in the diagram) will lie on an appropriate sphere and will diffract. The unit cell size in the figure corresponds to 12.5 Å. Typical unit cell sizes are 100 Å in a protein crystal. The reciprocal lattice points are very much closer together and consequently a very much larger number of reflections will be recorded than can be illustrated in the figure.

Monochromatic Data Collection

Monochromatic data collection is the most widely used method in protein crystallography (Fig. 3a). With monochromatic X-rays, only a small proportion of the lattice planes diffract at any particular orientation of the crystal and the crystal has to be rocked through a small angle (1–2 degrees) in order to record the full reflections (Fig. 4a). To bring other planes into the diffracting position, the crystal has to be rotated to a new setting and the procedure repeated with a new film. A full data set to a given resolution is one containing reflections from all possible arrangements of the lattice planes with interplanar spacing greater than or equal to the 'resolution' of the data set. In the case of glycogen phosphorylase b, for example, every structure shown in

Fig. 5 required 32 exposures, each with 1.5 degree oscillation (Fig. 4a) to produce a data set to 2.7 Å resolution. The use of a high-intensity synchrotron source sped up data collection from one week on our standard laboratory equipment to 25 minutes for our fastest monochromatic data set at the SRS. This meant an overall data rate of 50 reflections/second, due solely to the high intensity of the X-ray beam.

Case study: catalysis in crystals of glycogen phosphorylase b

Phosphorylase is active in the crystal[14–15]. Activity is not confined to the surface molecules but is a property of the whole crystal. Using a sequence of monochromatic data sets to record the diffraction images, we have been able to monitor the interconversion of substrates into products in the crystal[15–16].

In muscle, phosphorylase (EC 2.4.1.1) catalyses the phosphorylytic breakdown of glycogen to produce glucose 1-phosphate (for a recent

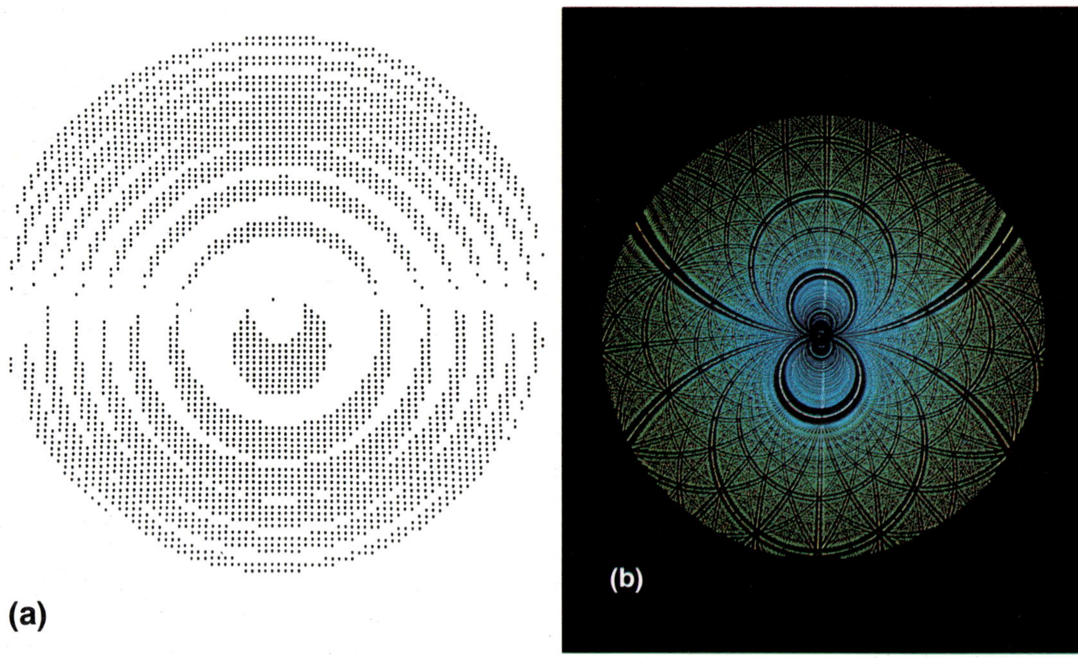

(a) **(b)**

FIGURE 4

Computer generated images of (a) monochromatic and (b) Laue diffraction patterns. (**a**) Simulated oscillation photograph using monochromatic X rays (wavelength: 0.88 Å, crystal to film distance: 174 mm, oscillation range: 0–1.5 degree) of a phosphorylase b crystal ($P4_32_12$, a=b=128.8 Å, c=116.2 Å) for one of the 32 oscillation photographs used to calculate the map shown in Fig. 6. (**b**) Computer-generated Laue diffraction pattern of phosphorylase with the crystal rotated to 22.5 degrees from the position with a* anti-parallel to the beam and c* coinciding with the spindle of an oscillation camera. (a* and c* = reciprocal cell axes.) Crystal to film distance=133.8 mm. Film radius=59 mm. Effective wavelength range contributing to reflections: 0.2–2.1 Å. Each spot on the photograph has been given a colour code to indicate the wavelength of the contributing X-ray beam (blue:0.2 Å; red:2.1 Å). With overlapping reflections, the colour corresponding to the longest component is shown. *(Programmes by Dr I. J. Clifton, SERC Daresbury Laboratory.)*

review see Ref. 17). The enzyme is also capable of slowly transforming the glycosylic substrate analogue, heptenitol (2,6-anhydro-1-deoxy-D-gluco hept-1-enitol) into heptulose 2-phosphate (Fig. 6) (Ref. 18–20). What makes these reactions so promising for time resolved analysis is the fact that the rate-limiting step is the interconversion of the respective ternary enzyme–substrate complexes to products. This means that the active enzyme–substrate complex will transiently accumulate in the crystal similarly to compound B of Fig. 2b. Hence, providing that data collection is fast compared to the half-life of the complex, it may be possible to observe the ternary enzyme–substrate complex before it is converted into products. During the reaction, some active sites will be occupied by substrates, and some will be occupied by products. The relative amount of these species will vary with time.

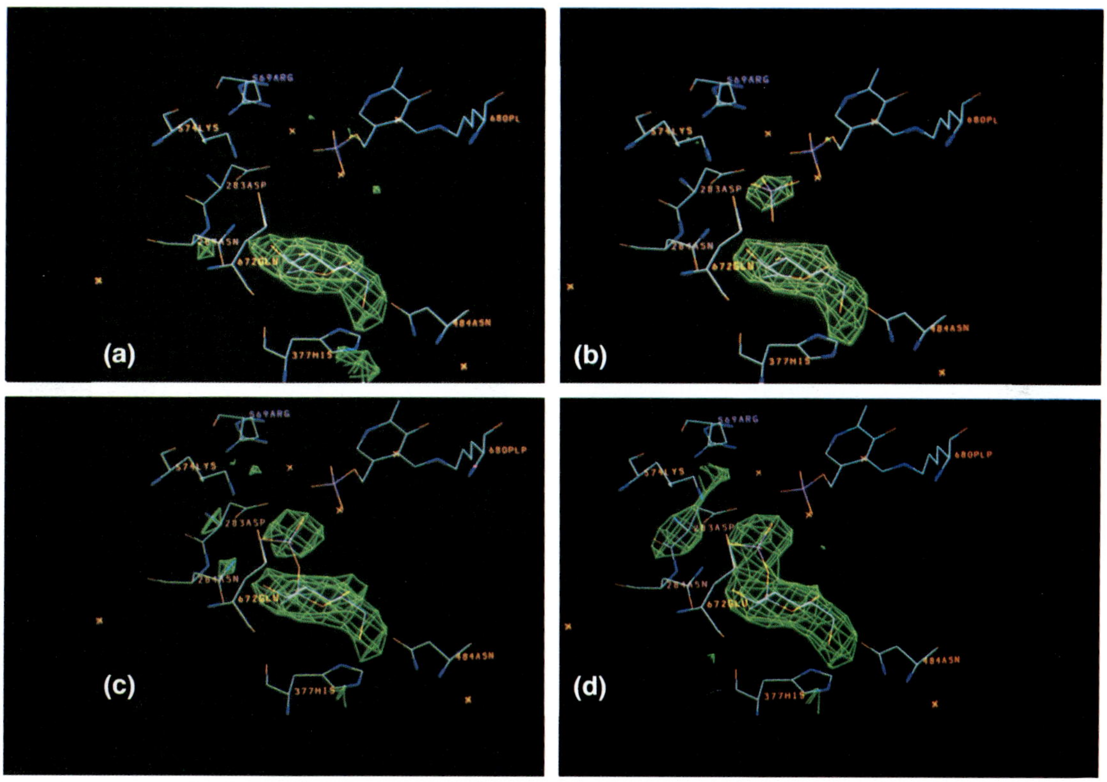

FIGURE 5

Four snapshots of the heptenitol + P_i to heptulose 2-phosphate reaction in crystals of glycogen phosphorylase b: the active site and its vicinity is shown. A single contour level is displayed at a third of the maximal peak height. Residues are shown as they appear in the unliganded native enzyme. Water molecules in the native structure are represented as crosses. A new crystal was used for each data set. (**a**) 100 mM heptenitol (no reaction). (**b**) 100 mM heptenitol + 50 mM phosphate. 10 min soak followed by rapid data collection (60 min) at 13°C. (**c**) 100 mM heptenitol + 50 mM P_i + 50 mM maltoheptaose, 15 min soak followed by rapid data collection (45 min) at 20°C. (**d**) 50 h soak in 100 mM heptenitol + 50 mM P_i + 50 mM maltoheptaose followed by 2.5 h of data collection. All solutions contained 2.5 mM AMP, 10 mM *N*,*N*-bis(2-hydroxyethyl)-2-aminoethane sulfonic acid and 10 mM magnesium acetate. The pH was set to 6.7 (20°C).

In the phosphorylase b experiments the crystal was mounted in a flow cell[21,22] and the reaction was initiated by flowing substrate over the crystal. Diffusion and saturation binding was over in about 10 min. Data collection took place either immediately after this or after various resting times to allow the reaction to proceed. Fig. 5 shows snapshots of the

$$\text{heptenitol} + P_i \rightarrow \text{heptulose 2-phosphate}$$

reaction depicting the formation and transformation of the enzyme–substrate complex into the product, heptulose 2-phosphate (Fig. 5d).

An obvious drawback to the time resolved studies with monochromatic X-rays is that a new crystal is required for each experiment and that the current time resolution is relatively coarse (~ 0.5–1 h). All of the crystal is needed for each data set and radiation damage renders the exposed crystal useless for further work.

THE LAUE METHOD

This method was used by Friedrich, Knipping and von Laue to record the first X-ray diffraction from a crystal of copper sulphate in 1912 (Fig. 3a). The method, which utilizes the complete white spectrum of the radiation, fell into disuse because conventional X-ray sources did not give a satisfactory spectrum and because of difficulties in unravelling the complicated diffraction patterns. Synchrotron sources, modern computers and developments in computing software have resolved these problems. With the Laue method (see for example, Refs 23 and 24) it is possible to record almost all of the data much more quickly from only a few orientations of the crystal. The crystal may then be translated so that a fresh portion is irradiated. As many as 20 Laue photographs may be recorded from a phosphorylase crystal 3 mm long.

The broad effective spectrum (0.2–2.5 Å) and the high intensity of the white X-radiation generated on the SRS wiggler magnet[6] allows the recording of Laue diffraction photographs on the millisecond time scale. Using polychromatic X rays and a stationary crystal, a large number of lattice planes diffract simultaneously as the Bragg condition is satisfied for each of these planes by at least one wavelength of the spectrum (Fig. 3b). The diffracting position of the lattice planes is determined by their orientation, spacing and the wavelength range applied; the wider the range the greater the number of planes in diffracting position. Thus, many reflections can be recorded in a short time with a single exposure. With crystals of high symmetry, a high

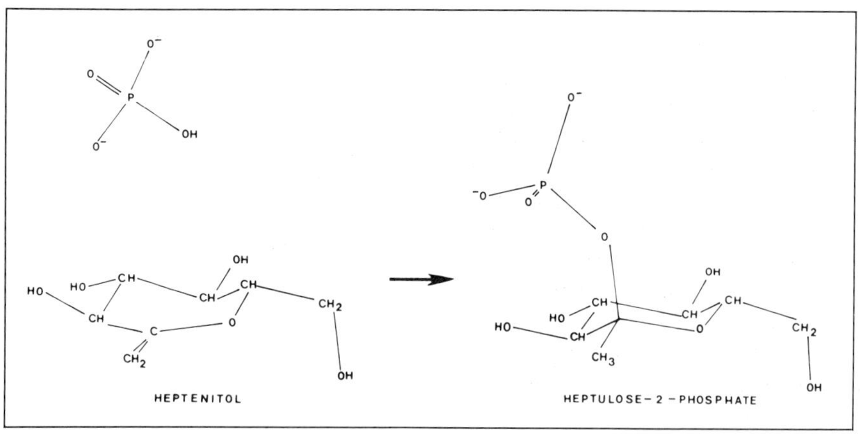

FIGURE 6

The phosphorylase catalysed heptenitol → heptulose 2-phosphate reaction.

proportion of the unique data set may be recorded on a single photograph. With crystals of lower symmetry, a few more photographs with different crystal orientations may be required. Using the high intensity white X-radiation of synchrotron sources to record Laue photographs, data rates well in excess of 100000 reflections per second can be achieved. This is due to the high intensity of the source and the large number of reciprocal lattice points that occupy diffracting positions simultaneously.

There are two ways of extracting information from Laue photographs of protein crystals. The difference methods are designed to analyse structural changes relative to a known starting structure[24] whereas wavelength normalization techniques are being developed to extract structural information without the knowledge of a starting 'native' structure[25–27]. With proteins, the difference method was the first to produce interpretable electron-density maps on the binding of a ligand to the enzyme phosphorylase b (Ref. 24). With this method, a sequence of (nearly) identical Laue data sets are taken before, during and after the initiation of the reaction. Data sets in the sequence are scaled to the starting (native) Laue data set to compensate for factors such as variations in exposure times, beam decay and radiation damage. The advantages of this method are that wavelength and position-dependent corrections to the intensities are not needed.

Case study: binding of oligosaccharides to phosphorylase b

Previous experiments with monochromatic radiation had established the binding site for the oligosaccharide, maltoheptaose (mol.wt 1152), at the surface glycogen storage site of phosphorylase b (Ref. 17). Comparative studies with the Laue method were undertaken in order to test the feasibility of the technique. In the Laue experiments[24], data sets of three exposures (1 s each) were taken from the same crystal at different angular settings before, during and after the addition of ligand. The intermediate data set indicated a transient disorder in the crystal and the data could not be processed. Efforts are under way to develop methods which will enable structural information to be revealed from such photos. The comparison of the two data sets for the native and the liganded enzyme resulted in a data set that contained approximately 58% of the total number of unique reflections to 2.4 Å resolution. When restricted to the highest quality measurements (i.e. intensities greater than three standard deviations, and greater than a quarter of the mean intensity) the total number of reflections was 9029, representing 25% of the complete data set to 2.4 Å resolution. The resulting difference Fourier synthesis clearly showed the oligosaccharide-binding site on the enzyme and compared favourably with the corresponding difference Fourier synthesis calculated with the equivalent measurements from the monochromatic data set.

With the wavelength normalization technique, it is possible to extract structural information for Laue data without the knowledge of a starting native structure. The success of this approach depends on a satisfactory treatment of wavelength-dependent effects such as incident intensity, absorption, Lorentz and polarization factors, obliquity and detector response. The determination of the so called 'wavelength normalization curve' to compensate for these effects is a crucial step in the extraction of independent structural information from Laue photographs. The first successful use of the wavelength normalization technique was in the determination of a small molecule structure by Wood *et al.* in 1982[27]. Lately the method has been used to determine the unknown structure of a small organometallic compound[28] where only microcrystals were available. With proteins, the first successful use of this technique was the localization of heavy atom sites in a crystal of xylose isomerase[26].

The detailed structural information available in the protein Laue maps calculated to date is limited by the use of only partial data. Theoretical calculations show that with a slightly different recording geometry a much greater proportion of the unique data can be recorded than with the present normal beam geometry. In principle therefore it should be possible to obtain as detailed structural information with the Laue method as with the more conventional monochromatic method. These developments are in hand.

DIFFUSION AND BINDING LIMITATIONS

With the very swift data collection times it is important to ensure that neither diffusion of ligands into the crystals nor binding of the ligand to the enzyme form the rate-limiting steps in the reaction. Diffusion times will depend on the size of the ligand and its free diffusion coefficient, the pore size in the crystal, the dimensions of the crystal, and temperature. For most protein crystals, diffusion times are in the range of seconds and minutes. Thus in many instances it will be necessary to pre-diffuse the ligand into the crystal under inactive conditions and to initiate the reaction by (say) pH jump, pressure jump, photodissociation or photo activation. The latter method appears to have considerable promise. The substrate (or enzyme) is blocked with a photoactive group that can be released rapidly to initiate the reaction. Crystallographic experiments with phosphorylase b have been performed using 'caged' phosphate compounds similar to those that have been developed in the study of ATP utilization in muscle contraction[29]. A problem arises with saturating concentrations of caged substrates. The photo-label when released (a nitroso ketone in this case) reacts with the protein, extensively modifying the nucleophilic side chains of the enzyme. This, in our experiment, may have lead to an inhibition of the enzyme. An alternative route that may alleviate this problem would be to cage the enzyme. Glycogen phosphorylase contains the essential cofactor, pyridoxal phosphate, which plays an obligatory role in catalysis through its phosphate group. Caged pyridoxal phosphate has been synthesized and kinetic, reconstruction and crystallization studies are under way. Many other approaches have been taken to photolabel enzymes including the synthesis of photolabile inhibitors for thrombin and other serine proteinases[30].

Towards a picosecond time resolution

The pulsed nature of the synchrotron radiation offers the possibility for stroboscopic experiments with time resolution in the picosecond range. In the Daresbury storage ring, the circulating electron bunches (160 of them) produce a short (120 ps) and intense burst of X rays whenever they pass by the window of the experimental station. Each pulse is followed by a pause of 2 ns. The length of the pause can be altered to some extent: In the so called 'single bunch' mode, there is only one electron bunch (120 ps) circulating in the whole storage ring with a pulse separation of about 320 ns. Other storage rings have different time structures (see, for example, Ref. 9). Let us consider a photoexcitable crystal which returns to the ground state before the next X-ray pulse arrives at the crystal. If we use a high intensity pulsed light source (a laser or the storage ring itself) to excite the sample crystal with the same frequency but with a fixed phase shift relative to the phase and frequency of the X rays of the storage ring, then we have everything in our hands to perform a stroboscopic experiment and to record a structure which has a lifetime in the time range set by the length, phase and frequency of the light and X-ray pulses. First, the exciter light pulse hits the crystal followed by the analyser X-ray pulse. Every X-ray photon arriving at the crystal and, consequently, at the detector will then report on a structure whose age is determined by the phase shift between the exciter light and the analyser X-ray pulses. A single X-ray pulse is usually not sufficient to produce a strong enough diffraction image at the detector (film) but the process can be repeated as many times as necessary to build up a diffraction picture (seconds, minutes). This picture will contain information on a structure that is only a few hundred picoseconds old. To record another picture at another time after excitation, the phase shift has to be changed and the whole procedure repeated. A system of mechanical beam choppers may extend the time range for crystals with longer relaxation times.

In a stroboscopic experiment at the SRS a crystal will have to survive 500 million excitations during a one second long exposure. With high-energy accelerators used in today's physics and available for tomorrow's synchrotron radiation experiments it may be possible to record a complete Laue pattern with a single X-ray pulse from a single electron bunch. This, in fact, would mean an overall exposure time of 10–100 ps and could lead to a new era in structural chemistry.

Conclusions

With existing technology a variety of methods is available for the three-dimensional study of transient structures: (1) Collection of a sequence of complete monochromatic data sets if the lifetime of the intermediate is longer than about 30 min. (2) Collection of a sequence of Laue photographs if the lifetime of the intermediate is longer than about 3 s. (3) Collection of streaked Laue photographs if the lifetime is of the order of 0.1 s. (The film is translated so that each diffraction spot is drawn into a streak in which the intensity at a particular position corresponds to a particular time interval. This saves the time needed to change the film.) (4) Stroboscopic experiments utilizing the pulsed nature of the synchrotron radiation for intermediates with lifetimes of a few hundred picoseconds (the distance light travels in a picosecond is 0.3 mm).

Fast data collection methods have only recently become accessible to the crystallographer. Developments in milliseconds and sub-millisecond data collection methods are of vital importance, offering a route to the dissection and detailed understanding of molecular events associated with catalysis, ligand binding and structural transitions in crystals. Equally important developments are needed at the software side to enable mixtures of structural states to be analysed and in the biochemistry and chemistry for the design of suitable substrates and ligands.

Acknowledgements

This work was supported by the Medical Research Council and the Science and Engineering Research Council. We thank our colleagues N. G. Oikonomakos, P. J. McLaughlin and K. Varvill for their contributions to the phosphorylase work and D. R. Trentham for collaboration with the caged phosphates. We also wish to show our appreciation to the staff of the Daresbury Laboratory for the facilities and service and to P. A. Machin, J. W. Campbell, T. J. Greenough, I. J. Cliftron, S. Zurek, S. Gover, J. R. Helliwell, R. C. Liddington and M. Elder for their contributions to the Laue work.

References

1 Phillips, D. C. (1967) *Proc. Natl Acad. Sci. USA* 57, 484–495
2 Alber, T., Petsko, G. A. and Tsernoglou, D. (1976) *Nature* 263, 297–300
3 Walter, J., Steigemann, W., Singh, T. P., Bartunik, H. D., Bode, W. and Huber, R. (1982) *Acta Crystallogr. Sect B* 38, 1462–1472
4 Fink, A. L. and Petsko, G. A. (1981) *Adv. Enzymol.* 52, 177–246
5 Makinen, M. and Fink, A. L. (1977) *Annu. Rev. Biophys. Bioeng.* 6, 301–342
6 Greaves, G. N., Bennett, R., Duke, P. J., Holt, R. and Suller, V. P. (1983) *Nucl. Instrum. Methods* 208, 139–142
7 Pool, M. W., Munro, I. H., Taylor, D. G. and Walker, R. P. (1983) *Nucl. Instrum. Methods Phys. Res.* 208, 143–145
8 Helliwell, J. R. (1985) *J. Mol. Struct.* 130, 63–91
9 Mills, D. M. (1984) *Physics Today*, April 1984
10 Huxley, H. E., Faruqi, A. R., Bordas, J., Koch, M. H. J. and Milch, J. R. (1980) *Nature* 284, 140–143
11 Kress, M., Huxley, H. E. and Faruqi, A. R. (1986) *J. Mol. Biol.* 188, 325–342
12 Mandelkow, E. M., Harmsen, A., Mandelkow, E. and Bordas, J. (1980) *Nature* 287, 595–596
13 Caffrey, M. (1985) *Biochemistry* 24, 4826–4844
14 Kasvinsky, P. J. and Madsen, N. B. (1976) *J. Biol. Chem.* 251, 6852–6859
15 McLaughlin, P. J., Stuart, D. I., Klein, H. W., Oikonomakos, N. G. and Johnson, L. N. (1984) *Biochemistry* 23, 5862–5873
16 Hajdu, J., Acharya, K. R., Stuart, D. I., McLaughlin, P. J., Barford, D., Klein, H. W., Oikonomakos, N. G. and Johnson, L. N. (1987) *EMBO J.* 6, 539–546
17 Johnson, L. N., Hajdu, J., Acharya, K. R., Stuart, D. I., McLaughlin, P. J., Oikonomakos, N. G. and Barford D. (1989) in *Allosteric Enzymes* (Herve, G., ed.), pp. 81–127, CRC Press
18 Hehre, E. J., Brewer, C. F., Uchiyama, T., Schlesselman, P. and Lehmann, J. (1980) *Biochemistry* 19, 3557–3564
19 Klein, H. W., Im, M. J., Palm, D. and Helmreich, E. J. M. (1984) *Biochemistry* 23, 5853–5861

20 Klein, H. W., Im, M-J. and Palm, D. (1986) *Eur. J. Biochem.* 157, 107–114
21 Wyckoff, H. W., Doscher, M., Tsernoglou, D., Inagami, T., Johnson, L. N., Hardman, K. D., Allewell, N. N., Kelly, D. M. and Richards, F. M. (1967) *J. Mol. Biol.* 27, 5372–5382
22 Hajdu, J., McLaughlin, P. J., Halliwell, J. R., Sheldon, J. and Thompson, A. W. (1986) *J. Appl. Crystallogr.* 18 528–532
23 Moffat, K., Szebenyi, D. M. E. and Bilderback, D. H. (1984) *Science* 223, 1423–1425
24 Hadju, J., Machin, P. A., Campbell, J. W., Greenough, T. J., Clifton, I. J., Zurek, S., Gover, S., Johnson, L. N. and Elder, M. (1987) *Nature* 329, 178–181
25 Campbell, J., Clifton, I., Machin, P. A., Zurek, S., Helliwell, J. R., Habash, J., Hajdu, J. and Harding, M. M. (1987) in *Biophysics and Synchrotron Radiation Research* (Bianconi, A. and Congiu Castellano, A., eds), pp. 52–60, Springer-Verlag
26 Farber, G. K., Machin, P. A., Almo, S., Petsko, G. A. and Hajdu, J. (1988) *Proc. Natl Acad. Sci. USA* 85, 112–115
27 Wood, I. G., Thompson, P. and Matthewman, J. C. (1983) *Acta Crystallogr.* B39, 543–547
28 Harding, M. M., Maginn, S. J., Campbell, J. W., Clifton, I. and Machin, P. A. (1988) *Acta Crystallogr., Sect. B* 44, 142–146
29 Hibberd, M. G., Dantzig, J. A., Trentham, D. R. and Goldman, T. E. (1985) *Science* 228, 1317–1319
30 Turner, A. D., Pizzo, S. V., Rozakis, G. and Porter, N. A. *J. Am. Chem. Soc.* (in press)
31 Blundell, T. L. and Johnson, L. N. (1976) *Protein Crystallography*, Academic Press

J. Hajdu, K. R. Acharya, D. I. Stuart, D. Barford and L. N. Johnson are at the Laboratory of Molecular Biophysics, Oxford University, Oxford OX1 3QU, UK.

What Can Two-dimensional NMR Tell us About Proteins?

Peter E. Wright

As a result of spectacular advances in recent years, nuclear magnetic resonance (NMR) is now firmly established as an essential tool in protein research, providing both a unique method for three-dimensional structure determination and powerful new approaches for studies of protein dynamics and folding.

In the last decade, nuclear magnetic resonance (NMR) has emerged as an immensely powerful tool for studies of the structure and dynamics of proteins. As a result of the great progress which has been made in NMR methodology, we now have a second approach (alongside X-ray crystallography) for determination of the three-dimensional structures of small proteins (up to ~20 kDa). What makes NMR unique, of course, is that it is the only method that can provide high resolution structural information from the solution state. However, its usefulness extends far beyond solution structure determination for globular proteins.

It has long been recognized that the NMR spectrum of a protein is a fingerprint of its three-dimensional structure. However, the potential of NMR as a tool for protein structure determination could not be realized until advances had been made in several areas: in superconducting magnet technology, eventually leading to today's 500 and 600 MHz NMR spectrometers; in 2D NMR methodology for resolution and assignment of the crowded ^{1}H spectra of proteins; and in computational methodology for calculation of protein conformations from the rather loose structural constraints available from NMR spectra.

2D NMR spectroscopy

The fundamentals of 2D NMR are treated in detail in any number of excellent texts, a selection of which is given in Refs 1–6. A general scheme for 2D NMR is shown in Fig 1(a). All 2D NMR experiments involve four time periods: a *preparation* period; an *evolution* period (t_1) during which the spins are frequency labeled; a *mixing period* or *mixing pulse* to transfer coherence between spins; and a *detection* period (t_2) during which the nuclear magnetization is sampled. A series of experiments is recorded with successive small increments in the evolution time t_1 to generate a time-domain matrix $S(t_1,t_2)$. Two-dimensional Fourier transformation then yields the frequency-domain matrix $S(\omega_1,\omega_2)$, i.e. the desired two-dimensional spectrum. The pulse sequence for the simplest 2D NMR experiment, homonuclear-correlated spectroscopy (COSY), is shown in Fig. 1(c) and a schematic spectrum is illustrated in Fig. 2. In the COSY spectrum, the diagonal corresponds to the conventional 1D NMR spectrum and the symmetrical cross peaks on either side of the diagonal arise from through-bond scalar correlations. Figure 2 illustrates the network of COSY connectivities that would be expected in a polypeptide for an amino

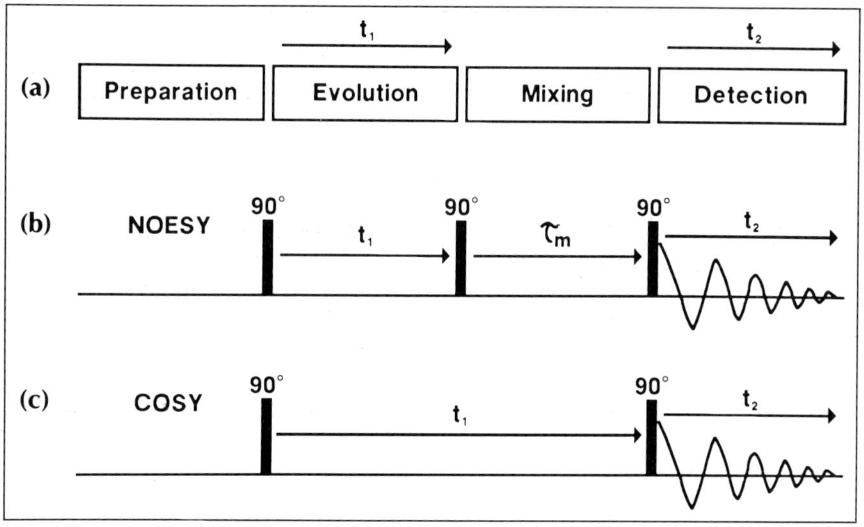

FIGURE 1

(a) General scheme for the 2D NMR experiment. (b) Pulse sequence for NOESY experiment. (c) Pulse sequence for COSY experiment.

acid, such as Asp, that has backbone NH, $C^{\alpha}H$ and two $C^{\beta}H$ methylene protons. Note that in practice, phase-sensitive NMR spectra are usually recorded in which the coupling patterns of the diagonal and cross peaks (not illustrated in Fig. 2) can be resolved. A large number of ^1H homonuclear scalar correlated 2D NMR experiments are available to supplement the basic COSY experiment. These include various forms of relayed coherence transfer spectroscopy, total correlated spectroscopy (TOCSY or HOHAHA), multiple quantum spectroscopy, and multiple quantum filtered COSY experiments.

The principal structural constraints used to determine three-dimensional structures of proteins by NMR spectroscopy are obtained from the through-space dipole–dipole interactions manifest in the nuclear Overhauser effect (NOE)[7,8]. The pulse sequence for ^1H homonuclear 2D NOE spectroscopy (NOESY) is shown in Fig. 1(b). The evolution time t_1 is incremented during the experiment; the mixing time τ_m is kept constant. The cross peaks in the NOESY spectrum arise from through-space correlations and provide information about the spatial proximity of pairs of protons. For short mixing times (less than 50 ms for small proteins of fewer than about 100 amino acid residues), only relatively short inter-proton distances (less than 3–3.5 Å), will give rise to observable NOE cross peaks. At longer mixing times, NOEs between protons up to 5 Å apart can be observed, but must be interpreted cautiously due to the possibility of spin diffusion.

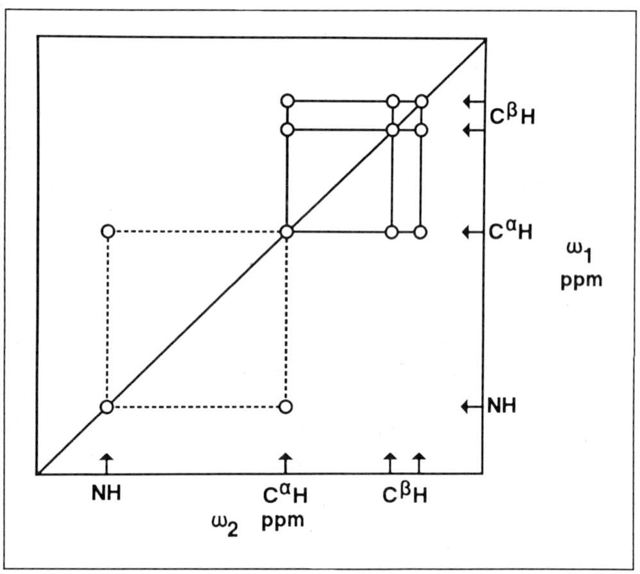

FIGURE 2

Schematic representation of a COSY spectrum for a residue such as Asp, containing backbone NH, $C^{\alpha}H$ and two $C^{\beta}H$ methylene protons. The hatched lines indicate that connectivities to the backbone NH proton are often observed only in spectra acquired in H_2O (rather than D_2O) solution. For simplicity, the splitting patterns observed in the cross peaks and diagonal peaks in phase-sensitive spectra have not been shown.

RESONANCE ASSIGNMENTS

Assignment of resonances to specific amino acids in the sequence represents the requisite first step in any NMR study of a protein. Much effort has been directed towards the assignment problem and using a battery of ^1H homonuclear 2D NMR methods it is now possible to obtain complete proton assignments for proteins of around 100 amino acid residues[9] and nearly complete backbone resonance assignments and extensive side chain assignments for somewhat larger proteins[10]. The first stage of assignments involves identification of the amino acid spin systems. This is achieved by using various scalar correlated NMR experiments to identify the characteristic networks of spin–spin coupled protons associated with the different amino acids. At the next stage, resonances are assigned to specific amino acids in the sequence. Central to the sequence-specific assignment of proton resonances is the powerful sequential assignment procedure proposed by Wüthrich and coworkers[11]. In essence, this method utilizes the through-space ^1H–^1H interactions manifest in the nuclear Overhauser effect (NOE) to identify sequentially neighbouring amino acid residues (Fig. 3). The sequential assignment method represented a major step in the evolution of protein NMR spectroscopy because it allowed reliable assignments to be made based only on knowledge of the amino acid sequence, without recourse to any prior knowledge of tertiary structure. Information about secondary structure and its location in the amino acid sequence is obtained essentially for free in the process of making sequential assignments. The NMR parameters diagnostic of secondary structure are summarized in Fig. 4.

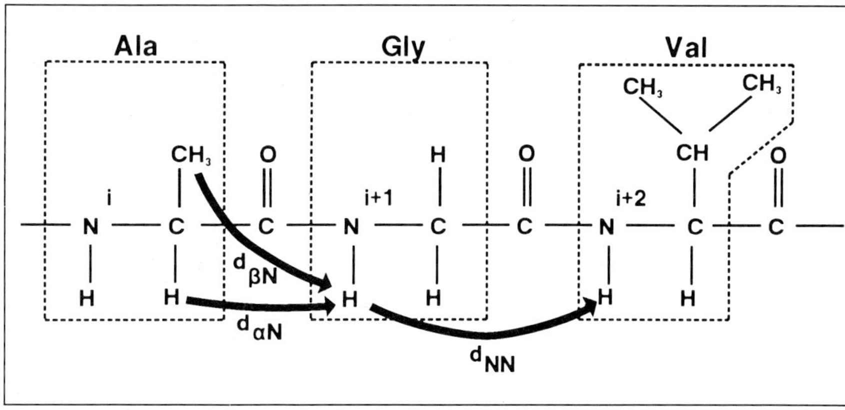

FIGURE 3

Schematic illustration of the sequential assignment procedure[11]. The amino acid spin systems (enclosed within the hatched lines) are first assigned using various through-bond scalar correlated 2D NMR experiments. Sequentially adjacent amino acid residues are then identified by way of characteristic NOE connectivities (solid arrows) involving backbone and $C^{\beta}H$ protons. In this way, the local amino acid sequence can be determined purely from NMR experiments. Comparison with the amino acid sequence of the entire protein then allows sequence-specific assignments to be made.

Complete assignment of the ^1H NMR spectra of proteins larger than ~15 kDa is difficult using homonuclear two-dimensional methods alone. However, several recent advances may allow sufficiently detailed assignments to be made for proteins as large as ~30 kDa such that three-dimensional structure determination will eventually be possible. Central to these methods is the use of extensive isotope labeling with ^2H, ^{13}C and ^{15}N. Thanks to modern molecular biology, this is relatively straightforward for proteins which have been expressed in bacterial or other cells. Selective isotopic labeling provides a powerful method for editing the complex overlapped spectra of large proteins. In an impressive demonstration of the capabilities of this method, complete backbone reso-

β-strand:	$d_{\alpha N}$ = 2.2Å	$^3J_{HN\alpha}$ > 8 Hz
α-helix:	d_{NN} = 2.8Å $d_{\alpha N}$ (i,i+3) = 3.4Å $d_{\alpha\beta}$ (i,i+3) = 2.5–4.4Å	$^3J_{HN\alpha}$ < 5.5 Hz
β-turn:	d_{NN}(3,4) = 2.4Å $d_{\alpha N}$(2,4) = 3.3Å	

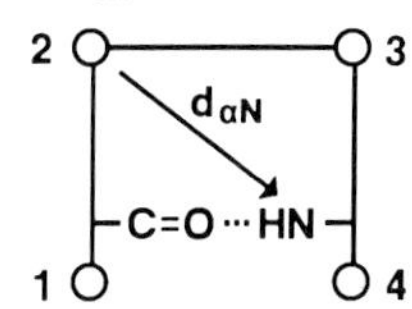

FIGURE 4

Summary of NMR parameters used to identify elements of secondary structure[54]. In the case of β-turns, detailed consideration of HN–$C^{\alpha}H$ coupling constants and patterns of NOEs frequently allows classification according to turn type[55].

(1) Distance:	NOE α $1/r^6$
(2) Dihedral angle:	$^3J_{HN\alpha} \rightarrow \phi$
	$^3J_{\alpha\beta} \rightarrow \chi_1$
(3) Hydrogen bond:	amide exchange

FIGURE 5

Summary of the protein structural constraints which can be derived from ^1H NMR spectroscopy.

nance assignments have been obtained for two proteins of 18 kDa, phage T4 lysozyme (F. W. Dahlquist, pers. commun.) and staphylococcal nuclease[12], and certain key residues have been assigned for the even larger human N-*ras* oncogene encoded p21 protein[13]. Another promising strategy utilizes uniform ^{13}C and ^{15}N labeling to facilitate concerted sequence-specific assignments for ^1H, ^{13}C and ^{15}N resonances and appears to be applicable to proteins of more than 20 kDa[14].

A major problem in studies of larger proteins is overlap of cross peaks in 2D NMR spectra. A logical way of increasing the resolution is to extend the NMR experiments into a third dimension. Preliminary reports of the application of homonuclear three-dimensional NMR spectroscopy to proteins have appeared[15,16] and the method clearly holds great promise. Heteronuclear 3D NMR experiments carried out on isotopically labeled protein samples are likely to be even more useful. Applications to proteins of molecular mass as high as 27.5 kDa have been described recently[17,18] and demonstrate the enormously increased resolution attainable. The method is sufficiently sensitive that data can be acquired in a reasonable time for samples of concentration approximately 1 mM.

CALCULATION OF THREE-DIMENSIONAL STRUCTURES

Assignment of the ^1H resonances is a necessary labour which must be completed before detailed structural studies can be undertaken. NMR methods for protein structure determination have been reviewed in depth recently by several authors[19–23] and only a brief overview will be given here. A number of NMR parameters provide structural constraints (Fig. 5), the most important being the NOE which gives information about inter-proton distances up to a maximum of about 5 Å. Distances are usually calibrated by an empirical procedure, using known inter-proton distances in regular secondary structure as a yardstick. Constraints on both backbone ϕ and sidechain χ_1 dihedral angles can be obtained from coupling constants. For a protein of 100 residues, the number of structural constraints which can be derived from 2D NMR data is very large, of the order of 1000 constraints.

Although no single method for protein structure determination has been adopted universally, a general computational strategy which is used widely has emerged and is illustrated in Fig. 6. Distance geometry methods are applied first to generate structures purely on the basis of geometric constraints derived from the NMR data and from knowledge of covalent architecture. Two quite different algorithms are employed; one operates in distance space (metric matrix approach)[24], while the other utilizes a variable target function approach with minimization in dihedral angle space[25]. Using either approach, the programs are run repeatedly to generate a large number of structures that are consistent with the NMR constraints. The structures resulting from distance geometry calculations frequently show significant violations of the constraints inferred from NMR measurements and also have large strain energies that generally arise from nonbonded contacts that are shorter than van der Waals distances. They can be improved by using 'refinement' procedures that seek to minimize the strain energy and reduce the residual violations of NMR constraints. Restrained molecular dynamics calculations provide a robust method for carrying out this optimization process. In this approach, structures resulting from distance geometry calculations (or sometimes from model building or from the X-ray structure of a homologous protein) are subjected to a molecular dynamics simulation with the conventional potential energy terms plus a penalty function for distances and angles that violate the NMR constraints[26–28]. A modification of this basic approach has been reported in which a simple repulsion term replaces the dihedral, van der Waals, electrostatic and hydrogen bonding potentials of the energy function used in conventional molecular dynamics. This modified approach has been termed dynamical simulated annealing[29]. Structures refined by restrained molecular dynamics generally have much lower 'potential' energy and fewer violations of the NMR constraints than the corresponding distance geometry structures.

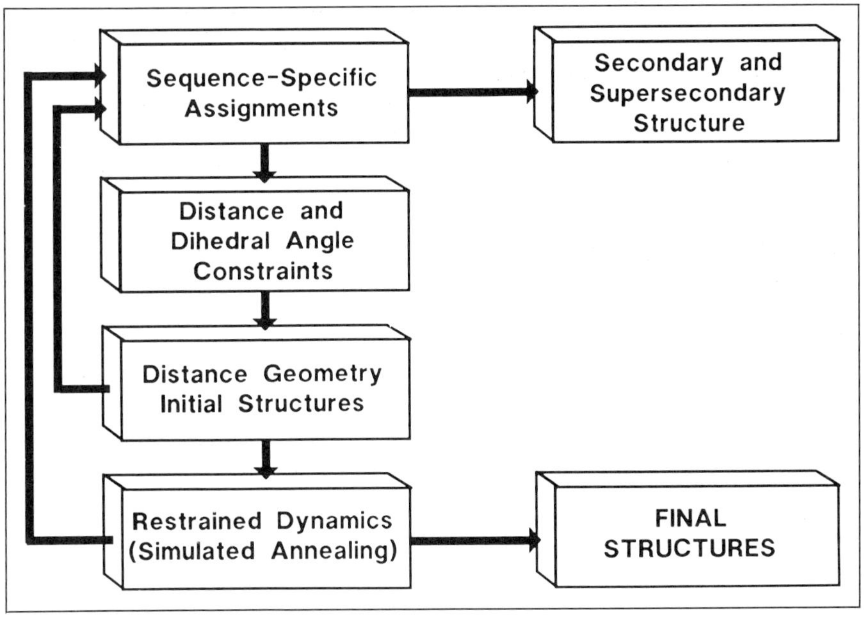

FIGURE 6

Flow chart showing the various steps involved in determination of the three-dimensional structure of a protein. The structures generated at intermediate stages of the calculations are often used to help resolve spectral ambiguities and extend the assignments.

QUALITY OF NMR-DERIVED PROTEIN STRUCTURES

The NMR structure determination method does not generate a unique structure of a protein but rather a family of structures consistent with the NMR constraints (Fig. 7). This is largely because only upper and lower bounds on interproton distances, rather than precise distances, are used as input for the structure calculations. In addition, some regions of a protein may be relatively unconstrained by the NMR data and hence their structures remain poorly defined. This could arise either from a failure to resolve certain NOEs or from internal mobility of the protein in solution.

At this time there exists no parameter analogous to the crystallographic R-factor (a measure of the agreement of a structural model with experimental diffraction data) by which the overall quality of structures derived from NMR data can be described. Rather, the precision of structure determination is usually assessed by the rather subjective measure of the root mean square (rms) deviation among the structures in the family. Not all regions of a protein are equally well constrained by

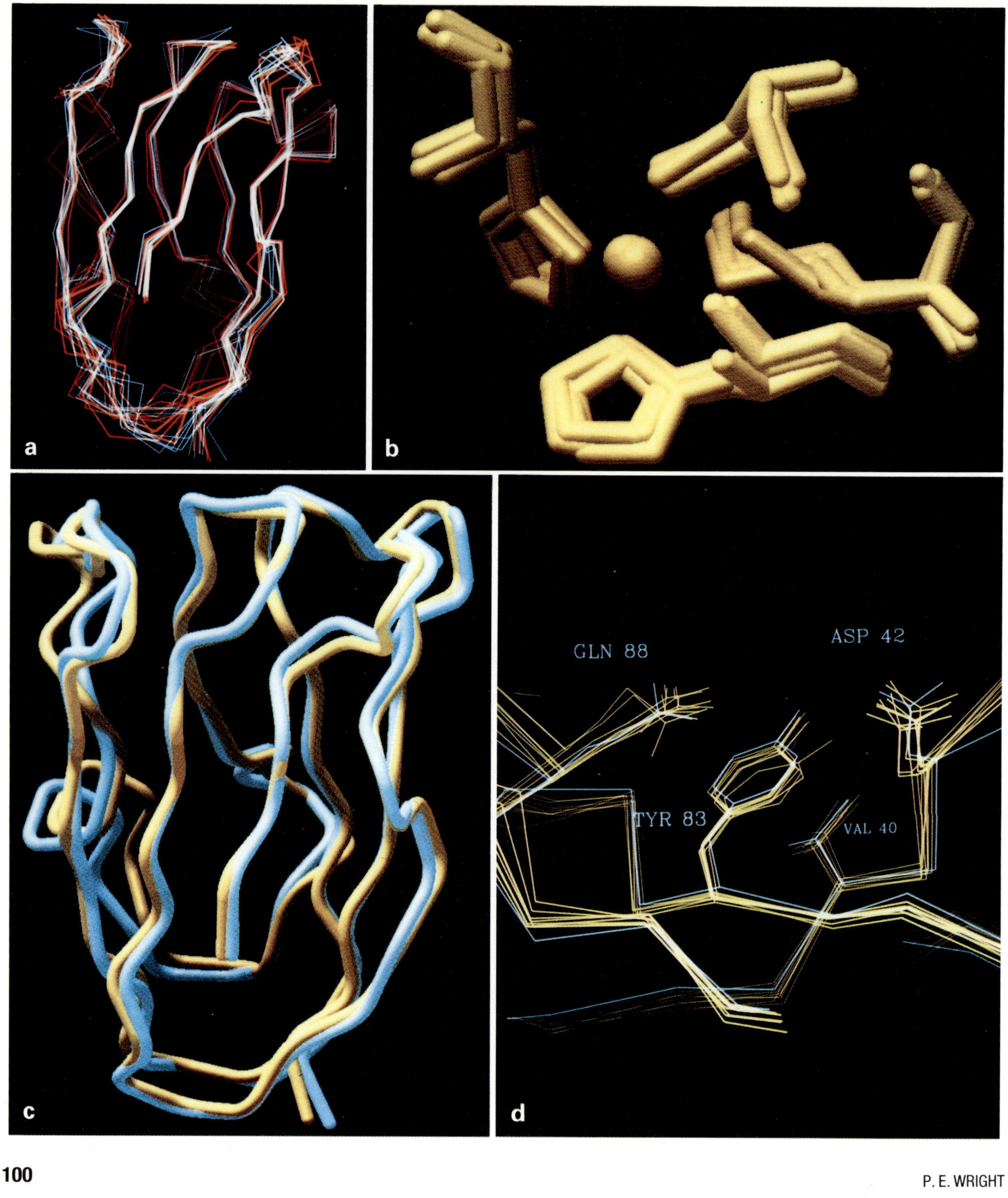
a
b
c
d
GLN 88
ASP 42
TYR 83
VAL 40

the NMR data and hence the local rms deviations may differ significantly in different parts of the polypeptide chain. This is evident in Fig. 7a, where the family of structures is seen to be tighter in the β-strands than in the connecting loops. In general, it has been observed that residues in the interior of a protein tend to be more highly constrained than surface residues. In the best-defined structures reported to date[30,31], the average pairwise backbone rms deviations among the structures is of the order of 0.9 Å. (This is frequently reported as the average backbone rms deviation from the mean, which is then of the order of 0.6 Å.) It should be noted that this level of precision in structure determination is usually only achieved when a very high density of NMR constraints is available and when stereospecific proton assignments are made. The high precision structures of the α-amylase inhibitor tendamistat[30] and the protein BDS-I from sea anemone[31] for example, are derived from data sets containing almost 13 NMR constraints (on average) per residue.

ARE PROTEIN NMR STRUCTURES CORRECT?

Careful consideration needs to be given to the reliability and limitations of the NMR method of protein structure determination. There are now several proteins whose structures have been determined by both NMR and X-ray crystallography. Early comparisons of the solution and crystal structures of bovine pancreatic trypsin inhibitor[32], barley serine proteinase inhibitor[29] and potato carboxypeptidase inhibitor[29] showed clearly that the global fold established by X-ray crystallography is reproduced in solution. Some small differences, especially on the protein surface, were observed. As a test of the NMR method for structure determination, X-ray and NMR structures of tendamistat were determined independently in different laboratories[33,34]: for the polypeptide backbone and internal side chains, the solution and crystal structures are nearly identical and the precision of the NMR structures is comparable to a 2.0 Å resolution X-ray structure[22]. Again, significant structural differences are observed at the protein surface. Finally, high precision NMR structures have been determined for reduced French bean plastocyanin (J. M. Moore, G. P. Gippert, D. A. Case and P. E. Wright, unpublished) which can be compared with the X-ray structure of poplar plastocyanin[35]. Despite the substitution of 21 out of 99 amino acid residues, the NMR and X-ray structures are remarkably similar with respect to overall shape, size and polypeptide chain fold (Fig. 7c). In fact

FIGURE 7

NMR structures of French bean plastocyanin. (**a**) Superposition of eight structures showing the well-defined fold of the polypeptide backbone. The structure is that of a Greek key β-barrel. The Cu atom (not shown) would be in the upper middle of the structure. (**b**) Superposition of five molecular dynamics refined structures showing Cu coordination geometry. Proceeding clockwise from the bottom, the ligands are His87, His37, Cys84 and Met92. (**c**) Superposition of one of the NMR structures of French bean plastocyanin (yellow) and the X-ray structure of poplar plastocyanin (blue). The backbone is represented as a tube of diameter 0.6 Å. (**d**) View of plastocyanin showing conserved surface side chains. Eight refined NMR structures of French bean plastocyanin (yellow) are superimposed on the X-ray structure of the poplar protein (blue). *The tube structures in (b) and (c) were generated using the program MCS written by Dr Michael Connolly.*

the average rms difference for backbone heavy atoms between eight French bean NMR structures and the X-ray structure of poplar plastocyanin is only 1.1 Å. For many conserved side chains, even at the protein surface, the NMR and X-ray structures are observed to correspond closely (Fig. 7d).

Since the two different structural techniques, X-ray crystallography and NMR, independently give the same structures for a number of proteins, we conclude that the NMR method is capable of correctly determining the structure of globular proteins with high precision. However, not all structures determined in solution and crystal are the same. There are substantial differences between the crystallographic and solution structures of rat metallothionein, both in the polypeptide fold and in the coordination within the metal–cysteine clusters[36]. Major differences have also been reported in the global fold of the histidine-containing phosphocarrier protein (HPr) of *E. coli* determined by NMR[37] and by X-ray crystallography[38]. Finally, independent NMR studies of the human anaphylatoxin C3a have established significant differences between the solution and crystal structures that appear to be attributable at least in part to effects of crystal packing[39,40].

Clearly, the strength of the NMR method is not merely as a check on structures determined by X-ray crystallography. Rather, NMR provides a completely independent method for determination of the three-dimensional structures of small proteins, with the added advantage that crystallization is not required. To date, solution structures have been determined for more than 20 different globular proteins containing 33–108 residues; for approximately half of these, no X-ray structures are available. The NMR structures of several proteins have been determined independently in different laboratories using different computational methods; the close similarity of the structures obtained gives further confidence in the NMR method for structure determination. In addition to globular proteins, NMR structures have been determined for more than 20 different polypeptides (e.g. hormones), often in solutions containing organic solvents to induce structure formation.

Despite its obvious successes, the NMR method for solution structure determination is not without its dangers. The most serious of these is the possibility that incorrect *local* structures could be generated in relatively unconstrained regions as a result of errors in assignments or in evaluation of structural constraints. One case has already arisen in which incorrect assignments resulted in errors in a three-dimensional structure[41]. With the adoption of rigorous iterative back-calculation procedures in the future, it is likely that such errors in structures can be avoided and that even more precise NMR structures will be obtained.

FUTURE PROSPECTS FOR NMR STRUCTURE DETERMINATION

The largest proteins for which three-dimensional solution structures have been determined to date have been around 12 kDa. However, the advent of 3D NMR methods and the powerful combination of NMR with the techniques of molecular biology which facilitate isotopic labeling make it probable that structures of proteins in the 20–30 kDa range might ultimately be determined. There are excellent prospects for determination of the conformations of small isotopically labeled ligands bound to large proteins. This approach has been used successfully to study the structure of a 35 kDa pepsin–inhibitor complex[42] and preliminary studies of antibody Fab′–peptide complexes[43] of around 60 kDa suggest that analysis of quite large systems will be possible. An alternative strategy applicable to many large multi-domain proteins involves isolation of independently folded domains small enough for study by NMR. This approach has been used successfully to determine three-dimensional structures of the DNA-binding domain of *lac* repressor[26] and of one of the 37 zinc finger DNA-binding domains of the regulatory protein *Xfin*[44] (Fig. 8). A similar approach may be applicable to fragments and subdomains of many integral membrane proteins, using organic solvents to simulate the properties of the membrane[45,46]. High resolution, solid state NMR methods are also beginning to play an increasingly important role in studies of large globular proteins and membrane proteins[47].

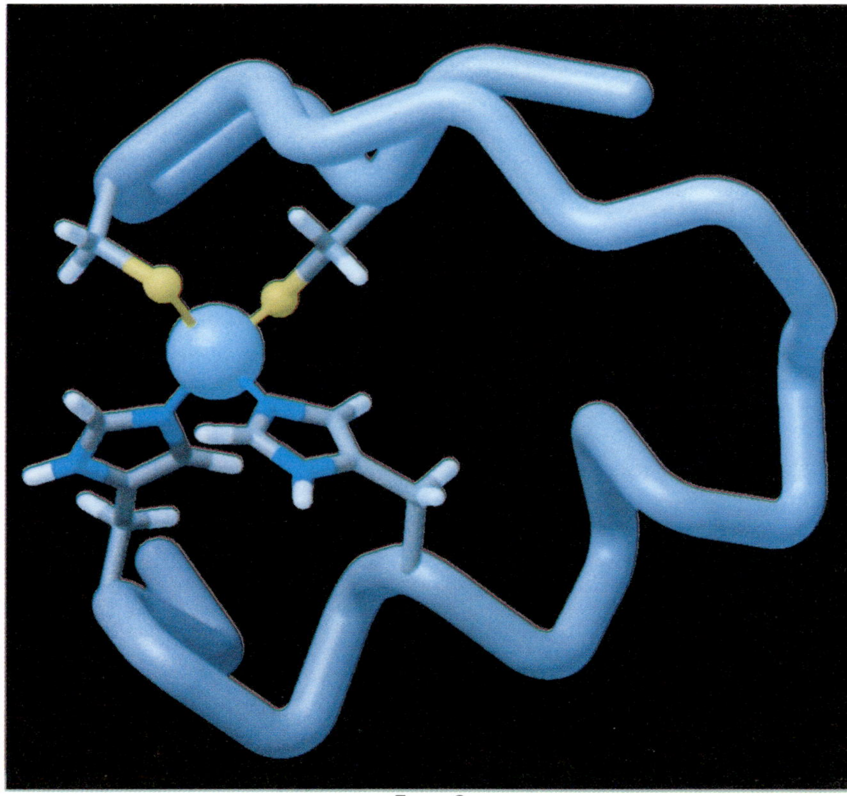

FIGURE 8

NMR structure of a single zinc finger DNA-binding motif from *Xfin*[44]. The C^{α} backbone chain tracing is represented as a tube of diameter 1 Å. The zinc is coordinated by two Cys and two His ligands. The structure shown is that with lowest residual constraint violations in a family of 37 structures generated using distance geometry and molecular dynamics calculations.

PROTEIN DYNAMICS AND PROTEIN FOLDING

While, understandably, much attention has been focused in recent years on the development of methodologies for solution structure determination, NMR continues to have a major impact in other areas of protein research. It has long been recognized that NMR can potentially provide new insights into protein dynamics on a time scale ranging from nanoseconds (relaxation measurements) to thousands of seconds (amide exchange) but these capabilities have not yet been exploited fully. With complete proton assignments available for many small proteins, there is now an opportunity for detailed investigations of side chain and backbone dynamics at individual sites throughout each protein. A highly promising approach is through measurement of natural abundance ^{13}C relaxation times using novel two-dimensional methods[48] to provide the resolution necessary to obtain information at individual sites. Amide proton exchange rates also provide important information on structural fluctuations and can be measured for essentially all residues using 2D NMR methods[49]. NMR studies of protein dynamics are in their infancy but should in the future lead to a new level of understanding of the structures and dynamics of proteins in solution and the relationship of these properties to biological function.

Finally, NMR holds great promise for investigations of the fundamental molecular mechanisms of protein folding. The combination of antipeptide antibody technology with 2D NMR spectroscopy is proving to be a powerful approach for identification and structural characterization of peptide fragments of proteins that have a high propensity to fold in aqueous solution and are likely to play a major role in initiation of protein folding (reviewed in Ref. 50). Elegant 2D NMR methods have recently been devised which provide the first detailed structural insights into that most elusive of entities, the folding intermediate[51–53]. Understanding the mechanisms of protein folding remains a major challenge (see article by Baldwin on p. 125). However, with the powerful NMR techniques that are now available, we can anticipate significant progress in the next few years.

Acknowledgements

I thank Garry Gippert and Michael Pique for assistance with computer graphics and Lonnie Harvey for preparation of the manuscript. This work is supported by the National Institutes of Health (Grant numbers GM36643 and GM38794).

References

1 Bax, A (1982) *Two-dimensional Nuclear Magnetic Resonance*, Reidel
2 Ernst, R. R., Bodenhausen, G. and Wokaun, A. (1987) *Principles of Nuclear Magnetic Resonance in One and Two Dimensions*, Clarendon
3 Croasmun, W. R. and Carlson, R. M. K., eds (1987) *Two-dimensional NMR Spectroscopy*, VCH
4 Derome, A. E. (1987) *Modern NMR Techniques for Chemistry Research*, Pergamon
5 Schraml, J. and Bellama, J. M. (1988) *Two-dimensional NMR Spectroscopy*, Wiley
6 Martin, G. E. and Zekter, A. S. (1988) *Two-dimensional NMR Methods for Establishing Molecular Connectivity*, VCH
7 Noggle, J. M. and Schirmer, R. E. (1971) *The Nuclear Overhauser Effect. Chemical Applications*, Academic Press
8 Neuhaus, D. and Williamson, M. P. (1989) *The Nuclear Overhauser Effect in Structural and Conformational Analysis*, VCH
9 Chazin, W. J. and Wright, P. E. (1988) *J. Mol. Biol.* 202, 623–636
10 Redfield, C. and Dobson, C. M. (1988) *Biochemistry* 27, 122–136
11 Billeter, M., Braun, W. and Wüthrich, K. (1982) *J. Mol. Biol.* 155, 321–346
12 Torchia, D. A., Sparks, S. W. and Bax, A. *Biochemistry* (in press)
13 Campbell Burk, S., Papastauros, M. Z., McCormick, F. and Redfield, A. G. (1989) *Proc. Natl Acad. Sci. USA* 86, 817–820
14 Stockman, B. J., Reily, M. D., Westler, W. M., Ulrich, E. L. and Markley J. L. (1989) *Biochemistry* 28, 230–236
15 Oschkinat, H., Griesinger, C., Kraulis, P. J., Sørenson, O. W., Ernst, R. R., Gronenborn, A. M. and Clore, G. M. (1988) *Nature* 332, 374–376
16 Vuister, G. W., Boelens, R. and Kaptein, R. (1988) *J. Magn. Reson.* 80, 176–185

17 Marion, D., Kay, L. E., Sparks, S. W., Torchia, D. A. and Bax, A. (1989) *J. Am. Chem. Soc.* 111, 1515–1517
18 Fesik, S. W., Gampe, R. T., Zuiderweg, E. R. P., Kohlbrenner, W. E. and Weigl, D. (1989) *Biochem, Biophys. Res. Commun.* 159, 842–847
19 Clore, G. M. and Gronenborn, A. M. (1987) *Protein Eng.* 1, 275–288
20 Braun, W. (1987) *Q. Rev. Biophys.* 19, 115–157
21 Kaptein, R., Boelens, R., Scheek, R. M. and van Gunsteren, W. F. (1988) *Biochemistry* 27, 5389–5395
22 Wüthrich, K. (1989) *Science* 243, 45–50
23 Wüthrich, K. (1986) *NMR of Proteins and Nucleic Acids*, Wiley
24 Havel, T. F. and Wüthrich, K. (1984) *Bull. Math. Biol.* 46, 673–698
25 Braun, W. and Gó, N. (1985) *J. Mol. Biol.* 186, 611–626
26 Kaptein, R., Zuiderweg, E. R. P., Scheek, R. M., Boelens, R. and van Gunsteren, W. F. (1985) *J. Mol. Biol.* 182, 179–182
27 Clore, G. M., Brünger, A. T., Karplus, M. and Gronenborn, A. M. (1986) *J. Mol. Biol.* 191, 523–551
28 Moore, J. M., Case, D. A., Chazin, W. J., Gippert, G. P., Havel, T. F., Powls, R. and Wright, P. E. (1988) *Science* 140, 314–317
29 Nilges, M., Gronenborn, A. M., Brünger, A. T. and Clore, G. M. (1988) *Protein Eng.* 2, 27–38
30 Kline, A. D., Braun, W. and Wüthrich, K. (1988) *J. Mol. Biol.* 204, 675–724
31 Driscoll, P. C., Gronenborn, A. M., Beress, L. and Clore, G. M. (1989) *Biochemistry* 28, 2188–2198
32 Wagner, G., Braun, W., Havel, T. F., Schaumann, T., Gó, N. and Wüthrich, K. (1987) *J. Mol. Biol.* 196, 611–639
33 Kline, A. D., Braun, W. and Wüthrich, K. (1986) *J. Mol. Biol.* 189, 377–382
34 Pflugrath, J. W., Wiegand, G. and Huber, R. (1986) *J. Mol. Biol.* 189, 383–386
35 Guss, J. M. and Freeman, H. C. (1983) *J. Mol. Biol.* 169, 521–563
36 Schultze, P., Wörgötter, E., Braun, W., Wagner, G., Vasák, M., Kägi, J. H. R. and Wüthrich, K. (1988) *J. Mol. Biol.* 203, 251–268
37 Klevit, R. E. and Waygood, E. B. (1986) *Biochemistry* 25, 7774–7781
38 El-Kabbani, O. A. L., Waygood, E. B. and Delbaere, L. T. J. (1987) *J. Biol. Chem.* 262, 12926–12929
39 Chazin, W. J., Hugli, T. E. and Wright, P. E. (1988) *Biochemistry* 27, 9139–9148
40 Nettesheim, D. G., Edalji, R. P., Mollison, K. W., Greer, J. and Zuiderweg, E. R. P. (1988) *Proc. Natl Acad. Sci. USA* 85, 5036–5040
41 Folkers, P. J. M., Clore, G. M., Driscoll, P. C., Dodt, J., Köhler, S. and Gronenborn, A. M. (1989) *Biochemistry* 28, 2601–2617
42 Fesik, S. W., Luly, J. R., Erikson, J. W. and Abad-Zapatero, C. (1988) *Biochemistry* 27, 8297–8301
43 Tsang, P., Fieser, T. M., Ostresh, J. M., Lerner, R. A. and Wright, P. E. (1988) *Peptide Research* 1, 87–92
44 Lee, M. S., Gippert, G. P., Soman, K. V., Case, D. A. and Wright, P. G. (1989) *Science* 245, 635–637
45 Arseniev, A. S., Maslennikov, I. V., Bystrov, V. F., Kozhich, A. T., Ivanov, V. T. and Ovchinnikov, Y. A. (1988) *FEBS Lett.* 231, 81–88
46 Moody, M. F., Jones, P. T., Carver, J. A., Boyd, J. and Campbell, I. D. (1987) *J. Mol. Biol.* 193, 759–774
47 Smith, S. O. and Griffin, R. G. (1988) *Annu. Rev. Phys. Chem.* 39, 511–535
48 Nirmala, N. R. and Wagner, G. (1988) *J. Am. Chem. Soc.* 110, 7557–7558
49 Wagner, G. and Wüthrich, K. (1982) *J. Mol. Biol.* 160, 343–361
50 Wright, P. E., Dyson, H. J. and Lerner, R. A. (1988) *Biochemistry* 27, 7167–7175
51 Roder, H., Elöve, G. A. and Englander, S. W. (1988) *Nature* 335, 700–704
52 Udgaonkar, J. B. and Baldwin, R. L. (1988) *Nature* 335, 694–699
53 Baum, J., Dobson, C. M., Evans, P. A. and Hanley, C. (1989) *Biochemistry* 28, 7–13
54 Wüthrich, K., Billeter, M. and Braun, W. (1984) *J. Mol. Biol.* 180, 715–740
55 Wagner, G., Neuhaus, D., Wörgötter, E., Vašák, M., Kägi, J. H. R. and Wüthrich, K. (1986) *J. Mol. Biol.* 187, 131–135

P. E. Wright is at the Department of Molecular Biology, Research Institute of Scripps Clinic, 10666 North Torrey Pines Road, La Jolla, CA 92037, USA.

A whole arsenal of physical methods can be used to elucidate the secondary and tertiary structure of polypeptides and proteins, of which the spectroscopic methods form an important part.

The amide moiety gives rise to several intensive bands in the infrared spectrum, whose fine structure is characterized by the secondary structure. UV spectroscopy is helpful in identifying the presence of aromatic chromophores (Phe, Tyr, Trp and also many proteids) and in characterizing them as, for example, in hemoglobin. NMR spectroscopy can now be applied even to relatively large proteins and polypeptides by making use of sophisticated two-dimensional methods (see article by P. Wright on p. 95). In addition to information about the secondary structure, the relative arrangement of elements of secondary structure can also be obtained by 2D-NMR. It should be noted, however, that the determination of a full structure may take months!

X-ray crystallography has the great advantage that the structure of the whole protein can be seen. However, the protein first has to be crystallized and good resolution is only obtained if one can make a whole series of measurements in which the protein is attached to different sample substrates, or where there exist solvates with several different solvents. Two other articles in this book (pp. 75 and 83) examine X-ray crystallography in some depth.

Circular Dichroism of Proteins

N. Neumann and G. Snatzke

This article reviews the use of circular dichroism for the elucidation of protein structure. In addition to a short introduction on the technique of circular dichroism measurements, the CD of peptide units and other important chromophores, and the use of these CD effects for the determination of the secondary structure of proteins is described.

Circular dichroism (CD) is one of several chiroptical methods which can differentiate between two enantiomers and are based on the interaction of incident polarized electromagnetic radiation with optically active matter. Other chiroptic methods include polarimetry (optical rotatory dispersion, ORD) and measurement of ellipticity[1–5].

CD spectroscopy, although in principle similar to UV spectroscopy, can provide much more information than the latter, since each band can be positive or negative. Thus frequently even small absorption bands, which may amalgamate in the UV-spectrum, may be nicely resolved in the CD spectrum. The CD spectrum is determined by the spatial arrangement of structural elements around a chromophore but for a given sign, very often two ranges of torsional angles are possible. A measured NMR-coupling constant is also compatible with several ranges of torsional angles, but these generally only partly overlap with those obtained from the CD measurement. Thus, these spectroscopic methods can be used to complement each other since often only one definite torsional angle fits data from both spectra. Beyond this, the absolute configuration determines the sign of each CD band, which thus, in turn, gives us the absolute configuration if the appropriate rule is known.

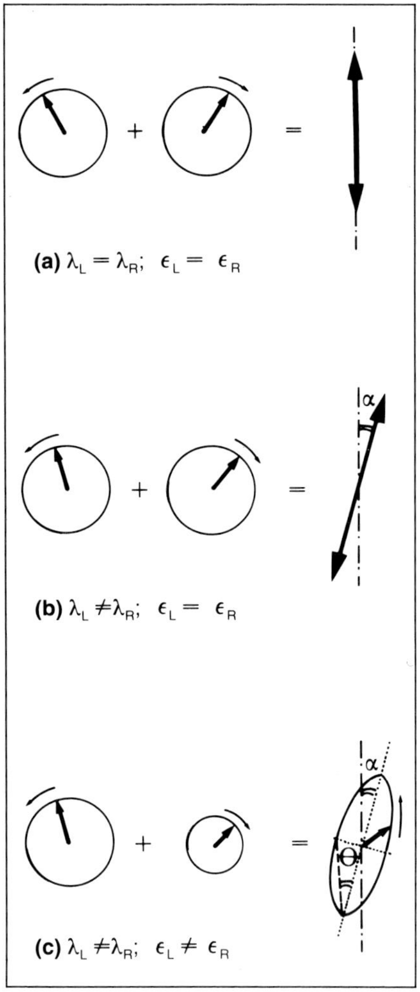

Basic concepts of circular dichroism

Polarimetric measurements determine the extent to which a beam of linearly polarized light is rotated on transmission through an optically active medium. This rotation is wavelength-dependent, leading to an ORD curve when measured over a distinct wavelength range.

The theory of CD had been developed by Biot and Fresnel. A linearly polarized light beam can be thought to consist of (and can be resolved into) two circularly polarized components, a right- (R) and a left-handed (L) one, which are in phase along the lightpath and of same amplitude. By passing through an optically active medium these two beams experience different refractive indices, $n_L \neq n_R$. This causes a phase difference between the two beams ($c_L \neq c_R$) leading to a rotation of their sum vector ('plane of rotation') by an angle α (see diagram). The two beams may also be absorbed by the optically active medium to different extents, causing a difference in amplitude between the two beams which results in elliptical polarization; i.e. the tip of the sum vector describes an ellipse instead of a straight line. The shape of any ellipse is fully characterized by its ellipticity (ζ) defined by $\zeta = \tan^{-1}(b/a)$, a being the major and b the minor axis of the ellipse. If the pathlength (d) within the solution of a substance is measured in cm, and the concentration (c) in mol l^{-1}, its absorption is characterized by

$$\varepsilon = (1/cd)A$$

If I_0 is the intensity of the light impinging onto the cell of length d, I that of the leaving beam, the absorbance $A = \log(I_0/I)$ (Beer–Lambert–Bouguier law). CD, in strict sense, is defined as the difference between the absorption coefficients for right (R) and left (L) circularly polarized light

$$\Delta\varepsilon = \varepsilon_L - \varepsilon_R$$

ε has units of 10^3 $cm^2\,mol^{-1}$. In general, a CD spectrometer is organized in the same fashion as a single-beam UV-spectrometer. The light source is either a xenon lamp or a combination of a deuterium lamp (<350 nm) and a tungsten lamp (for longer wavelengths). The cell that contains the solution should have quartz windows specifically selected for good UV transparency and minimal linear dichroism. As a detector, a photomultiplier can be used, or, at long wavelengths, a solid-state phototransistor.

The light is first linearly polarized using two appropriately cut quartz prisms which act together with a dispersion element that selects the wavelength. In addition to these elements, the modulator is the 'heart' of the CD instrument. This produces alternately right and left circularly polarized light by either the Pockels effect (two electrodes, e.g. two very thin layers of moist glycerol or two gold grids are applied to both sides of a plate cut perpendicularly to the long axis of an ammonium dideuterium phosphate crystal) or the electro-optical effect (a block of quartz periodically contracted and elongated by application of an AC voltage to two disks made from piezoelectric ceramics, which are glued to two steel blocks that in turn hold the quartz in between). The latter modulator is now more frequently used, since it allows a modulation frequency of 18 kHz while the Pockels cell operates only up to 75 Hz (Ref. 6).

The ratio of AC and DC signal on the detector is proportional to the CD and gives an electric signal which is usually calibrated to measure the change in absorbance ($\Delta A = A_L - A_R$), although some machines are calibrated in ellipticities. The specific ellipticity [ζ] and the molar ellipticity [Θ] can be calculated from the measured angle of ellipticity in the same way as specific rotation [α] and molar rotation [ϕ] are obtained from the angle of rotation, α:

$$[\alpha] = (100\alpha)/(lc_p) \text{ and } [\zeta] = (100\zeta/(lc_p)$$

in which α stands for the angle of rotation and ζ for ellipticity (both measured in degrees). For historical reasons, l is measured in dm and the concentration, c_p, is expressed as a percentage. Note that specific rotation [α] and specific ellipticity [ζ] are both measured in deg cm^2 dag^{-7} and not degrees[7]. The respective molar units are: $[\phi] = M_r[\alpha]/100$ and $[\Theta] = M_r[\zeta]/100$ (M_r=relative molar mass). For all practical cases $[\Theta] = 3300 \cdot \Delta\varepsilon$. With polymers it is the custom to use the M_r of an 'average monomer' in this formula instead.

During the last two decades it has proved possible to make CD measurements using wavelengths as short as 150 nm. Since air, quartz and all usual solvents absorb strongly below 179 nm, vacuum spectrographs and prisms and/or modulators made of CaF_2 or MgF_2 have to be used, and all solvents avoided. CD measurements of solvents at these shorter wavelengths provide much additional information although some of this refers to moieties such as C–C, C–O, etc.[8,9] which have usually not been considered as chromophores before.

NONAMIDE CHROMOPHORES

Once the CD spectrum of a protein is obtained, the structure of the protein is deduced by fitting together spectra of other 'model peptides' until their sum approximates the spectrum of the protein. Much of the structural information is contributed by absorption of the peptide bonds, but other chromophores, e.g. aromatic side chains and disulphide bonds, will affect the observed CD curve. We will first focus on the contribution of these non-amide chromophores[10–12] before a more general discussion of how structural information may be obtained by CD.

Aromatic rings and disulfide groups, are the main contributors to the near-UV CD of proteins. Tyrosine, tryptophan, phenylalanine, histidine and cystine absorb between 240 and 300 nm as well as in the typical peptide region. This may result in greater uncertainty when determining α-helical and β-form content of such a protein. The treatment by perturbation theory as developed by Tinoco *et al.*[13] is too crude because the perturbation introduced by chromophores in close proximity is not always so small that it can be neglected. For example, tryptophan residues 63 and 64 of chicken egg lysozyme are so close to each other that neither can move without a corresponding motion of the other: such proximity effects are certainly reflected in CD.

L-Phenylalanine in aqueous solution shows positive Cotton effects at 264 and 258 nm, whereas in polyamides or proteins often a weaker CD at 268 and 262 nm is observed. (These Cotton effects correspond to the α-band absorption of the phenyl chromophore.) Small peptides of phenylalanine or *N*-acetyl-L-phenylalanine amide commonly serve as model compounds. Various oligomers show additional positive or weakly negative CD effects in the peptide region, which are altered by a change of solvent. This is often erroneously interpreted as a change of α-helical content.

Below pH 4, the imidazole moiety of histidine gives rise to a positive CD at 222 nm. Nevertheless, under these conditions poly-L-histidine is often used for calibration purposes with the assumption that its CD is characteristic for a random coil or extended chain.

The tyrosine chromophore shows the ^1L_b-, ^1L_a-, and ^1B-bands of the phenolic group besides those of the n–π*- and π–π*-bands of the peptide chromophore, of which the latter occur below 240 nm (224/194 nm). The ^1L_a-band is redshifted upon alkaline titration and the ^1L_b-band also moves from 275 to 295 nm. Since the ^1L_b-transition of the phenyl chromophore is polarized perpendicular to its long axis, the sign and magnitude of its CD are most sensitive to any rotation around the phenolic pivot bond. The same is valid for any interchromophoral interaction of the exciton type between the perpendicularly polarized component of the ^1B-transition and other strongly absorbing chromophores nearby. This type of interaction leads to much stronger couplets than those of the polypeptide backbone, and they may be of opposite sign. Thus, poly-L-tyrosine has long been assumed to form a left-handed instead of the right-handed helix! Furthermore, the reported studies on ribonuclease and on insulin show that CD allows a clear differentiation between exposed and buried tyrosine residues (Fig. 1). It also allows one to follow assembly reactions as, for example, in the case of insulin subunits in different media.

The tryptophan chromophore shows its absorption bands around 277, 218 and 196 nm. Its CD as well as those of its esters and of *N*-acetyl-L-tryptophan amide reveal several positive Cotton effects between 260 and 300 nm. The 290 nm band has been assigned to the 0–0 line of the ^1L_b-transition by low temperature measurements; this is followed by the first overtone vibration line at 283 nm ($0+850\ cm^{-1}$ transition). The position of these bands is not affected by solvent changes and can therefore be differentiated from the broad, intense ^1L_a-band between 265 and 297 nm. ^1L_b- and ^1L_a-transitions are polarized at right angles to each other, but their exact orientation has not been established. Like its *N*-acetyl-derivative at neutral pH, tryptophan itself shows a positive Cotton effect around 222 nm and a negative one around 195 nm. From studies of poly-L-tryptophan and several of its co-polymers with γ-ethyl-L-glutamate, it was proposed that it forms a right-handed helix in some solvents, and that the CD in the short-wavelength region of known α-helical proteins is strongly influenced upon incorporation of L-tryptophan.

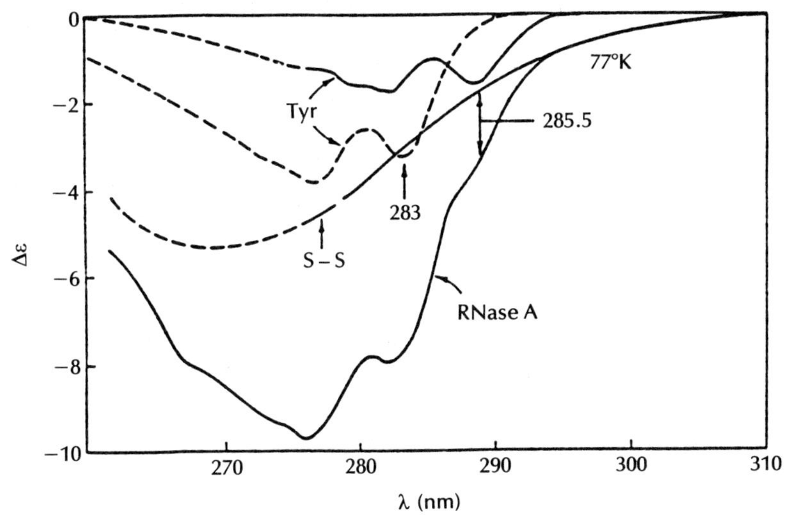

FIGURE 1

Analysis of Ribonuclease A CD at 77K (——) and contributions of the exposed (----) tyrosines with wavelengths of their 0–0 lines indicated. SS indicates the contribution by disulfide bonds. The sum of the component spectra yield the experimental spectrum of RNase A. *Redrawn, with permission, from Ref. 32.*

Because of the different behavior of the ^1L_a- and ^1L_b-CD with changes of the environment, it is possible to resolve spectra of tryptophan-containing proteins by superposition of ^1L_a- and ^1L_b-type curves from model compounds in the same way as described for the differentiation of buried and exposed tyrosine residues.

The disulfide chromophore[4–16] differs from the aforementioned ones as it is inherently chiral with either a left-handed (M) or right-handed (P) screw sense. For cystine itself the CSSC torsional angle ω is close to 90° (minimum overlap of the nonbonding electrons on the S_2-unit). By definition, ω=0 describes the planar Z-form and ω=180° the E-form. The magnitude of its Cotton effect depends strongly on the preferred torsional angle and on the ratio of conformers with positive and negative ω. From theory and many experiments it has been proved that P-helicity gives rise to a positive CD, and M-helicity to a negative one (Carmack–Neubert rule). The rotational equilibrium of the disulphide chromophore has been investigated by temperature- and pH-dependence of L-cystine CD spectra. The 250 nm Cotton effect is red-shifted upon reducing the dihedral angle from 90° (Ref. 12). The broad featureless appearance of this CD band is often of great help in assigning it to the disulfide chromophore. Ribonuclease, lysozyme and insulin are among the natural proteins containing cystine bridges that have been studied. Schwyer *et al.*[17] have used the sign of this disulphide Cotton effect to prove that the torsional angle of the S-S bond is bigger than 90° since the sign of this CD band changes and then corresponds to the sign of this torsional angle.

THE PEPTIDE CHROMOPHORE AND THE DETERMINATION OF SECONDARY STRUCTURES

A careful and critical examination of the peptide bond absorption (≤240 nm) of a particular protein allows estimation of the amount of

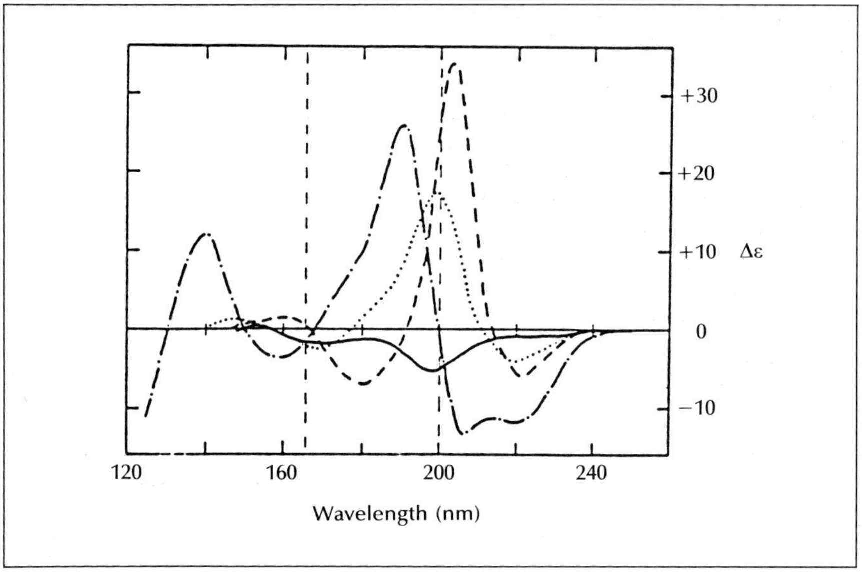

FIGURE 2

The vacuum UV-CD of model polypeptides believed to be in a single secondary structure. That of the α-helix (-·-) is an average of the solution CD-spectrum of poly-γ-L-glutamate in F_6iPrOH according to Johnson and Tinoco (1972), and the film CD-spectrum of poly-L-alanine according to Young and Pysh (1973). The CD-curve of the antiparallel β-sheet (···) is that of a film CD-spectrum of BOC(L-alanine)$_7$OMe, redrawn from Balcerski *et al.* (1976). The parallel β-sheet-CD (---) is the film CD-spectrum of BOC(L-valine)$_7$OMe, redrawn from Balcerski *et al.* (1976). The random coil-CD (——) is the solution CD-spectrum of unordered collagen in F_6iPrOH, redrawn from Jenness *et al.* (1976). *Reproduced, with permission, from Ref. 8.*

secondary structures it contains, (α-helix, β-sheets, β-turns and random structure), which are the main contributors to its tertiary structure. For all globular proteins, the first three structural types are highly ordered (α- and β-form periodically, β-turns aperiodically) with torsional angles independent of the nature of the structure of the amino acid, whereas the 'random coil' is differently arranged in individual proteins[18]. Since the Cotton effects of the ordered structures are rather strong and those of the 'random coil' rather weak, the CD of a protein may be determined mainly by the ordered structures if the percentage of the 'random coil' is not too high (say less than 30%). With the help of standard curves for these secondary structures (obtained from model peptides) the conformational composition can thus be easily estimated. Uncertainties, which can never be ruled out, can arise, for example, from the interaction of peptide chromophores in different structural units coming close to each other in globular proteins or from the CD contribution of other chromophores as discussed above. In general, it is therefore unrealistic to cite exact percentages.

The characteristic Cotton effects of the peptide bond are around 222 (n–π*-transition), 208–210 (π–$\pi^*_{(\parallel)}$-transition) and 191–193 nm (π–$\pi^*_{(\perp)}$-transition)[12,18–21] (Fig. 2). The mechanism of chiral interaction between intense bands is called 'exciton-type splitting', leading to intense positive and negative CD bands in close proximity. A simple treatment of this interaction may be found in literature[22,23].

Sperm whale myoglobin (0.1 M NaF, pH 7), poly-L-lysine (pH 10) or poly (γ-methyl-L-glutamate) in hexafluoroisopropanol have been used as typical models for an α-helix although their CD curves do not coincide. In addition in the far-UV region there is a positive shoulder at 178 nm and a negative band near 150 nm, followed by a positive one.

A typical β-pleated sheet reveals a negative Cotton effect in the 220 nm region and a positive one between 195 and 200 nm. In the vacuum-CD, an additional negative band below 170 nm can be found. Poly(L-glutamic acid) or poly(L-lysine), transformed from their α-helical form into the β-conformation upon heating, serve model compounds.

The typical β-structure is characterized by co-polypeptides, composed of alternating hydrophobic and hydrophilic amino acid residues which form a bilayer structure with one face exposed to the solvent. They serve as model compounds for the antiparallel β-pleated sheet, found, for example, in superoxide dismutase (SOD). Parallel β-pleated sheets, on the other hand, are almost always buried on both sides, so that their central amino acid sequence is the hydrophobic one, and the amino acids towards the end form the hydrophilic ones, as found, for example, in flavodoxin (FXN). Furthermore, the bending of β-sheets into different barrel-like shapes may contribute to the CD spectrum. CD does not differentiate with certainty between parallel and antiparallel β-structure.

By using naturally occurring proteins whose structures are known from X-ray diffraction as reference compounds, one automatically takes into account interactions between different secondary structures, although only in a statistical way. The approach developed by Brahms and Brahms[24] also includes contributions of β-turns neglected completely until the middle of the 1970s. Venkatachalam originally defined 15 possible types of β-turns in globular proteins, which were later divided into ten and then into three fundamental types, covering about 80% of all β-turns occurring in proteins. According to Woody's theoretical analysis, all three types should have similar spectra with a weak negative band between 220 and 230 nm, a positive band between 200 and 210 nm, and a strong negative band between 180 and 190 nm. Poly(Ala_2–Gly_2) is believed to give a type I β-turn and its CD closely resembles that found for practically all β-turns in proteins[24–27].

Poly(Pro–Lys^+–Leu–Lys^+–Leu) in salt-free solution gives a CD curve similar to that one of the charged poly(L-lysine)·HCl, representing

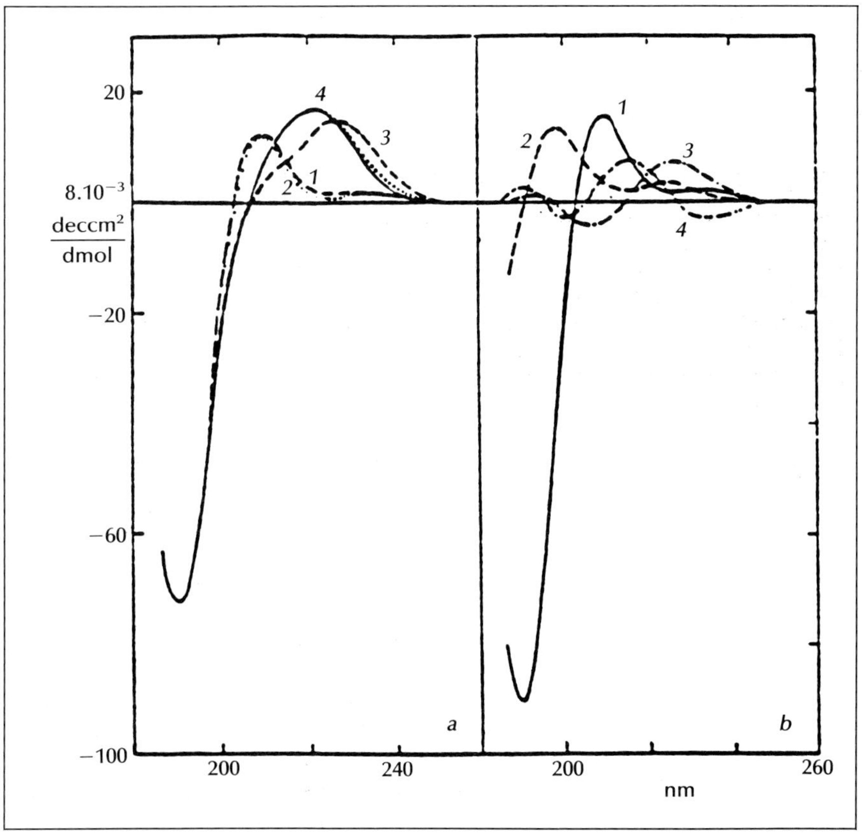

FIGURE 3

(a) Comparison of the experimental curve of cyclo(L-Pro-L-Val) in TFE (full curve) with the curve of the first subspectrum (---- 1), the sum of the first two CD-subspectra (······ 2), that of the first three subspectra (---- 3) and the sum of the first four subspectra (······ 4). **(b)** Decomposition of the experimental curve of cyclo(L-Pro-L-Ala) in TFE into subspectra 1–4. *Reproduced with permission from Ref. 33.*

the random coil form of proteins, with a negative band near 200 nm and a very weak band near 220 nm, which can be either positive or negative.

As mentioned previously, all chiroptical effects of the various conformations contribute to the observed CD spectrum in an additive way. One tries, therefore, to match the experimental curve by a sum of 'standard' curves, which may be either constant for each type of structure or may be adjusted to a given class of protein (termed direct analysis by Provencher and Glöckner[28]). The more general methods use experimental curves of a group of related proteins and apply a mathematical procedure called 'factor analysis' to it. In this way, one obtains a main curve and several others, with progressively smaller CD values. Each measured CD curve is then built up as a sum of these curves, multiplied by a factor obtained from the curve analysis. Although it is tempting to identify each of these independent spectra with a known structure (e.g. α-helix, random coil, etc.), this is not possible, since any of the standard curves contains much more information in cryptic form[29,30] (Fig. 3).

The direct analysis mentioned above is only possible for globular proteins, because one standard curve is used for each of the mentioned secondary structures (α-helix, β-sheets, β-turns, random coil) which are obtained from the CD spectra of different globular proteins by curve fitting. Hennessey and Johnson (see Ref. 27) have developed an eigenvector method of multicomponent analysis with an unconstrained least-square method. Each of these methods has its own advantage, but both only give reliable results, when the basic reference spectra included into the set reflect all the structural characteristics of the protein in question[19,30].

The most reliable, and at the same time simplest procedure, has been outlined by Curtis Johnson and colleagues[31]. From a set of related protein CD spectra (typically 15 proteins and one polypeptide) a matrix is constructed, each line of which consists of 42 CD values measured from 178 to 260 nm at an interval of 2nm. This matrix is inverted yielding five generalized inverse CD spectra (Fig. 4) from which a second inversion yields five typical secondary-structure spectra (α-helix, the parallel- and antiparallel β-sheet, the β-turn and the average random structure) (Fig. 5), following the Moore–Penrose theorem (see Ref.

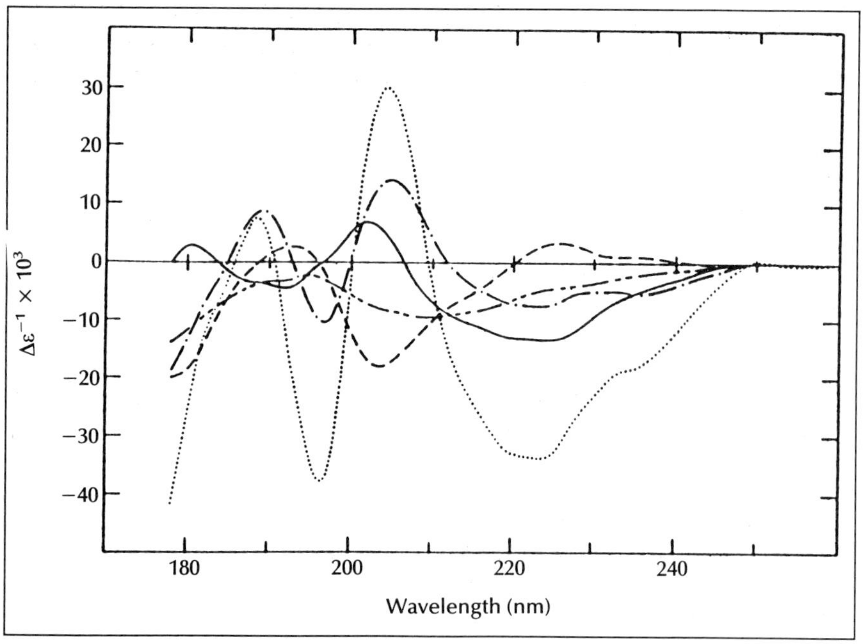

FIGURE 4

Generalized inverse CD-spectra for five major secondary structures from 178 to 260 nm: α-helix (———), antiparallel β-sheet (---), parallel β-sheet (-·-), β-turn (-··-), other (random) structure (···). *Reproduced, with permission, from Ref. 34.*

29). From these tabulated data and the corresponding digitalized CD spectrum of any investigated protein a CD curve can be constructed which provides the relative contribution of each type of the aforementioned secondary structures. The whole procedure involves only multiplication and addition and can be done with a simple pocket calculator. Small negative percentages are tolerated, and the mathematical synthesis is believed to be successful if the sum of the fine structures add up to values between 0.9 and 1.1.

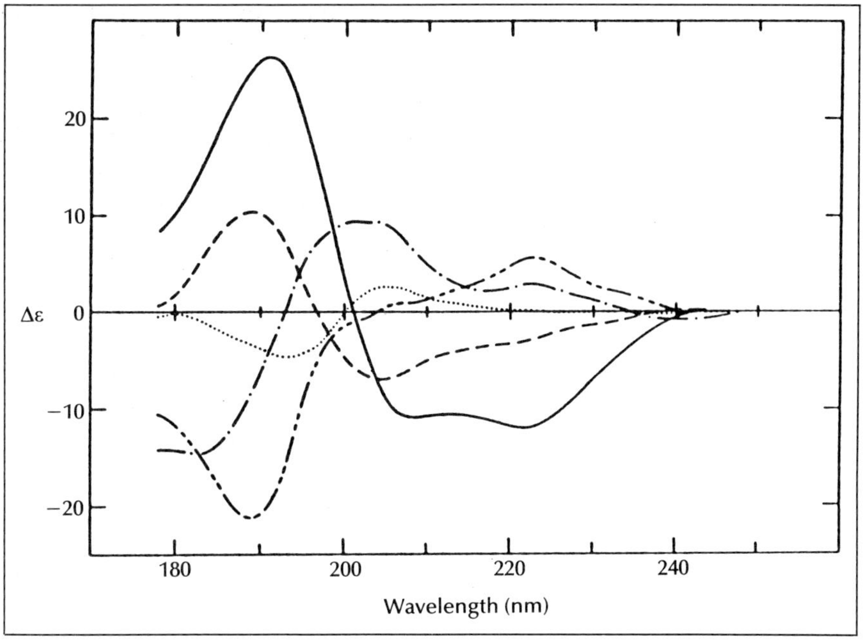

Figure 5

Secondary structure spectra for five major secondary structures from 178 to 260 nm as obtained from the curves of Fig. 4: α-helix (——), antiparallel β-sheet (---), parallel β-sheet (-·-), β-turn (-··-), other (random structure) (···). *Reproduced, with permission, from Ref. 34.*

Summary

It is beyond the scope of this review to give more than a few specific examples of the use of circular dichroism for elucidation of protein structure, but each year there are published several hundred relevant papers. With the expected accumulation of much more CD data on protein structure, we should be able to improve our understanding of the changes of Cotton effects with changes of the environment and experimental conditions (solvents, salts, prosthetic groups, substrates, etc.). When the long-term stability of the instruments is improved the kinetics of enzyme reactions can be monitored more precisely by CD spectroscopy, and once data banks are available then more reliable 'standard curves' can be constructed. A wealth of principally new information is to be expected when CD spectrometers, which will allow routine measurements down to around 130–150 nm, are on the market. Several laboratories are working towards this goal and with the newly developed electronics, moving mechanical parts will be eliminated from the optical pathway. Diode arrays can be substituted for conventional detectors, so that one measurement can be done within seconds: this technique also allows miniaturization of the instrument. Such measurements have to be performed in vacuum, however, so thin films must be used instead of solutions. The price of such new equipment is expected to be much lower than that of today's devices, and the necessary data processing can be done on any available laboratory computer. CD measurements are thus expected to be even more important in the future than today, and chiroptical data will be recorded routinely to characterize proteins and to follow the kinetics of enzyme reactions.

References

1 Caldwell, D. J. and Eyring, H. (1971) *The Theory of Optical Activity*, John Wiley & Sons
2 Buckingham, A. D. and Stiles, P. J. (1974) *Acc. Chem. Res.* 7, 258–264
3 Schellman, J. A. (1975) *Chem. Rev.* 75, 323–331
4 Kauzmann, W. (1957) *Quantum Chemistry*, Academic Press
5 Snatzke, G. (1979) *Angew. Chem.* 91, 380–393
6 Snatzke, G. (1967) *Z. Instrumentenkd.* 3/4, 11–124
7 McGlashan, M. L., ed. (1968) *Manual of Symbols and Terminology for Physicochemical Quantities and Units* ('The Green Book'), R. Inst. Chem. Monograph No. 15; IUPAC Monograph (1979) Second Revision (D. H. Whiffen, ed.), Pergamon Press; see also (1979) *Pure Appl. Chem.* 51, 1–41
8 Johnson, C. W., Jr (1986) *Photochem. Photobiol.* 44, 307–313
9 Bustamate, C., Maestre, M. F. and Wells, K. S. (1986) *Photochem. Photobiol.* 44, 331–341
10 Kahn, P. C. (1979) *Methods Enzymol.* 61, 339–378
11 Adler, A. J., Greenfield, N. J. and Fasman, G. D. (1973) *Methods Enzymol.* 27, 675–735
12 Beychok, S. (1968) *Annu. Rev. Biochem.* 37, 437–462
13 Tinoco, I., Jr (1973) in *Fundamental Aspects and Recent Developments in Optical Rotatory Dispersion and Circular Dichroism* (Ciardelli, F. and Salvadori, P., eds), pp. 66–86, Heyden & Son
14 Carmack, M. and Neubert, L. A. (1967) *J. Am. Chem. Soc.* 89, 7134–7135
15 Linderberg, J. and Michl, J. (1970) *J. Am. Chem. Soc.* 92, 2619–2625
16 Woody, R. W. (1973) *Tetrahedron* 29, 1273–1283
17 Ludescher, U. and Schwyzer, R. (1971) *Helv. Chim. Acta* 54, 1637–1644
18 Richardson, J. S. and Richardson, D. C. (1989) in *Prediction of Protein Structure and the Principles of Protein Conformation* (Fasman, G. D., ed.), pp. 1–89, Plenum Press
19 Johnson, C. W., Jr (1988) *Annu. Rev. Biophys. Biophys. Chem.* 17, 145–166
20 Fasman, G. D. (1987) *Biopolymers* 26, 59–79
21 Rosenkranz, H. (1974) *Z. Klin. Chem. Klin. Biochem.* 12, 415–422
22 Schellman, J. A. (1969) *Acc. Chem. Res.* 1, 144–151
23 Harada, N. and Nakanishi, K. (1983) *Circular Dichroic Spectroscopy – Exciton Coupling in Organic Chemistry*, University Science Books
24 Brahms, S. and Brahms, J. (1980) *J. Mol. Biol.* 138, 149–179
25 Chang, C. T., Wu, C. S. C. and Yang, T. J. (1978) *Anal. Biochem.* 91, 13–31
26 Chou, P. Y. and Fasman, G. D. (1977) *J. Mol. Biol.* 115, 135–175
27 Lewis, P. N., Momany, F. A. and Scheraga, H. A. (1973) *Biochim. Biophys. Acta* 303, 211–229
28 Provencher, S. W. and Glöckner, J. (1981) *Biochemistry* 20, 33–37
29 Prevelige, P., Jr and Fasman, G. D. (1989) in *Prediction of Protein Structure and the Principles of Protein Conformation* (Fasman, G. D., ed.), pp. 391–416, Plenum Press
30 Yang, J. T., Wu, C. S. C. and Martinez, H. M. (1986) *Methods Enzymol.* 130, 208–269
31 Compton, L. A. and Johnson, W. C., Jr (1986) *Anal. Biochem.* 155, 155–167
32 Horwitz, J., Strickland, E. H. and Billups, C. (1970) *J. Am. Chem. Soc.* 92, 2119
33 Malŏn, P., Pančoška, P. and Bláha, K. (1985) in *Proceedings of the First International Conference on Circular Dichroism*, Vol. 6, p. 181, Bulgarian Academy of Sciences
34 Compton, L. A. and Johnson, W. C., Jr (1986) *Anal. Biochem.* 155, 155–167

N. Neumann and G. Snatzke are at the Department of Structural Chemistry, Ruhr-University Bochum, PO Box 10 21 48, FRG.

Conformational Stability of Globular Proteins

C. Nick Pace

The conformational stability of ribonuclease T_1 has been measured as a function of the variables of most interest to biochemists: temperature, pH, salt concentration, disulfide-bond content and amino acid sequence. The results provide insight into the forces that stabilize globular proteins.

Ribonuclease T_1 (RNase T_1) has proven to be an excellent model for investigating the conformational stability of globular proteins. The three-dimensional structure of RNase T_1 has been determined at high resolution[1] and shows that the 104 amino acid residues fold to a compact globular conformation in which the hydrophobic core is sandwiched between a 4.5-turn α-helix, and an extended antiparallel β-sheet (Fig. 1). The folded conformation is stabilized in part by two disulfide bonds, by 87 intramolecular hydrogen bonds, and by hydrophobic interactions in which 85% of the nonpolar residues are removed from contact with water. Thus, RNase T_1 provides a good model for studies of the folded conformation. When the disulfide bonds are broken, RNase T_1 is unfolded under physiological conditions, also providing a good model for studies of the unfolded conformations[2]. Recent studies of staphylococcal nuclease suggest that the unfolded conformations of a protein make more important contributions to the net conformational stability than previously thought[3]. Here we summarize our studies of the conformational stability of RNase T_1. They show that the conformational stability of this small globular protein can be varied experimentally over a range of almost 20 kcal mol^{-1} (84 kJ mol^{-1}).

The conformational stability of globular proteins is remarkably low

We define the conformational stability of a globular protein as the difference in free energy between the folded and unfolded conformations under physiological conditions, and denote this as $\Delta G(H_2O)$ (Fig. 1). Estimates of the conformational stability of RNase T_1 and several other proteins that are under active experimental study at present are given in Table I. The conformational stability of almost all naturally occurring globular proteins is between 5 and 15 kcal mol^{-1} (21–63 kJ mol^{-1}). It appears to be advantageous to living organisms to have proteins for which the folded, biologically active conformation is only marginally more stable than unfolded, inactive conformations. The reasons for this are still not clear, but it may have to do with the protein's turnover[15] or function[16]. Since a protein based on a four-helix bundle structure has been constructed with a conformational stability of 22.5 kcal mol^{-1} (94.5 kJ mol^{-1})[17], it seems likely that evolution could have generated more stable proteins if they were needed.

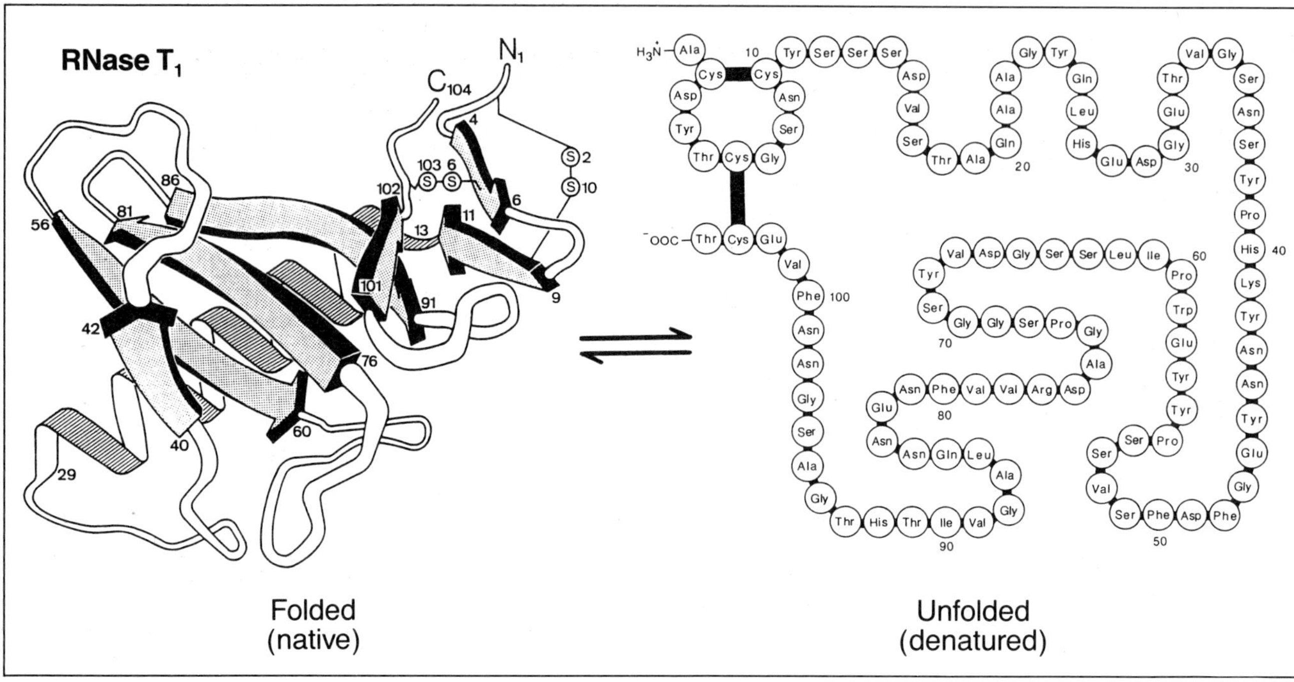

FIGURE 1

ΔG for this reaction under physiological conditions is the conformational stability, i.e. $\Delta G(H_2O)$. The three-dimensional structure of RNase T_1 was determined in Saenger's laboratory[1] and is reproduced here with permission.

GLOBULAR PROTEINS CAN BE UNFOLDED BY HEATING OR COOLING

At pH 7, RNase T_1 is most stable near $-5°C$ and the stability decreases both at higher and at lower temperatures (Fig. 2). Becktel and Schellman[16] refer to plots of ΔG vs temperature as protein stability curves and have discussed them in detail. This interesting temperature dependence is described by a form of the Gibbs–Helmholtz equation (Eqn 1):

$$\Delta G(T) = \Delta H_m(1 - T/T_m) - \Delta C_p[(T_m - T) + T\ln(T/T_m)] \qquad (1)$$

where $\Delta G(T)$ is ΔG at temperature T, T_m is the midpoint of the thermal unfolding curve, ΔH_m is the enthalpy change for unfolding measured at T_m, and ΔC_p is the difference in heat capacity between the unfolded and folded conformations. The temperature of maximum stability (T_S) occurs at the temperature where $\Delta S = 0$, and is given by

$$T_s = T_m \exp(-\Delta H_m/[T_m \Delta C_p]).$$

T_s is between $-10°C$ and $35°C$ for the proteins that have been studied to date. T4 lysozyme[13] (T_S = 12.5°C, and T_m = −3 and 28°C), and staph nuclease[18] (T_S = 18°C, and T_m = −19 and 57°C) provide interesting examples where T_s occurs above 0°C so that both low and high temperature unfolding can be observed directly.

When a protein unfolds, the buried nonpolar side chains are brought into contact with water. In order to accommodate these side chains, they are surrounded by cages of water molecules so that the extent of hydrogen bonding is increased. The heat capacity of the unfolded protein is greater than that of the folded protein because now these cages must be melted in order to raise the temperature of the solution. Thus, the same effect that gives rise to hydrophobic interactions also gives rise to the larger ΔC_p that dominates the thermodynamics of protein folding. (See Baldwin[19], and Privalov and Gill[20] for interesting discussions.)

Table I
Conformational stability [$\Delta G(H_2O)$] of RNase T_1 and six other proteins[a]

Protein	$\Delta G(H_2O)$ (kcal mol^{-1})	$\Delta G(H_2O)$ (kJ mol^{-1})	Ref.
RNase T_1[b]			
pH 7.0, 25°C	5.6	23.52	4
+0.2 M Na_2HPO_4	10.0	42.00	7
pH 7.0, −6°C	8.3	34.86	6
pH 4.5, 25°C	8.9	37.38	5
Gln25→Lys	6.5	27.30	8
Glu58→Ala	4.8	20.16	8
BPTI[c]	14.3	60.06	9
Barnase	9.4	39.48	10
RNase A	9.0	37.80	5
Nuclease	6.1	25.62	11
DHFR	5.9	24.78	12
T4 Lysozyme	12.8	53.76	13

[a] Methods used to measure conformational stability are described in Ref. 14.
[b] For wild-type RNase T_1, and for the two mutants, 30 mM MOPS buffer (pH 7.0, 25°C) was used.
[c] Abbreviations: BPTI, bovine pancreas trypsin inhibitor; DHFR, dihydrofolate reductase; MOPS, 3-(*N*-morpholino) propanesulfonic acid.

Globular proteins are most stable near their isoelectric points

At 25°C, RNase T_1 is most stable at pH 4.5, and this is near the pH where the net charge is zero (Fig. 3). At zero net charge, RNase T_1 will have six positive charges (3 His, 1 Lys, 1 Arg and 1 α-amino) and a total of six negative charges from the 13 carboxyl groups. The charges on the surface of globular proteins are generally arranged so that there are more favorable than unfavorable electrostatic interactions among the charged groups; thus they should contribute favorably to the conformational stability[21]. When the protein unfolds, the charges will be further apart and the effective dielectric constant increased; consequently, electrostatic interactions will be substantially reduced, especially at moderate salt concentrations. Note that RNase T_1 is only about 6 kcal mol^{-1} (25.2 kJ mol^{-1}) less stable at pH 10 where the net charge on the molecule is about −12, and repulsive electrostatic interactions among the charges will predominate. This suggests that electrostatic interactions among the charges on the surface of a globular protein make only a small contribution to the conformational stability, probably much smaller than the contributions from hydrophobic interactions and hydrogen bonding.

Another important contribution to the pH dependence of $\Delta G(H_2O)$ results from differences between the p*K*s of groups in the folded and unfolded conformations. The 6 kcal mol^{-1} (25.2 kJ mol^{-1}) increase in $\Delta G(H_2O)$ between pH 10 and 5 results in part from the presence of groups that bind protons more tightly (higher p*K*s) in the folded than in

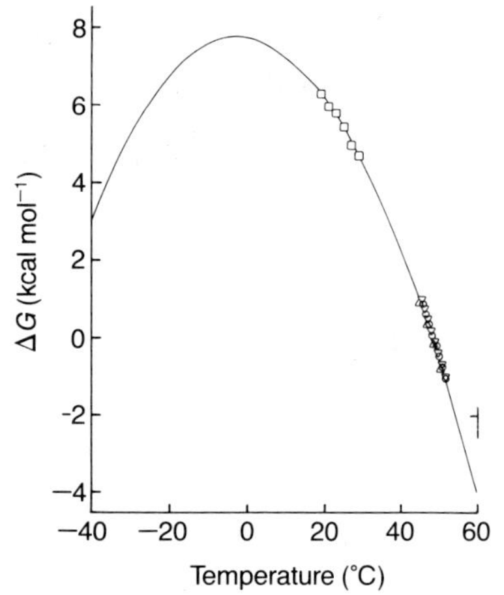

Figure 2

ΔG as a function of temperature for the unfolding of RNase T_1 at pH 7. The data points below 30°C are from urea unfolding curves, and those above 40°C are from thermal unfolding curves. The solid line was calculated with Eqn 1 using the thermodynamic parameters given here; T_m = 48.25°C, ΔH_m = 95.2 kcal mol^{-1} (399.84 kJ mol^{-1}), ΔC_p = 1.72 kcal mol^{-1} deg^{-1} (7.22 kJ mol^{-1} deg^{-1}). *Adapted from Ref. 6.*

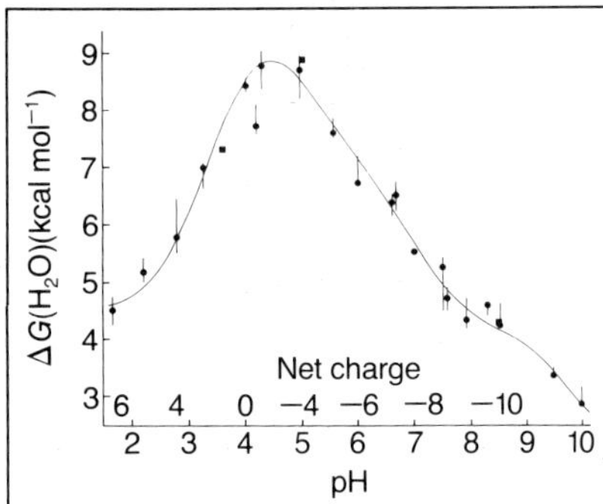

Figure 3

$\Delta G(H_2O)$ as a function of pH for the unfolding RNase T_1 at 25°C. The approximate net charge given is based on the titration curve. The solid curve was calculated assuming that certain groups have different p*K*s in the folded and unfolded conformations as described in Ref. 5. *Reproduced, with permission, from Ref. 5.*

the unfolded conformations, and thereby shift the conformational equilibrium toward the folded conformation. NMR titrations have shown that the p*K*s of the histidine residues in folded RNase T_1 are 7.9 (His92), 7.8 (His40), and 7.2 (His27)[22], and a p*K* of about 6.4 would be expected for the histidine residues in the unfolded conformations. This alone would account for a change of over 3 kcal mol^{-1} (12.6 kJ mol^{-1}) in $\Delta G(H_2O)$ with pH[5]. Higher p*K*s are expected when the net charge on the molecule is negative, and lower p*K*s are expected when the net charge is positive. Thus, the entire decrease in $\Delta G(H_2O)$ at low pH can be explained by the presence of three aspartic acid residues with p*K*s of 2.7 in the folded and 4.1 in the unfolded conformations. So, both p*K* differences and electrostatic interactions among the charged groups will tend to favor the maximum conformational stability occurring near the isoelectric pH, as is generally observed.

Proteins are stabilized by compounds that bind specifically to the folded conformation

The conformational stability of RNase T_1 can be almost doubled by adding 0.2 M Na_2HPO_4 (Table I). This remarkable increase in $\Delta G(H_2O)$ also occurs in the presence of other salts, and has been shown to result primarily from the preferential binding to the folded protein of one divalent or two monovalent cations at a cation binding site and the binding of one HPO_4^{2-} ion at an anion binding site (the active site)[7]. The structure of both of these sites is now known through X-ray diffraction studies[1]. The creation of specific cation and anion binding sites on the surface of a protein through genetic engineering might be a generally useful way of increasing the conformational stability. An obvious advantage of using ion binding for stabilization is that it need not interfere with the functioning of the active site of an enzyme. The use of specific binding to stabilize a protein is not a new idea. One hundred years ago in a remarkable 98 page paper, O'Sullivan and Tompson reported[23]: '. . . we have shown that invertase when in the presence of cane sugar will stand without injury a temperature fully 25°C higher than in its absence. This is a striking fact, and, as far as we can see, there is only one explanation of it, namely, the invertase enters into combination with the sugar'.

Disulfide bonds stabilize proteins by raising the free energy of the unfolded state

In Table I, the three proteins that contain disulfide bonds unfold when the disulfide bonds are broken. When the temperature is lowered, however, RNase T_1 will refold and retains its enzyme activity even when the SH groups are masked with bulky blocking groups (Fig. 4 and Table II). This is surprising because one disulfide bond is completely buried and only about 12 Å from the active site. It will be interesting to see how the carboxymethyl groups are accommodated in the folded protein without disrupting the active site. Note that breaking the 2 to 10 disulfide bond lowers the conformational stability by over 3 kcal mol^{-1} (12.6 kJ mol^{-1}), and breaking both disulfide bonds lowers the stability by from 7 to 9 kcal mol^{-1} (29.4–37.8 kJ mol^{-1}) depending on the blocking group. Since the melting temperature is below 25°C for both RCM- and

RCAM-RNase T_1, the unfolded states existing under physiological conditions can be compared with the unfolded states at higher temperatures and in the presence of urea and GdnHCl. These studies are under way at present, and should improve our understanding of the unfolded states of proteins in the absence of denaturing conditions.

Disulfide bonds increase the conformational stability mainly by constraining the unfolded conformations of the protein and thereby decreasing their conformational entropy. Based on our results and other experimental studies in the literature, we have proposed a new equation for predicting the effect of a crosslink on the conformational entropy (ΔS_{conf}) of a protein:

$$\Delta S_{conf} = -2.1 - (3/2)R\ln n \quad (2)$$

where n is the number of residues in the loop forming the disulfide bond[2]. This equation predicts that the conformational stability of a protein will be raised by 3, 4 and 5 kcal mol^{-1} (12.6, 16.8 and 21 kJ mol^{-1}) by adding loops of 15, 45 and 135 residues, respectively. Thornton[24] has shown that 49% of the disulfide bonds in globular proteins are separated by less than 24 residues and that 15 residues is the average separation. However, it has proven difficult to stabilize subtilisin BPN′ by introducing disulfide bonds[25]. Seven different disulfide bonds have been tried and none significantly enhances the stability to irreversible inactivation. In part, this indicates that it is difficult to add a disulfide bond that does not increase the free energy of the folded state due to strain or other factors.

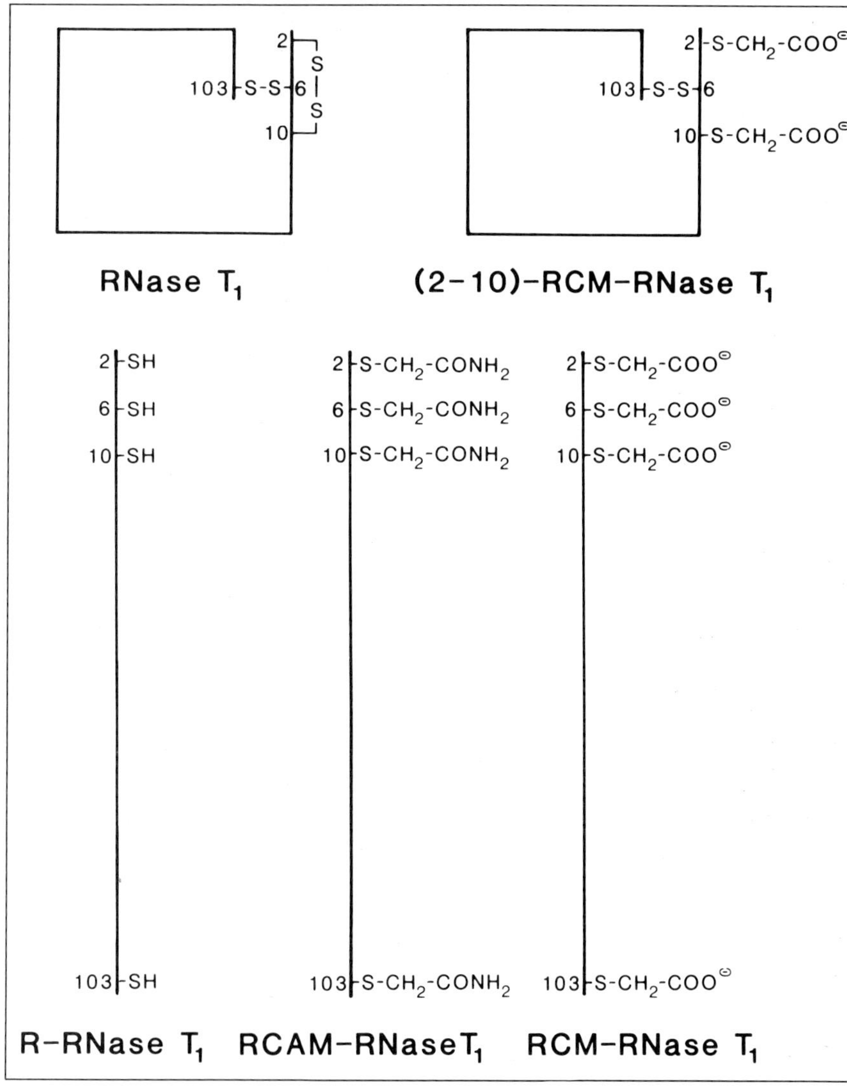

FIGURE 4

Schematic representation of RNase T_1 and the derivatives with one or both disulfide bonds broken that are characterized in Table II.

PROTEINS CAN BE STABILIZED OR DESTABILIZED BY CHANGES IN AMINO ACID SEQUENCE

The new found ability to construct proteins to order has generated great interest in many areas of biochemistry[26]. In the protein folding

TABLE II

Activity, melting temperature and conformational stability of RNase T_1 and derivatives (see Fig. 4) with one or both disulfide bonds broken[a]

Protein	Activity[b]	T_m[c]	$\Delta G(H_2O)$[d]		$\Delta(\Delta G)$	
	(%)	(°C)	(kcal mol^{-1})	(kJ mol^{-1})	(kcal mol^{-1})	(kJ mol^{-1})
RNase T_1	100	59.3	10.2	42.84	–	–
(2–10)-RCM-T_1	86	53.3	6.8	28.56	3.4	14.28
R-T_1	53	26.7	3.0	12.60	7.2	30.24
RCAM-T_1	43	21.2	1.5	6.30	8.7	36.54
RCM-T_1	35	16.6	0.9	3.78	9.3	39.06

[a] 0.1 M sodium formate, pH 5.0, 0.25 M NaCl.
[b] RNA hydrolysis at 5°C.
[c] Midpoint of thermal denaturation curve.
[d] Conformational stability at 12.5°C.

field, these new approaches should allow us to gain a better understanding of the forces that stabilize globular proteins[27]. On the basis of studies of barnase mutants created by site-directed mutagenesis, for example, Kellis *et al.*[10] have suggested that the contribution of hydrophobic interactions to protein stability '... exceeds by severalfold the values obtained from model experiments ...' and studies with λ repressor by Lim and Sauer[28] have given us a much better understanding of the 'packing rules' for a hydrophobic core. Results with a mutant of RNase T_1 that increases the conformational stability by about 0.9 kcal mol^{-1} (3.78 kJ mol^{-1}), and one that decreases the stability by about 0.8 kcal mol^{-1} (3.36 kJ mol^{-1}) are shown in Table I. Much larger changes in the conformational stability are possible. In the Gln16→Leu mutant of λ repressor, T_m is increased by 14°C, and a $\Delta G(H_2O)$ by 2.8 kcal mol^{-1} (11.76 kJ mol^{-1})[29]. To date, the largest increase in conformational stability resulting from a single change in amino acid sequence is the Asn57→Ile mutant of yeast iso-1-cytochrome *c* where T_m is increased by 17°C, and $\Delta G(H_2O)$ by ≈4.2 kcal mol^{-1} (17.64 kJ mol^{-1})[30]. So far, nature and random mutagenesis have proven much better at finding stable mutants than biochemists working from the three-dimensional structures of folded proteins, and, despite considerable progress, this is likely to remain the case for some time to come. This is surely an exciting time to be studying proteins, and especially the reaction that is the 'secret of life', protein folding, which is the subject of the next article in this book.

ACKNOWLEDGEMENTS

I am grateful to NIH (GM 37039), the Robert A. Welch Foundation (A 1060), and the Texas Agricultural Experiment Station for financial support.

REFERENCES

1 Kostrewa, D., Choe, H-W., Heinemann, U. and Saenger, W. (1989) *Biochemistry* 28, 7592–7599
2 Pace, C. N., Grimsley, G. R., Thomson, J. A. and Barnett, B. J. (1988) *J. Biol. Chem.* 263, 11820–11825
3 Shortle, D. (1989) *J. Biol. Chem.* 264, 5315–5318
4 Thomson, J. A., Shirley, B. A., Grimsley, G. R. and Pace, C. N. (1989) *J. Biol. Chem.* 264, 11614–11620
5 Pace, C. N., Laurents, D. V. and Thomson, J. A. *Biochemistry* (in press)
6 Pace, C. N. and Laurents, D. V. (1989) *Biochemistry* 28, 2520–2525
7 Pace, C. N. and Grimsley, G. R. (1988) *Biochemistry* 27, 3242–3246
8 Shirley, B. A., Stanssens, P., Steyaert, J. and Pace, C. N. (1989) *J. Biol. Chem.* 264, 11621–11625

9 Schwarz, H., Hinz, H-J., Mehlich, A., Tschesche, H. and Wenzel, H. R. (1987) *Biochemistry* 26, 3544–3551
10 Kellis, J. T., Nyberg, K. and Fersht, A. R. (1989) *Biochemistry* 28, 4914–4922
11 Shortle, D. and Meeker, A. K. (1986) *Proteins: Struc. Func. Gen.* 1, 81–89
12 Garvey, E. P. and Matthews, C. R. (1989) *Biochemistry* 28, 2083–2093
13 Chen, B. and Schellman, J. A. (1989) *Biochemistry* 28, 685–691
14 Pace, C. N., Shirley, B. A. and Thomson, J. A. (1989) in *Protein Structure: A Practical Approach* (Creighton, T. E., ed.), pp. 311–330, IRL Press
15 Arfin, S. M. and Bradshaw, R. A. (1988) *Biochemistry* 27, 7979–7984
16 Becktel, W. J. and Schellman, J. A. (1987) *Biopolymers* 26, 1859–1877
17 Regan, L. and Degrado, W. F. (1988) *Science* 241, 976–981
18 Griko, Y. V., Privalov, P. L., Sturtevant, J. M. and Venyaminov, S. Y. (1988) *Proc. Natl Acad. Sci. USA* 85, 3343–3347
19 Baldwin, R. L. (1986) *Proc. Natl Acad. Sci. USA* 83, 8069–8072
20 Privalov, P. L. and Gill, S. J. (1988) *Adv. Protein Chem.* 39, 191–234
21 Wada, A. and Nakamura, H. (1981) *Nature* 293, 757–758
22 Inagaki, F., Kawano, T., Shimada, I., Takahashi, K. and Miyazawa, T. (1981) *J. Biochem. (Tokyo)* 89, 1185–1195
23 O'Sullivan, C. and Tompson, F. W. (1890) *J. Chem. Soc.* 57, 834–931
24 Thornton, J. M. (1981) *J. Mol. Biol.* 151, 261–287
25 Mitchinson, C. and Wells, J. A. (1989) *Biochemistry* 28, 4807–4815
26 Mutter, M. and Vuilleumier, S. (1989) *Angew. Chem. Int. Ed. Engl.* 28, 535–554
27 Alber, T. (1989) *Annu. Rev. Biochem.* **58**, 765–798
28 Lim, W. A. and Sauer, R. T. (1989) *Nature* 339, 31–36
29 Pakula, A. A. and Sauer, R. T. (1989) *Proteins: Struc. Func. Gen.* 5, 202–210
30 Das, G., Hickey, D. R., McLendon, G. and Sherman, F. (1989) *Proc. Natl Acad. Sci. USA* 86, 496–499

C. N. Pace is at the Biochemistry Department, Texas A&M University, College Station, TX 77843, USA.

How Does Protein Folding Get Started?

Robert L. Baldwin

Three models for the initiation of protein folding are considered in the light of recent experiments. The three models emphasize the possible roles of: (1) a hydrophobic collapse, (2) formation of secondary structure, or (3) formation of one or more specific interactions. Whether these models are likely to be contradictory or complementary is discussed.

Work on the initiation of protein folding has reached an interesting stage: fundamental discoveries have been made but the results don't fit a single picture. On the one hand, there is growing evidence that denatured proteins undergo a hydrophobic collapse at low denaturant concentrations, like the conditions used to start refolding. On the other hand, it is now clear that secondary structures such as α-helices and reverse turns can be formed in peptide fragments more readily than was believed in the past. A decade ago, it was difficult to argue that any structure could be detected at the start of folding. Now there is good evidence for some structure of two quite different kinds. Nevertheless, a careful analysis of the role of folding in initiating the formation of disulfide bonds in bovine pancreatic trypsin inhibitor (BPTI)[1] concludes that, although structure may be present in the reduced protein, it plays no significant role in directing the pathway of disulfide bond formation until the first correct disulfide bond is formed.

In this review I discuss protein folding *in vitro* and limit discussion to those small, single-domain proteins which unfold and refold in the absence of ribosomes, enzymes and other factors. The central question for these small proteins is how the amino acid sequence determines the three-dimensional structure and the pathway of folding. Folding is a fast reaction (e.g. 0.1 s) for small proteins unless the *cis–trans* isomerization of proline residues or the formation of disulfide bonds is involved.

With one clearly demonstrated exception, the folding reactions of small proteins are reversible: provided that artifacts are avoided such as aggregation and precipitation of the unfolded protein, or autolysis, or chemical damage, the unfolded protein refolds readily in suitable conditions. The interesting exception is the bacterial protease subtilisin, which presumably is just the first of this class of proteins to be studied. It is synthesized with an N-terminal 'pro'sequence which is removed by self cleavage after folding, as the proenzyme becomes mature, active subtilisin. Folding of the proenzyme can be achieved *in vitro* but mature subtilisin, once unfolded, cannot be refolded in the same conditions used to fold the proenzyme[2].

Current work on the initiation of folding is focused on three topics: (1) hydrophobic collapse and its relation to the molten globule state, (2) stable secondary structures that are observable in peptide fragments, and (3) specific interactions that stabilize folding. Each of these topics has its own model for folding. The molten globule model proposes that folding begins with a hydrophobic collapse to the molten globule state and that properties of the molten globule state are critical for later steps in folding. The model does not specify exactly what these steps are. One possibility is exploration of the available compact conformations, since this number is quite small[3] compared to the astronomically large number of conformations of the unfolded polypeptide chain. The term 'molten globule state'[4] refers to a rarely observed state which nevertheless may occur commonly during folding (see below). The name is based on the unusual properties of this conformation: it is highly compact and has a high content of secondary structure as estimated by circular dichroism (CD), but the side chains do not have fixed conformations, unlike buried side chains in native proteins.

The framework model of folding postulates that secondary structures play the central role in determining the folding pathway. Folding begins with the formation of individual, transient secondary structures ('flickering clusters') that are stabilized by packing against each other. Folding is a hierarchical process in which simple structures are formed first and these interact to give more complex structures. A stable framework of secondary structure is made on which the final tertiary structure is assembled by squeezing out H_2O from the hydrophobic core, as the side chains are fitted together like pieces of a jigsaw puzzle. The structure of myoglobin has been used to argue that its folding might easily proceed by a framework model[5,6]. Its individual helices would form transiently and then be stabilized by fitting together at helix pairing sites[7], which are described by a simple and elegant model of 'ridge into groove' pairing[8]. Experiments on folding intermediates indicate that secondary structure is formed before the tertiary structure[9] and these observations have lent weight to the framework model: the experiments might, however, be explained by the molten globule model.

Specific interactions stabilize folding and vice versa, if the folded structure enables the specific interaction to be made. Thus, specific interactions could direct the folding pathway. Consider the linkage between the pathways of folding and disulfide bond formation in BPTI[1]. Formation of a correct disulfide bond in reduced, unfolded BPTI should stabilize a folded conformation favoring that disulfide bond by at least 100-fold, as estimated[1] from theory for the effect of a crosslink on the conformational entropy of a flexible chain polymer. After the first disulfide bond is made at random, disulfide interchange allows rapid equilibration of all possible 1-disulfide intermediates, so long as the folding is weak enough not to interfere seriously with disulfide equilibration. A predominant species with a correct disulfide bond (the 30–51 bond) is found in the absence of urea, together with a nearly random distribution of other 1-disulfide species, but in 8 M urea the 30–51 species drops to a low level. The change between 0 M and 8 M urea can be used to estimate to what extent the 30–51 bond is stabilized by the folded conformation present in 0 M urea. In considering purely noncovalent folding reactions, it is also possible to suppose that specific interactions direct folding[1].

I describe next some recent experiments that lend support to each of these models for folding. In older work, considerable reliance was placed on analogies with model compounds (such as the soap micelle discussed next) to deduce which structures might be formed first in folding. It was always recognized that these analogies are only suggestive, and appeals to different models sometimes give conflicting results. Recently, peptide fragments have been used to model folding intermediates: finding native-like structure in a fragment strongly suggests that the fragment structure has a place on the folding pathway (see Fig. 1). Even with the use of these more realistic peptide models, long-range effects may be missing that are important in determining the folding pathway of the whole protein.

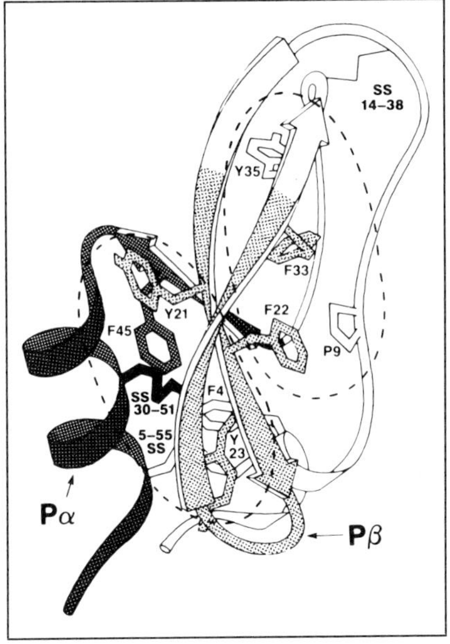

FIGURE 1

Schematic drawing showing the regions of bovine pancreatic trypsin inhibitor (BPTI) present in a peptide model of a folding intermediate[23]. The two peptides, Pα and Pβ are connected by the 30–51 disulfide bond. *Reproduced, with permission, from Ref. 23.*

HYDROPHOBIC COLLAPSE: STRUCTURE OF THE MOLTEN GLOBULE STATE*

Analogy between a folded protein and a soap micelle has been used to argue in favor of a hydrophobic collapse at the start of folding. A globular protein has its ionizing groups on the outside and its hydrophobic side chains buried on the inside: so does a soap micelle. The major force driving both protein folding and soap micelle formation is the hydrophobic interaction. The folded protein was designed by evolution to have a closely packed, tightly fitting hydrophobic core[10,11] unlike the soap micelle whose hydrophobic side chains are packed loosely. Thus, a protein might undergo a non-specific collapse at the start of folding, resulting in a loosely packed structure. Synthetic co-polymers of hydrophilic and hydrophobic amino acids do show fairly compact conformations[12].

Convincing evidence for a hydrophobic collapse of denatured staphylococcal nuclease came by a curious route. Some point mutants were found that show reduced stability to unfolding and the mutations apparently affect the denatured protein, not the native protein, as judged by the surprisingly large changes in the slopes of their denaturation curves[13]. Since these changes in slope might have other explanations, C-terminal deletions of staphylococcal nuclease were made[14] that cause the native protein to unfold: the properties of the denatured polypeptide chain could be studied in the absence of denaturants in these long fragments. The mutations studied earlier were transferred into the fragments, where they were found to stabilize a compact form, as measured by size exclusion chromatography. Structure could also be detected by CD. The mutations introduce side chains that are more hydrophobic, consistent with the notion of a hydrophobic collapse, and

* **Addendum**. Because it is not yet known if the model for the molten globule form of a protein proposed by Ptitsyn[16] describes the experimentally observed form, it is desirable to distinguish between them. In a current review[36], the term 'molten globule' is reserved for Ptitsyn's model and 'collapsed form' for the experimentally observed form. The term 'form' of the protein is used in preference to 'state' because the collapsed form of a protein may not be a unique state like the native state.

the mutations do not stabilize the native state. In fact, when the mutations are put in rank order according to their effect on stability, this order is opposite in the native state to that of the collapsed, denatured state. The compactness of the collapsed denatured protein is remarkable: as estimated by hydrodynamic radius, it is nearly as compact as the native protein[14].

Earlier, a kinetic study of the urea unfolding transitions of several small proteins near 0°C, using urea-gradient gel electrophoresis[15], had shown that the denatured form typically becomes compact before reaching the rate-limiting intermediate in refolding. Thus, a plausible connection exists between a hydrophobic collapse of the denatured protein and molten globule intermediates in refolding. Molten globule intermediates have been found in the equilibrium unfolding transitions of only a few small proteins (for review, see Ref. 16). High salt concentrations stabilize a molten globule form of the β-lactamase from *Bacillus cereus*, produced by unfolding either at low pH or high pH[17]: perhaps molten globule forms of other proteins can be found in this way. The best-studied molten globule intermediate is the A-form of bovine α-lactalbumin (α-LA), produced by unfolding at low pH. Intermediates with properties similar to the A-form are found at neutral pH both in the GuHCl- (Ref. 18) and heat-induced unfolding transitions and in the kinetics of refolding[19]. Rapidly formed folding intermediates, which appear to be molten globule intermediates, have been found in the refolding reactions of several small proteins[16,19,20]. Mutants of β-lactamase from *Staph. aureus* have been found that appear to be blocked in folding at a molten globule stage[21].

The technique of two-dimensional NMR (see also article by Wright on p. 95) has been used successfully[22] to investigate the structure and dynamics of the A-form of guinea pig α-LA. Perhaps the most striking result is the comparison of α-helix content found by NMR and estimated by CD. Only a small number (8–10) of slowly exchanging peptide NH protons are found by NMR, as compared to the high helix content estimated by CD (30–40%). The slowly exchanging NH protons have been assigned to residues 89–96, which form an α-helix in native α-LA. Thus the molten globule form of guinea pig α-LA appears to stabilize a single helix found in the native protein*, and the explanation for the high helix content estimated by CD remains to be found. It may be rapidly fluctuating secondary structure[16], which does not give rise to stably protected NH protons.

α-Helices, hairpin loops and reverse turns in peptides

The peptide hydrogen bond, -NH . . . OC- has at best marginal stability in H_2O because the NH and CO groups make competing hydrogen bonds to H_2O. Knowing this behavior, established with model compounds, protein chemists for a long time accepted at face

* **Addendum**. At least two of the α-helices present in native guinea pig α-lactalbumin have now been found in the 'collapsed' or 'molten globule' form of this protein, as judged by the presence of protected peptide NH protons (P. A. Evans, C. M. Dobson, pers. commun.).

value the early negative results of experiments testing for α-helix formation in peptide fragments of proteins. The generalization was accepted that short, individual, secondary structures can be stabilized only through interactions between subdomains that are present in the tertiary structures of proteins. In the last few years, however, it has been shown that α-helices, reverse turns and even a hairpin loop of antiparallel β-sheet[23] can be observed in aqueous solution in peptide fragments. The formation of secondary structures in peptides and its significance for the initiation of protein folding has been reviewed recently[24]. Finding that a secondary structure as small as a reverse turn can be detected in a peptide in aqueous solution shocked the scientific community when it was first announced.

The 'α-helix revolution' began with an investigation of why the 13-residue C-peptide of ribonuclease A (RNaseA) apparently shows partial helix formation in H_2O: this was predicted to be impossible, if the rules for forming the C-peptide helix are the same as for random sequence copolymers of amino acids. The latter studies had indicated that all short peptides (≤20 residues) should be unable to form the α-helix in H_2O. A major reason* for the unexpected stability of the C-peptide helix turned out to be specific interactions between pairs of side chains in the helix[25,26]: A $Glu2^-$ . . . $Arg10^+$ ion pair interaction, and a Phe8 . . . $His12^+$ ring interaction. The interaction of charges at either end of the helix with the α-helix macrodipole was also found to be important for helix stability. Since the $Glu2^-$ . . . $Arg10^+$ and Phe8 . . . $His12^+$ interactions are probably present in both the isolated C-peptide helix and intact RNaseA, the C-peptide helix is an *autonomous folding unit*[25]: the interactions that direct its folding are present within its own amino acid sequence. The $Glu2^-$ . . . $Arg10^+$ salt bridge in RNaseA terminates the N-terminal end of this helix since, in order to make the hydrogen-bonded ion pair (separated by eight residues in sequence), it is necessary that Glu2 be in a non-helical conformation.

Direction of folding by specific interactions

A straightforward thermodynamic argument, based on linkage[1], shows that a specific interaction should stabilize a folded structure favoring that interaction. Nevertheless, experimental confirmation using a peptide model[23] came as a major breakthrough. The 30–51 disulfide bond was used to join two peptides, each of which is structureless by itself (Fig. 1): a 16-residue C-terminal peptide from a helical region of BPTI was joined to a 14-residue peptide that forms a hairpin loop of antiparallel β-sheet in BPTI. CD and NMR data indicate that the two peptides assume native-like conformations once they are joined by the 30–51 bond (see Fig. 1). Thus, a specific interaction that joins two segments (covalently, in this case) can stabilize the folding of a subdomain; moreover, native-like structure is formed.

* **Addendum**. Another major reason for the unexpected stability of the C-peptide helix is the surprisingly high intrinsic helix-forming tendency of alanine residues[37].

The drawing in Fig. 1 emphasizes the secondary structures that are formed but the main major factor energetically is probably the formation of a small hydrophobic core, including the side chains of Tyr21, Phe45 and Tyr23. The effects of these and other side chains on the relative stabilities of the 1-disulfide and 2-disulfide intermediates in whole BPTI are being tested by site-directed mutagenesis[27]. A packing algorithm has been proposed[10] for predicting which mutations in the hydrophobic core of a protein are compatible with stable folding, and tests based on randomizing the core by mutagenesis[11] indicate that proper formation of the core is the central feature energetically of protein folding, but the problem of specifying how the side chains must be packed in the core is still not solved.

The formation of a disulfide bond always has a major effect on the folding of a trapped intermediate because the disulfide bond is covalent. It remains to be seen, in purely noncovalent folding reactions, whether any single noncovalent interaction can play a directing role in folding analogous to disulfide bond formation. The mechanism by which two subdomains can mutually stabilize their folding is commonly thought to be the mechanism proposed in the framework model of folding: close packing over an extended surface, thus gaining stability through the hydrophobic interaction (exclusion of H_2O from nonpolar surfaces) as well as through van der Waals and other local interactions.

Sequential and Alternative Pathways of Folding

A question of much current interest is whether protein folding follows a simple sequential pathway (Fig. 2a) or employs alternative pathways simultaneously (Fig. 2b). It has even been proposed[28] that protein folding may resemble the assembly of a jigsaw puzzle, with no fixed starting point and no unique pathway or pathways. An intriguing experiment which suggests that there is a single unique pathway is based on the folding of yeast iso-2 cytochrome *c* at mild alkaline pH[29]. An inactive form of the oxidoreduction protein is produced in these conditions, and it is formed with easily measurable kinetics when native cytochrome *c* is brought from neutral pH to alkaline pH. The intriguing result is that, when refolding is caused to occur at alkaline pH, a native-like, active form of cytochrome *c* appears transiently as an intermediate in the folding process, indicating that in refolding the native form must be produced first and the inactive form is then produced from the native form.

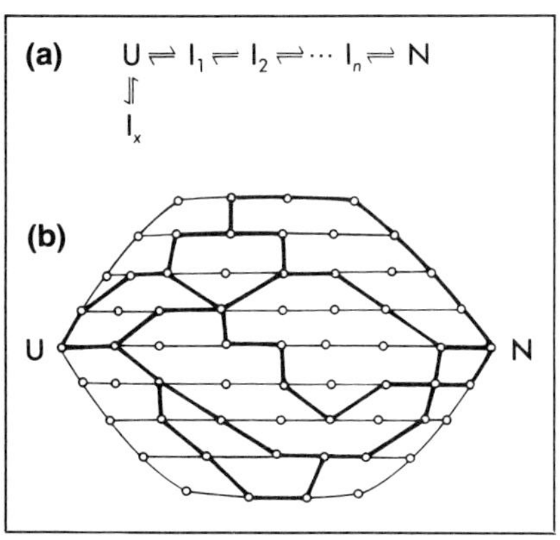

FIGURE 2

Two possible representations of a protein folding pathway from unfolded (U) to native (N). (**a**) Simple sequential model: the intermediates I_1, $I_2 \ldots I_n$ are on-pathway, and the intermediate I_x is off-pathway. (**b**) Alternative pathways, used simultaneously (*diagram after a suggestion of Dr David Shortle*).

In experiments with folding intermediates, how is it possible to decide whether an intermediate is on-pathway or off-pathway? If the folded structure of the intermediate is native-like, this obviously suggests it should be on the folding pathway. A test which may be employed is to ask whether the intermediate, when isolated by a trapping procedure, refolds more rapidly than U and unfolds more rapidly than N. This is the behavior expected of the on-pathway intermediates $I_1, \ldots I_n$ in the simple sequential mechanism of Fig. 2a, whereas the off-pathway intermediate I_x is expected to refold more slowly than U. The test has been applied to a pH 2 unfolding intermediate of RNaseA[30] and to the native-like refolding intermediate (I_N) of RNaseA[31]: in both cases the test indicates that the intermediates are on-pathway.

Pulse-labeling folding intermediates

As discussed above, one can test for structural intermediates at the start of folding by analysing the folding of fragments. This approach has produced evidence for a hydrophobic collapse in long fragments of staphylococcal nuclease[14], for autonomous helix formation by the C-peptide of RNaseA[25,26], for the presence of reverse turns in small peptides in aqueous solution[24], and for native-like structure in two subdomains of BPTI joined by a disulfide bond[23]. The measurements take advantage of the fact that individual fragments cannot fold completely to form the structure of the intact protein, although pairs of complementing fragments may do so. Thus, the fragment structures can be characterized at leisure by NMR and CD.

To determine the kinetic pathway of folding of the whole protein, a kinetic trapping method is needed. The 1-disulfide and 2-disulfide intermediates of BPTI have been trapped by covalent blocking of the remaining free -SH groups[1]. For noncovalent folding reactions a trapping method has been proposed[32,33] that allows pulse labeling of the H_2O-accessible peptide NH (amide) protons, followed by two-dimensional NMR determination of the ^1H label present in each amide resonance after folding is complete. The method gives a kinetic curve of availability for exchange versus time of folding for each amide proton that is stable to exchange with D_2O in the native protein. The limitation arises because the two-dimensional NMR experiment, which requires 8–10 hours, is made in D_2O. Thus, only about 40 of the 120 peptide NH protons in RNaseA can be studied.

Preliminary results with this new method have been published for horse heart cytochrome *c*[33] and RNaseA[32]. In each case, the results show that individual steps in folding can be resolved in which subdomains are protected against exchange with solvent. In the folding of cytochrome *c*, the pairing and mutual stabilization of the N-terminal- and C-terminal helices is an early event in folding. For RNaseA, the surprising result is that the entire β-sheet appears to be formed rapidly and cooperatively at an early stage in folding. The four disulfide bonds of RNaseA are already formed in this *in vitro* folding reaction, and they place sharp restrictions on the possible secondary structures that can be formed. The pulse-labeling method shows only which amide protons are protected and what is the extent of protection; inferences about structures are usually based on the assumption that native-like secondary structure are formed. The method is highly sensitive to early events in folding, and it should give detailed answers to questions about initial events in folding as well as the events that come immediately after.

Summary

Are these three models for the initiation of folding contradictory or complementary? Before taking up this question, we should consider a basic puzzle inherent in the properties of the molten globule state. Its compact conformation is probably produced by a hydrophobic collapse, yet circular dichroism measurements indicate that it has a high content of secondary structure. Why should a hydrophobic collapse induce the formation of hydrogen-bonded secondary structure? One possible explanation is that generation of a compact structure tends to produce secondary structures, because the number of ways that compact conformations can be formed is limited, and this limitation favors secondary structures[34] which are themselves compact. Another possible explanation is that a hydrophobic collapse tends to exclude H_2O from the vicinity of the peptide NH and CO groups, which then are driven energetically to form peptide hydrogen bonds[35]. A third possible explanation might be that the high content of secondary structure indicated by CD measurements is only apparent, and that some other structural property of the collapsed state is responsible for the CD spectrum. The fourth possible explanation is that there is a direct linkage between the hydrophobic collapse and the formation of secondary structures. If one side of an α-helix or β-strand is buried in the native protein, that side will typically be hydrophobic. Thus, formation of secondary structures generates hydrophobic surfaces which can interact in the collapsed state.

The molten globule state of guinea pig α-LA stabilizes an α-helix that is present in the native protein[22]. This result clearly demonstrates that there is no inherent contradiction between a hydrophobic collapse and the presence of some native secondary structure. Thus, both types of structure may coexist at the start of folding. The example of the C-peptide helix, which is stabilized by specific side chain interactions within the helix[25,26], shows that specific interactions may play an important role within the framework model of folding. Thus, these three models for the initiation of folding may fit together harmoniously. To understand the problem better, we need to learn what the next steps in folding are.

Acknowledgements

I thank Drs T. E. Creighton, P. S. Kim, H. Roder, F. X. Schmid, D. Shortle and J. A. Schellman for their discussion of these topics and Dr R. T. Sauer for a preprint of his manuscript[11]. This work was supported by grants from the US National Science Foundation (DMB 881 8398) and National Institutes of Health (GM 19988 and GM 31475).

References

1 Creighton, T. E. (1988) *Biophys. Chem.* 31, 155–162
2 Ikemura, H. and Inouye, M. (1988) *J. Biol. Chem.* 263, 12959–12963
3 Dill, K. A. (1985) *Biochemistry* 24, 1501–1509
4 Ohgushi, M. and Wada, A. (1983) *FEBS Lett.* 164, 21–24
5 Ptitsyn, O. B. and Rashin, A. A. (1975) *Biophys. Chem.* 3, 1–20
6 Cohen, F. E., Richmond, T. J. and Richards, F. M. (1979) *J. Mol. Biol.* 132, 275–288
7 Lesk, A. M. and Chothia, C. (1980) *J. Mol. Biol.* 136, 225–270

8 Chothia, C., Levitt, M. and Richardson, D. (1981) *J. Mol. Biol.* 145, 215–250
9 Kim, P. S. and Baldwin, R. L. (1982) *Annu. Rev. Biochem.* 51, 459–489
10 Ponder, J. W. and Richards, F. M. (1987) *J. Mol. Biol.* 193, 775–791
11 Lim, W. A. and Sauer, R. T. (1989) *Nature* 339, 31–36
12 Rao, S. P., Carlstrom, D. E. and Miller, W. G. (1974) *Biochemistry* 13, 943–952
13 Shortle, D. and Meeker, A. K. (1986) *Proteins* 1, 81–89
14 Shortle, D. and Meeker, A. K. (1989) *Biochemistry* 28, 936–944
15 Creighton, T. E. (1980) *J. Mol. Biol.* 137, 61–80
16 Ptitsyn, O. B. (1987) *J. Protein Chem.* 6, 273–293
17 Goto, Y. and Fink, A. L. (1989) *Biochemistry* 28, 945–952
18 Kuwajima, K., Nitta, K., Yoneyama, M. and Sugai, S. (1976) *J. Mol. Biol.* 106, 359–373
19 Kuwajima, K., Hiraoka, Y., Ikeguchi, M. and Sugai, S. (1985) *Biochemistry* 24, 874–881
20 Kuwajima, K., Yamaya, H., Miwa, S., Sugai, S. and Nagamura, T. (1987) *FEBS Lett.* 221, 115–118
21 Craig, S., Hollecker, M., Creighton, T. E. and Pain, R. H. (1985) *J. Mol. Biol.* 185, 681–687
22 Baum, J., Dobson, C. M., Evans, P. A. and Hanley, C. (1989) *Biochemistry* 28, 7–13
23 Oas, T. G. and Kim, P. S. (1988) *Nature* 336, 42–48
24 Wright, P. E., Dyson, H. J. and Lerner, R. A. (1988) *Biochemistry* 27, 7167–7175
25 Shoemaker, K. R., Fairman, R., Kim, P. S., York, E. J., Stewart, J. M. and Baldwin, R. L. (1987) *Cold Spring Harbor Symp. Quant. Biol.* LII, 391–398
26 Rico, M., Santoro, J., Bermejo, F. J., Herranz, J., Nieto, J. L., Gallego, E. and Jimenez, N. A. (1986) *Biopolymers* 24, 1031–1053
27 Goldenberg, D. P., Frieden, R. W., Haack, J. A. and Morrison, T. B. (1989) *Nature* 338, 127–132
28 Harrison, S. C. and Durbin, R. (1985) *Proc. Natl Acad. Sci. USA* 82, 4028–4030
29 Nall, B. T. (1986) *Biochemistry* 25, 2974–2978
30 Hagerman, P. J., Schmid, F. X. and Baldwin, R. L. (1979) *Biochemistry* 18, 293–297
31 Schmid, F. X. (1983) *Biochemistry* 22, 4690–4696
32 Udgaonkar, J. B. and Baldwin, R. L. (1988) *Nature* 335, 694–699
33 Roder, H., Elove, G. A. and Englander, S. W. (1988) *Nature* 335, 700–704
34 Chan, H. S. and Dill, K. A. (1989) *J. Chem. Phys.* 90, 492–509
35 Finkelstein, A. V. and Ptitsyn, O. B. (1987) *Prog. Biophys. Mol. Biol.* 50, 171–190
36 Kim, P. S. and Baldwin, R. L. (1990) *Annu. Rev. Biochem.* 59 (in press)
37 Marqusee, S., Robbins, V. H. and Baldwin, R. L. (1989) *Proc. Natl Acad. Sci. USA* 86, 5286–5290

R. L. Baldwin is at the Department of Biochemistry, Beckman Center, B400, Stanford University, Stanford, CA 94305, USA.

Protein Conformational Prediction

Gerald D. Fasman

The prediction of the secondary and tertiary structure of globular and membrane proteins is reviewed. Prospects are encouraging for future developments, but present algorithms require cautious interpretation.

The goal of the prediction of the conformation of polypeptides and proteins remains elusive, but is being actively pursued by a host of researchers using a variety of methodologies. As new primary structures appear daily, mainly through the use of DNA sequencing technologies, the desire to estimate their structures to aid in understanding structure and function remains a high priority.

Although understanding the role of structure–function relationships is today a topic of intense research, the aim of the prediction of protein structure from the primary sequence has been attracting the attention of researchers since 1959, when Anfinsen[1] published a monograph which focused on this topic. The classical experiments of Anfinsen *et al.*[2,3] had shown that ribonuclease could be denatured and refolded without loss of enzymatic activity. This implied that the amino acid sequences contained sufficient information to define the three-dimensional structure of a protein in a particular environment. Since that discovery both theoreticians and experimentalists have tackled the chain folding problem with limited success.

Aside from the intrinsic desire to understand the beauty of nature's architecture, there have been proposed many exceptionally important applications: technological developments of industrial, clinical and agricultural importance may be achieved in the coming years by imitation of the interactions between macromolecules and ligands that occur naturally in the living cell. Engineered hormone analogs could possibly have more desirable properties than their native counterparts.

Theoretical efforts can be categorized into three main areas: energetic, heuristic and statistical. The school that follows the assumption that a protein folds so as to minimize the free energy of the system has had many contributors (e.g. Refs 4–7). Although offering long-range solutions for the future, these methods are still in their infancy and will not be discussed in this abbreviated review. Work of this type can be traced back to 1965 when Liquori and coworkers[8] and Ramachandran *et al.*[9] first demonstrated that the peptide unit can adopt only certain allowed conformations. Energy calculations have been refined recently, but minimization schemes have failed to predict chain folding accurately[10,11]. The first attempted prediction of the secondary structure of protein conformation was pioneered by Guzzo[12] in 1965. The Schiffer and Edmundson[13] method of locating hydrophobic arcs by the helical wheel method is still widely used.

Today there are at least between 15–20 proposals for the prediction of the secondary structure of protein conformation from the amino acid sequence. All of these cannot be discussed herein because of space limitations, but are discussed in detail in Ref. 14.

A problem arises in evaluating X-ray determined structures because various criteria have been used by X-ray crystallographers to define various secondary structures[15]. However compilation of secondary structures published by Kabsch and Sander[16], a pattern-recognition process of hydrogen-bonded and geometrical features, extracted from X-ray coordinates, is widely used.

Proteins may be classified into three main categories: globular, membrane and fibrous proteins. The latter class will not be discussed herein. Globular proteins tightly pack to form distinct tertiary conformations, by association of various secondary structures, α-helices, β-strands, with intervening β-turns, into various domains. These domains were first discussed by Wetlaufer[17] and Liljas and Rossman[18]. Earlier work on defining turns, laid the foundation for the concept of domains. Domains have been classified into five classes according to their secondary structural arrangement[19,20]: (1) all α, with α-helices and no β-strands; (2) all β, with only β-strands and no α-helices; (3) α/β, in which the chain alternates between α-helices and β-strands; (4) α and β, in which α- and β-regions tend to segregate into separated regions; and (5) coil, in which there is little or no secondary structure. An excellent recent review by Richardson and Richardson (in Ref. 14) illustrated with schematic drawings, can be highly recommended for additional reading.

The folding of proteins has been attributed mainly to hydrophobic interactions and hydrogen bonding networks. However other factors are emerging as playing important roles, for example electrostatics (see Rogers in Ref. 14). Ion pairs were shown[21] to play a role, when on the average, one-third of the charged residues in a protein are involved in ion pairs and 76% of these are concerned with stabilizing the tertiary structure. Sundaralingam *et al.*[22] surveyed 47 globular proteins and found many ion pairs of the type $i,i+3$ and $i,i+4$ in α-helices. Other weaker interactions, such as aromatic–aromatic interactions, have been discussed in a recent review[23].

Techniques of Protein Prediction

The first requirement is to have available the amino acid sequence. Protein sequences are collected and available from the Protein Information Resource (PIR) data bank. The second source is from DNA data banks which can be easily converted into primary amino acid structures.

One of the earliest attempts to determine whether the relationships existing between protein sequences were due to homology or by chance was by the sequence alignment algorithm of Needleman and Wunsch[24]. The use of dynamic programming methods to carry out global comparison of two sequences was soon amplified by many techniques. Because insertions and deletions often occur at the loop regions of protein secondary structures, improvements can be made by introducing penalties for insertions/deletions in α-helices or β-strands[25,26]. There are over 20 algorithms based on various sequence alignment techniques available in the literature.

HYDROPHOBICITY

The hydrophobic effect, first expounded by Kauzman[27], has been the central focus in the studies of protein folding, self-assembly, and conformation. The most widely used method based on hydrophobicity is that of Kyte and Doolittle[28], which progressively evaluates the hydrophilicity and hydrophobicity of a protein along with amino acid sequence. They, however, emphasized that the extent to which residues are buried depends not only upon strict hydrophobicities but also upon stearic effects that determine packing between the secondary structure in the crowded interior of the macromolecule. Chothia[29] pointed out that hydrophobic indices of the amino acids have a poor correlation to the extent which a residue is buried in the native folded protein matrix. Eisenberg *et al.*[30] analysed the structure of protein in terms of the 'hydrophobic moment' of (1) the entire molecule and (2) the segment of secondary structure that makes up the polypeptide chain. The hydrophobic dipole can be used to measure the amphilicity of the structure. Recently Cornette *et al.*[31] compared 38 published hydrophobicity scales for their ability to identify the characteristic period of α-helices, and have computed an optimum scale for this purpose using a new Eigenvector method.

APPROACHES TO PROTEIN CONFORMATION

Many methods exist to evaluate protein conformation from different perspectives.

(1) *Solvent accessibility*. The solvent accessibility, a counterpart of the buried hydrophobic areas, usually water, has been used by Lee and Richards[32] to give a quantitative assessment of the probability of an amino acid being buried or exposed, which was found to be proportional to the surface area. Richards and coworkers have expanded on this idea and have examined the total volume group value distributions and packing densities of proteins.

(2) *Packing of residues*. The degree to which ideal packing occurs determines the degree of relative stability of a particular protein. Crippen and Kuntz[33] surveyed the atom packing in high resolution X-ray crystal structures of 21 proteins and determined that the atom density of a central atom is determined primarily by its covalent bonded neighbors and proximity to the surface of the protein.

(3) *Distance geometry*. The mathematics of distance constitutes the basis of a group of algorithms for revealing the structural consequences of diverse forms of information about the macromolecule's conformation.

(4) *Amino acid physico-chemical properties*. A large number of physico-chemical properties, manifest in the amino acid side chains have been thoroughly investigated by many investigators. Attempts

have been made to correlate these properties with their relatedness between protein sequences.

PREDICTION OF SECONDARY STRUCTURES, α-HELIX, β-STRANDS AND β-TURNS

The three most widely used, or misused methods for prediction of protein secondary structure are those of Chou and Fasman[34], Garnier, Osguthorpe and Robson[35–37] and Lim[38]. The first two methods use parameters obtained from the analysis of known sequences and structures, and the third method is based on stereochemical criteria. The P_α, P_β, P_t parameters of Chou and Fasman have been widely used to predict secondary structures, to determine sequences with potential for conformational change, to design analogs of smaller polypeptides with pharmacological action (drug design), to aid in site-directed mutagenesis, and to throw more heat than light on the discussion concerning the validity of this approach to prediction of secondary structures.

The method of Garnier *et al.*[35–37] is based on information theory and can be programmed easily and unambiguously. Recently Garnier *et al.* reevaluated this method[36,37] and concluded that the existing data base does not allow evaluation of parameters required for an exact treatment of the problem. The prediction of β-turns has been widely researched, as it was early proposed that these turns might be the sites of initiation of protein folding[39]. The method of Chou and Fasman[40] for the prediction of β-turns has received the widest application, although other methods are available with claims of much higher accuracy. Leszczynski and Rose[41] have examined looped structures, of which the β-turn is a member, in protein X-ray data and termed these 'omega loops'. These sequences were previously usually regarded as 'random coil', but have been proposed to form a non-regular secondary structure. β-Turns have also been identified as sites of carbohydrate attachment in glycoproteins[42] and of phosphorylation[43].

The three methods mentioned above have been scrutinized and evaluated carefully as to their accuracy and reliability. The exact percentage of accurate predictions is of course predicated on the definition one uses to define the secondary structures. As pointed out above, this varies widely, and thus so will the determination of accuracy. The first true trial of various predictive schemes was carried out by Schulz *et al.*[44] for adenylate kinase. Eleven predictive schemes were applied by their originators. The Chou–Fasman and Lim methods were by far the most accurate (note: that of Garnier *et al.* was not yet published). Since that time many analyses of accuracy have appeared (see, for example, Ref. 45). Kabsch and Sander[46] pointed out the lack of significance of short sequence homologies and this has been widely cited to discredit predictive schemes. However, the pentapeptides analysed by Kabsch and Sander which assumed various conformations in different proteins, confirmed the previous well-known adage, that a pentamer is not of sufficient length to lock into any one conformer, and it will be driven by the sequences on either side of it to assume a preferred conformer. The degree of accuracy has been estimated to be no better than 50%

(random guess) to as high as 70–75%. Thus judicious use of these algorithms is recommended. However the application to which predictions are to be used should dictate the relative confidence required. Thus Kaiser and coworkers[47], using the Chou–Fasman parameters, synthesized a helical cytotoxic peptide with mellitin-like activity.

OTHER METHODOLOGIES

(1) *Knowledge-based modelling and design of proteins.* The approach termed 'knowledge-based' prediction depends on the analogies between a protein that is to be modelled and other proteins of known three-dimensional structure at all levels in the hierarchy of protein organization: secondary structures, motifs, domains, and quaternary or ligand interactions. Blundell and colleagues are establishing a Relational Database of Structures (ORACLE) of sequences and structures, developing procedures for aligning and classification of three-dimensional structures, and modelling of globular proteins (see articles on pp. 153 and 163 of this book). These methods will be used in modelling targets of interest in drug design and protein engineering.

(2) *Minimum energy calculations.* Many research groups are concentrating on chain folding, which is simulated computationally to find the energy minimum. The assumption in all these calculations is that a protein folds so as to minimize the free energy of the system and many investigators have developed potential functions to describe the energy surface of the polypeptide chain. There are two major problems with this approach which have yet to be effectively overcome: (1) decisions of the appropriate parameterization of the energy, including derivation of the values of the parameters; and (2) the multiple minimum problem on the global energy surface. Substantial progress is being made but the approximations required to properly model the computationally complex electrostatic and solvent effects are major stumbling blocks. A recent relevant review can be found by Mackay, Cross and Hagler in Ref. 14.

(3) *Systematic search of possible conformations.* Moult and James[48] examined the feasibility of determining the conformation of segments of a polypeptide chain up to six residues in length in globular proteins by means of a systematic search through the possible conformations. The electrostatic energy, including a solvent screening term, and the exposed hydrophobic area are evaluated for each accepted conformation. It was found that there is a wide spread of energies among the accepted conformations, and the lowest energy ones have satisfactorily small root mean square deviations from the X-ray structure. The application of computer graphics for predictive purposes is receiving renewed attention.

(4) *Class prediction.* Chou[49,50] analysed the X-ray structure of 64 different proteins, in terms of their α-helical and β-sheet regions, as well as their amino acid compositions. The four distinct classes of proteins, α, β, α+β, α/β, had significantly different amino acid compositions.

Compared to the conformational parameters of Chou and Fasman[51], there were a few significant changes, and the relative order of the amino acids was relatively constant. A computerized algorithm was developed which could assign proteins to the correct structural class based on their amino acid compositions with 80% accuracy. Chou[49,50] developed a new set of conformational parameters, P_α and P_β for the amino acids in the four classes of proteins. Klein and DeLisi[52] used the multidimensional statistical technique of discriminant analysis to allocate amino acid sequences to one of the four structural classes: high α content, high β content, mixed α and β, low content or disordered structures. Nakashima *et al.*[53] analysed the folding type of 135 proteins, whose three-dimensional structure was known, in terms of the amino acid composition. The distribution pattern of proteins in this approach was examined in relation to five folding types, α, β, α+β, α/β and irregular. The assignment of the folding type to five classes by this method gave an accuracy of 70% and to four folding classes with a reliability of 79%. Klein[54] improved upon the Nakashima *et al.*[53] analysis by use of discriminant analysis. Analysis by class, α, β, mixed (α+β or α/β) and irregular (four classes) can be predicted with an 83% reliability and to one of five classes (separate α/β and α+β) with 78% reliability. It was shown that optional linear combinations of amino acid frequencies used as attributes give slightly better accuracy than attributes using secondary prediction methods for each residue. Deléage and Roux[56] have used the technique of Nakashima *et al.*[53] for class prediction and utilized this in a modified Chou and Fasman[35,55] algorithm, called a 'double prediction method' for the prediction of proteins from their amino acid sequence (see Ref. 14), and obtained 72% success in class prediction, with 61.3% of residues correctly predicted for three states (α, β and coil).

(5) *Neural network*. The most recent addition to the prediction methodologies is the application of neural networks. Neural networks learn from existing protein structures how to predict the secondary structure of local sequences of amino acids. Three publications applying this model have appeared to date. Qian and Sejnowski[57], using non-linear neural network models, attempted to predict the secondary structure of proteins for which no homologous structures were available. The average success rate was 64.3% on three types of secondary structures (α-helix, β-sheet and coil). It was concluded that no method based solely on local information in the protein sequence is likely to produce significantly better results for non-homologous proteins. Bohr *et al.*[58] applied neural networks to the prediction of the α-helices in rhodopsin and compared their results to other predictive methods. As the structure is unknown, no evaluation of the results is possible. Holley and Karplus' use of neural networks for the prediction of protein secondary structure was based on training using a sample set of 48 proteins of known structure[59]. Then, using 14 other proteins of known structure, the method achieved a maximum overall predictive accuracy of 63% for three states, helix, sheet and coil. These authors also conclude that introduction of information related to tertiary structure

will be necessary for any dramatic improvements in prediction. It has long been appreciated that without the use of tertiary structural data, the prediction of secondary structure could at best reach an accuracy approaching 70%. The caveat that tertiary structure stabilized secondary structure has been an unwritten maxim among the aficionados.

(6) *Antigenicity*. Three methods have been developed specifically to evaluate and determine sequences which will serve as epitopes. Intrinsic hydrophilicity parameters for individual amino acids is the basis for the most popular method of prediction to date, developed by Hopp and Woods[60,61]. Using similar algorithms, Karplus and Schulz[62] and Welling *et al.*[63] proposed alternate predictive schemes based, respectively, on flexibility parameters and calculated antigenicity values for each amino acid. Antigenicity values[63] were derived by analysing antigenic determinants from a number of proteins with respect to their amino acid compositions and by comparing this with the overall amino acid compositions of the proteins themselves. These methods are reviewed in an excellent article on antibody binding to protein antigens[64].

(7) *Expert systems*. Robson and coworkers[65] developed a suite of programs called LUCIFER for the conformational study of drugs, proteins and other biomolecules. These programs, called expert systems, are computer-aided designs of complex biomolecules. The suite explicitly seeks to carry out global energy minimization as rapidly as possible and puts great emphasis on the rational use of external data to overcome any deficiencies. One scans data bases for structural homologies and then molecular modelling is utilized to complete the synthesis of the data obtained.

PREDICTION OF THE SECONDARY STRUCTURE OF MEMBRANE PROTEINS

Membrane proteins are a diverse group of proteins which play an important role in modulating the activity of many of the cell's functions. The diversity of proteins residing in the lipid bilayer of the membrane is immense, as is their function from receptors to active enzymes. The method of prediction of membrane proteins, utilized most frequently is that of Kyte and Doolittle[28]. Using a program called SOAP, this simple method for displaying the hydropathic character of a protein has been used to define the sequences which have a high probability of being imbedded in the non-polar membrane environment. Kyte and Doolittle carefully avoided associating the hydrophobic regions with any particular structure, i.e. α-helix or β-sheet. Unfortunately, in the literature it has been generally accepted and stated that these segments represent α-helical segments. It is possible of course, that they *do* represent helical sections, and probably frequently are helical, but the Kyte and Doolittle method cannot be used to verify the presence of a helical segment in a bilayer. It is worth recalling that β-sheets can also be extremely hydrophobic, or amphipathic, and porin, which spans the outer membranes of *E. coli*, and forms voltage-dependent transmembrane channels, has a conformation dominated by β-structure. It has been frequently pointed out that the application of predictive methods, derived from data on globular proteins, is unlikely to give correct predictions of membrane imbedded proteins[66]. Argos[67] developed an algorithm based on physical characteristics of the 20 amino acids to predict membrane-buried regions. However, as only two membrane protein structures have been determined by X-ray diffraction methodologies, namely *Rhodopseudomonas viridis*[68] and *Rhodopseudomonas sphaeroides* R-26[69], and these two are closely related, one cannot evaluate the accuracy of the majority of predictions of membrane proteins. Thus faith is stronger than reason in their general acceptance. Eisenberg *et al.*[70] published an algorithm which identifies α-helices involved in the interactions with membrane proteins with lipid bilayers and which distinguishes them from helices in soluble proteins. The membrane-associated helices are classified with the aid of the *hydrophobic moment plot.* The magnitude of the hydrophobic moment measures the amphipathicity of the helix (and hence its tendency to seek a surface between hydrophobic and hydrophilic phases) and the hydrophobicity measures its affinity for the membrane interior. Segments of membrane proteins in α-helices tend to fall into one of three regions of a hydrophobic moment plot: (1) monomeric transmembrane anchors lie in regions of highest hydrophobicity and smallest hydrophobic moment; (2) helices presumed to be paired and helices which are bundled together in membranes (e.g. bacteriorhodopsin) fall in the adjacent region with a higher hydrophobic moment and a smaller hydrophobicity; and (3) helices from surface-seeking proteins (such as melittin) fall in the region with still higher hydrophobic moment. α-Helices from globular proteins mainly fall in a region of lower mean hydrophobicity and hydrophobic moment. This method shows great promise. The method of Finer-Moore and Stroud[71], based on Fourier analysis of hydrophobicities, using Eisenberg hydrophobicity analysis and the Garnier *et al.* algorithm for secondary structure prediction, can determine both amphipathic α-helices and amphipathic β-sheets. This method also has great potential and is discussed in detail with many examples of its application in a recent review by Finer-Moore *et al.* (see Ref. 14). Klein *et al.*[72] used discriminant analysis to precisely classify membrane proteins as integral or peripheral and estimated the probability that the prediction was correct. When used to analyse the photoreaction center, whose structure is known (see above) it yielded excellent results. Jähnig and coworkers[73] employed a structural predictive method for membrane proteins which takes into account the amphipathic α-helices and β-sheets. For the outer membrane proteins of *E. coli*, the porins, the model suggested mainly β-sheet, which agrees well with other data. Engelman *et al.*[74] suggest that the helical structure will be the dominant motif in integral membrane protein organization. These authors have developed the GSE scale (Goldman, Steitz, Engelman) for the characteristics of the amino acids and use it to compare other hydrophobicity scales. Another computational technique for detecting amphipathic structures, mainly α-helices but also useful for β-sheets in proteins has been proposed[31]. A discussion of the predictions of the membrane proteins found in bacteriorhodopsin, acetylcholine receptor, ATPases, T-cell receptor, and the sodium channel, can be found in an article by Fasman (see Ref. 18).

Prediction of Tertiary Structure

The future prospects for the prediction of the tertiary structure of proteins are unlimited. The opportunity for the design and construction of new proteins, e.g. vaccines, drugs, herbicides and pesticides, await a better understanding of the protein folding problem. The present status of the prediction of the tertiary structure of proteins cannot be called exemplary. There are three broad approaches utilized presently for the prediction of the tertiary structure of a polypeptide: (1) use of sequence homology with peptides of known three-dimensional structure; (2) prediction of secondary structure units followed by the assembly of these units into a compact structure; and (3) use of empirical energy functions *ab initio* to derive the tertiary structure of minimum potential energy. The first two rely heavily on the data base of structures determined by X-ray crystallography, while the third, in principle, does not. The first procedure uses rules of structure only by implication. The second uses them quantitatively, with the goal of a rough outline of the structure. The third relies on quantitative calculations for evaluation of trial structures. One should recall that in 1987, despite having over 300 structures in the Brookhaven data bank, there were only 61 structures that were solved to a resolution of 2.0 Å or better[75]. Thus, the majority of the analyses of the X-ray determined structures of proteins have included very ill-defined structures, and the results should be accepted with caution. As the accomplishments are modest in this area of tertiary structure prediction, only a brief summary will be offered. Kuntz[76] discussed the state of the art in 1975, when it was very rudimentary. By 1980, when approximately 80 globular structures were known to atomic resolution, it became evident that the tertiary fold is largely determined by packing of α-helices and/or β-strands[77]. Taylor and Thornton[78] devised a procedure which further assists in recognizing these supersecondary structures. Cohen, Sternberg and coworkers[79] have developed a stepwise method, termed the combinatorial approach, for predicting the three-dimensional structure of a protein from its amino acid sequence (for review see Ref. 80). They predict the secondary structure, pack the α-helices and β-strands into an approximate native fold, and use simplified energy calculations to refine the fold into the native structure. The results are variable. Cohen and coworkers[81] have applied their combinatorial approach to tertiary structural prediction and combined it with circular dichroism measurements to establish the class of the protein. When applied to the α-unit of tryptophan synthetase, the predicted structure did not agree with the subsequently obtained X-ray determined structure. Intensive research continues into the problem of tertiary structural prediction of proteins. It has been pointed out by Rooman and Wodak[82] that the identification of predictive motifs is limited by the size of the protein structure data base, and much more data will be necessary before the goal of predicting tertiary structures of proteins will be successfully resolved. However, new X-ray structures appear regularly and it may be in the not too distant future that the data base will satisfy the requirements to perfect predictive schemes.

References

1 Anfinsen, C. (1959) *The Molecular Basis of Evolution*, John Wiley & Sons
2 Anfinsen, C. B. (1973) *Science* 181, 233–239
3 Anfinsen, C. B., Haber, E., Sela, M. and White, F. H. (1961) *Proc. Natl Acad. Sci. USA* 47, 1309–1314
4 Levitt, M. and Warshel, A. (1975) *Nature* 253, 694–698
5 Nemethy, G. and Scheraga, H. A. (1977) *Q. Rev. Biophys.* 10, 239–352
6 McCammon, J. A., Gelin, B. R. and Karplus, M. (1977) *Nature* 267, 585–590
7 Weiner, S. J., Kollman, P. A., Case, D. A., Singh, U. C., Ghio, C., Alagona, G., Profeta, S. and Weiner, P. (1984) *J. Am. Chem. Soc.* 106, 765–784
8 De Santis, P., Giglio, E., Liquori, A. M. and Ripamonti, A. (1965) *Nature* 206, 456–458
9 Ramachandran, G. N., Ramakrishnan, C. and Sasisekharan, V. (1963) *J. Mol. Biol.* 7, 95–99
10 Hagler, A. T. and Honig, B. (1978) *Proc. Natl Acad. Sci. USA* 75, 554–558
11 Cohen, F. E. and Sternberg, M. J. E. (1980) *J. Mol. Biol.* 138, 321–333
12 Guzzo, A. V. (1965) *Biophys. J.* 5, 809–822
13 Schiffer, M. and Edmundson, A. B. (1967) *Biophys. J.* 7, 121–135
14 Fasman, G. D., ed. (1989) *Prediction of Protein Structure and the Principles of Protein Conformation*, Plenum Press
15 Blundell, T. L. and Johnson, L. N. (1976) *Protein Crystallography*, Academic Press
16 Kabsch, W. and Sander, C. (1983) *Biopolymers* 22, 2577–2637
17 Wetlaufer, D. B. (1973) *Proc. Natl Acad. Sci. USA* 70, 697–701
18 Liljas, A. and Rossman, M. G. (1974) *Annu. Rev. Biochem.* 43, 475–507
19 Levitt, M. and Chothia, C. (1976) *Nature* 261, 552–558
20 Richardson, J. S. (1981) *Adv. Prot. Chem.* 34, 167–339
21 Barlow, D. J. and Thornton, J. M. (1983) *J. Mol. Biol.* 168, 867–885
22 Sundaralingam, M., Sekharudu, Y. C., Yathindra, N. and Ravichandran, V. (1987) *Proteins* 2, 64–71
23 Burley, S. K. and Petsko, G. A. (1988) *Adv. Protein Chem.* 39, 125–189
24 Needleman, S. B. and Wunsch, C. D. (1970) *J. Mol. Biol.* 48, 443–453
25 Barton, G. J. and Sternberg, M. J. E. (1987) *Protein Eng.* 1, 89–94
26 Lesk, A., Levitt, M. and Chothia, C. (1986) *Protein Eng.* 1, 77–78
27 Kauzman, W. (1959) *Adv. Prot. Chem.* 14, 1–63
28 Kyte, J. and Doolittle, R. F. (1982) *J. Mol. Biol.* 157, 105–132
29 Chothia, C. (1976) *J. Mol. Biol.* 105, 1–14
30 Eisenberg, D., Weiss, R. M., Terwilliger, T. C. and Wilcox, W. (1982) *Faraday Symp. Chem. Soc.* 17, 109–120
31 Cornette, J. L., Cease, K. B., Margalit, J. H., Spouge, J. L., Berzofsky, J. A. and DeLisi, C. (1987) *J. Mol. Biol.* 195, 659–685
32 Lee, B. and Richards, F. M. (1971) *J. Mol. Biol.* 55, 379–400
33 Crippen, G. M. and Kuntz, I. D. (1978) *Int. J. Pept. Protein Res,* 12, 47–56
34 Chou, P. Y. and Fasman, G. D. (1974) *Biochemistry* 13, 211–222, 222–245
35 Garnier, J., Osguthorpe, D. J. and Robson, B. (1978) *J. Mol. Biol.* 120, 97–120
36 Gibrat, J-F., Garnier, J. and Robson, B. (1987) *J. Mol. Biol.* 198, 425–443
37 Biou, V., Gibrat, J-F., Levin, J. M., Robson, B. and Garnier, J. (1988) *Protein Eng.* 2, 185–191
38 Lim, V. I. (1974) *J. Mol. Biol.* 88, 857–872, 873–894
39 Kuntz, I. D. (1972) *J. Am. Chem. Soc.* 94, 4009–4012
40 Chou, P. Y. and Fasman, G. D. (1979) *Biophys. J.* 26, 367–384
41 Leszczynski, J. and Rose, G. D. (1986) *Science* 234, 849–855
42 [illegible] Lefebvre, M-H. (1976) *Arch. Biochem. Biophys.* 175, 400–409
43 Small, D., Chou, P. Y. and Fasman, G. D. (1977) *Biochem. Biophys. Res. Commun.* 79, 341–346
44 Schulz, G. E., Barry, C. D., Friedman, J., Chou, P. Y., Fasman, G. D., Finkelstein, A. V., Lim, V. I., Ptitsyn, O. B., Kabat, E. A., Wu, T. T., Levitt, M., Robson, B. and Nagano, K. (1974) *Nature* 250, 140–142

45 Matthews, B. W. (1975) *Biochim. Biophys. Acta* 405, 442–451
46 Kabsch, W. and Sander, C. (1984) *Proc. Natl Acad. Sci. USA* 81, 1075–1078
47 DeGrado, W. F., Kezdy, E. J. and Kaiser, E. T. (1981) *J. Am. Chem. Soc.* 103, 679–681
48 Moult, J. and James, M. N. G. (1987) *Proteins* 1, 146–163
49 Chou, P. Y. (1979) in *CECM Protein Folding Workshop*, Centre National de la Recherche Scientifique, Université de Paris-Sud, 91405 Orsay Cedex, France
50 Chou, P. Y. (1980) in *Abstracts, Second Chemical Congress of the North American Continent*
51 Chou, P. Y. and Fasman, G. D. (1978) *Adv. Enzymol.* 47, 45–148
52 Klein, P. and DeLisi, C. (1986) *Biopolymers* 25, 1659–1672
53 Nakashima, H., Nishikawa, K. and Ooi, T. (1986) *J. Biochem. (Tokyo)* 99, 153–162
54 Klein, P. (1986) *Biochim. Biophys. Acta* 874, 205–215
55 Chou, P. Y. and Fasman, G. D. (1978) *Annu. Rev. Biochem.* 47, 251–276
56 Deléage, G. and Roux, B. (1987) *Protein Eng.* 1, 289–294
57 Qian, N. and Sejnowski, T. J. (1988) *J. Mol. Biol.* 202, 865–884
58 Bohr, H., Bohr, J., Brunak, S. and Cotterill, R. M. J. (1988) *FEBS Lett.* 241, 223–228
59 Holley, L. H. and Karplus, M. (1989) *Proc. Natl Acad. Sci. USA* 86, 152–156
60 Hopp, T. P. and Woods, K. R. (1981) *Proc. Natl Acad. Sci. USA* 78, 3824–3828
61 Hopp, T. P. and Woods, K. R. (1983) *Mol. Immunol.* 20, 483–489
62 Karplus, M. and Schulz, G. E. (1985) *Naturwissenschaften* 72, 212–213
63 Welling, G. W., Weiser, W. J., van der Zee, R. and Welling-Webster, S. (1985) *FEBS Lett.* 188, 215–218
64 Getzoff, E. D., Tainer, J. A., Lerner, R. A. and Geysen, H. M. (1988) *Adv. Immunol.* 43, 1–98
65 Robson, B., Platt, E., Fishleigh, R. V., Marsden, A. and Millard, P. (1987) *J. Mol. Graphics* 5, 8–17
66 Paul, C. and Rosenbusch, J. P. (1985) *EMBO J.* 4, 1593–1597
67 Argos, P., Mohana-Rao, J. K. and Hargrave, P. A. (1982) *Eur. J. Biochem.* 128, 565–575
68 Deisenhofer, J., Epp. O., Miki, K., Huber, R. and Michel, H. (1985) *Nature* 318, 618–624
69 Allen, J. P., Feher, G., Teates, T. O., Rees, D. C., Deisenhofer, J., Michel, H. and Huber, R. (1986) *Proc. Natl Acad. Sci. USA* 83, 8589–8593
70 Eisenberg, D., Schwartz, E., Komaromy, M. and Wall, R. (1984) *J. Mol. Biol.* 179, 125–142
71 Finer-Moore, J. and Stroud, R. M. (1984) *Proc. Natl Acad. Sci. USA* 81, 155–159
72 Klein, P., Kanehisa, M. and DeLisi, C. (1985) *Biochim. Biophys. Acta* 815, 468–476
73 Vogel, H., Wright, J. K. and Jähnig, F. (1985) *EMBO J.* 4, 3625–3631
74 Engelman, D. M., Goldman, A. and Steitz, T. A. (1981) *Methods Enzymol.* 88, 81–88
75 McGregor, M. J., Islam, S. A. and Sternberg, J. E. (1987) *J. Mol. Biol.* 198, 295–310
76 Kuntz, I. D. (1975) *J. Am. Chem. Soc.* 97, 4362–4366
77 Levitt, M. and Chothia, C. (1976) *Nature* 261, 552–558
78 Taylor, W. R. and Thornton, J. M. (1984) *J. Mol. Biol.* 173, 487–514
79 Edwards, M. S., Sternberg, M. J. E. and Thornton, J. M. (1987) *Protein Eng.* 1, 173–181
80 Schulz, G. E. and Schirmer, R. H. (1979) in *Principles of Protein Structure*, Springer-Verlag
81 Hurle, M. R., Matthews, C. R., Cohen, F. E., Kuntz, I. D., Toumadje, A. and Johnson, W. C., Jr (1987) *Proteins* 2, 210–224
82 Rooman, M. J. and Wodak, S. J. (1988) *Nature* 335, 45–49

G. D. Fasman is in the Graduate Department of Biochemistry, Brandeis University, Waltham, MA 02254-9110, USA.

Membrane Protein Models: Possibilities and Probabilities

Jean-Luc Popot and Donald M. Engelman

Many integral membrane proteins show hydrophobic amino acid stretches that are interpreted as transbilayer helices, and experimental data often support such a contention. To what extent can hydrophobic analysis be trusted? Should one expect other folding motifs to have equal status?

In a previous note in *TIBS*[1] about the topography of the nicotinic acetylcholine receptor (nAChR), we pointed out some conflicts between predictions or conclusions derived from hydrophobicity analysis of the sequence, site-directed mutagenesis, biochemistry and immunology. Our main objective was to call attention to possible pitfalls in some of these methods. Reactions to this text pointed to another issue, namely the lack of universal agreement about the use of hydrophobicity analysis in the prediction of topology. We had noted that most models were based on identifying in the sequence stretches of residues long enough (about 20 residues) and hydrophobic enough (free energy of transfer from water to lipids about -20 kcal mol^{-1} [84 kJ mol^{-1}]) to form stable transmembrane α-helices. We only briefly alluded to the search for other structural elements, e.g. amphipathic helices. One reader reminded us that in the *E. coli* outer membrane there are integral proteins devoid of any hydrophobic helices[2]. Another pointed out that in large membrane proteins, not all transmembrane segments are in contact with the lipid phase and that different functions might entail different structural principles; he advocated the existence of shorter (10 residues) transmembrane helices, which hydrophobic analysis would not detect[3]. Although the point was not raised, we can add that voltage-gated channels most likely include voltage sensors in the form of charged transmembrane helices (see below).

We are not unaware of the above observations and arguments. So why, in the face of evidence or suggestions that other structures can or could exist, should one feel justified in building and testing topological models comprised exclusively of long hydrophobic transmembrane helices? We summarize below what we feel to be the main reasons why, with so many plausible models, some should be given priority. We also indicate some limits to the arguments.

Throughout this text, we use the term 'membrane protein' in a restricted sense, namely that of integral (membrane crossing) proteins. Our discussion centers on 'multispanning' (polytopic) integral membrane proteins, although the oligomeric interactions of proteins anchored by a single helix (e.g. many cell surface receptors) may involve similar factors.

Known structures

Strictly speaking, photosynthetic reaction centers are the only inte-

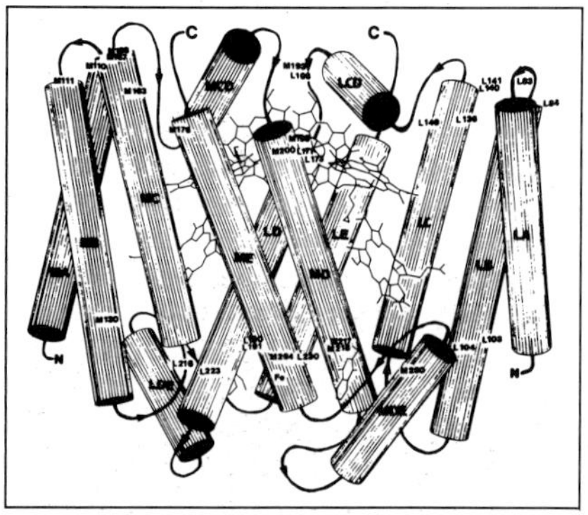

FIGURE 1

Scheme of the photosynthetic reaction centre L and M subunits. Helices are approximated by thin columns and helix connections are simplified. *Reproduced, with permission, from Ref. 4.*

gral membrane proteins in which transmembrane structure is known at high resolution. In these, the only polypeptide structures traversing the lipid bilayer are hydrophobic helices[4] (see Fig. 1).

The transmembrane secondary structure of bacteriorhodopsin (BR) can be inferred from spectroscopic data[5] and the 7Å resolution crystallographic structure[6]. X-Ray diffraction data on BR were, 15 years ago, the first evidence for α-helical structure in the transmembrane region of a membrane protein[7,8]. Henderson pointed out that, because of the absence of water in the center of lipid bilayers, transmembrane structures that satisfy main chain hydrogen bond formation should be strongly stabilized[7,9]. While the spectroscopic evidence is difficult to reconcile with ideal α_I-helices[10,11], the proposal that BR contains transmembrane β-strands[12] is not compatible with either electron[13] or neutron[14] diffraction data.

Porin is probably the third best known membrane protein, even though its structure has not yet been established to better than about 25 Å resolution. Spectroscopic and crystallographic evidence indicates that porin contains practically no α-helix, but a large proportion of β-strand (see Ref. 2 for references). Its sequence is very hydrophilic and contains few clues as to how it folds to create large pores in the outer membrane of *E. coli*. Porin and its congeners from the bacterial outer membrane are the only membrane proteins for which such strong evidence against an α-helical structure is available.

The structure of other membrane proteins is generally very poorly known. Except for bacterial outer membrane proteins and a mitochondrial protein, proteins classified as integral on the basis of their biochemical behavior are found either to be soluble proteins anchored by covalent binding to a lipid, or to have at least one segment that could form a hydrophobic transmembrane helix. Few of these segments have actually been identified as transmembrane by biochemical approaches; more importantly, most current approaches cannot exclude the possibility that other types of secondary structure are also present. Although it is still speculative, one case that deserves special mention is the family of voltage-sensitive ion channels[15–17]. Examination of the sequences suggests a common organization in groups of six transmembrane helices, each group comprised of five hydrophobic helices and a sixth helix that is hydrophobic but for the presence of several basic residues. Both symmetry considerations (that require an even number of crossings per group if all groups are to be arranged with the same transmembrane orientation) and functional requirements (that imply the existence of a voltage sensor) can be satisfied if the charged segment crosses the membrane.

FOLDING

If anything is more obscure than the final structure of membrane proteins, it is the way they acquire it. However, examination of the few known structures and some experimental data suggest that hydrophobic transmembrane α-helices behave as individual folding domains. This view has been discussed elsewhere[18,19] and will be only sketchily summarized here. Each of the membrane spanning helices of the photosynthetic reaction centers and of BR can be predicted by hydrophobicity analysis, and would be expected to be stable as an isolated transmembrane entity in a lipid bilayer. Such a situation could be the outcome of a two-stage folding process in which the final structure would form by accretion of preformed isolated helices, with minor adjustments (see Fig. 2).

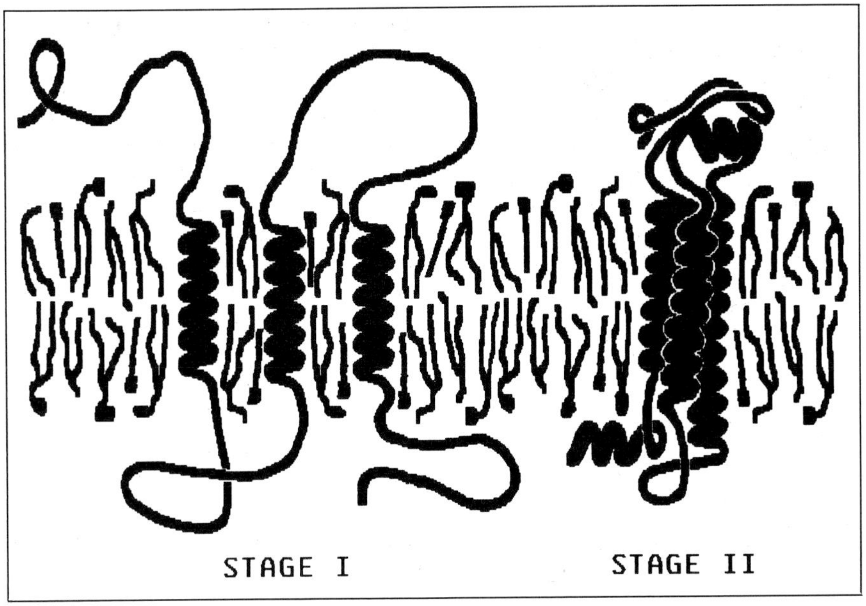

FIGURE 2

Two-stage model for the folding of an α-helical integral membrane protein. The first stage is the formation of independently stable trans-bilayer helices, principally in response to the hydrophobic effect and the formation of main-chain hydrogen bonds in the non-aqueous environment. The second stage is the interaction of the helices to form the tertiary fold of the polypeptide. Factors that could contribute to the energetics of stage II are the links between the helices, packaging of helices and lipid molecules, polar interactions between helices, and, when applicable, association with prosthetic groups or with other proteins. *Reproduced, with permission, from Ref. 19.*

Direct experimental evidence in favor of two-stage folding comes from renaturation studies with BR. In particular, it has been shown that denatured BR fragments reform helical structures when inserted individually into lipid and that they can reassemble and reform BR when the vesicles are fused. The refolded structure is spectroscopically and crystallographically indistinguishable from the native one (see Ref. 18 and references therein). Two-stage folding may not be limited to electron or proton pumps. The best known of the channel-forming receptors, the nicotinic acetylcholine receptor (nAChR), forms a large, cation-specific aqueous channel through the muscles or nerve plasma membrane[20,21]. Recent data strongly suggest that its transmembrane region is entirely comprised of hydrophobic transmembrane helices[1,19].

Altogether, these experiments and observations suggest that the building block of many membrane proteins is the hydrophobic α-helix. As with domains and subunits in soluble globular proteins, the packing of helices to give an integral monomer would be driven by the same forces as the packing of monomers into a complex. This view is supported by an examination of the structure of complexes in organelles (see below).

A two-stage folding mechanism is implicitly postulated in attempts at predicting the transmembrane topology of proteins by hydrophobicity analysis and may explain the success of the approach. However, this should not be carried too far. Examination of hydrophobic fragments in soluble and integral membrane proteins makes it obvious that there are cases in which it is not easy to decide from hydrophobicity analysis alone whether a given segment is a transmembrane one or not. One of

BR's helices, for instance, can be located in the sequence when its existence is known, but might have remained dubious were it not for crystallographic and biochemical data. Since there must be hydrophobic regions inside large extramembrane domains of integral proteins, the status of some hydrophobic stretches can remain ambiguous.

The asymmetry of the process of insertion during biosynthesis suggests that the unit to be considered when calculating the thermodynamics of insertion might be pairs of segments (hairpins) rather than individual helices[22,23]. This is an obvious departure from the tentative rule that every transmembrane segment ought to be stable by itself (and therefore identifiable by hydrophobicity analysis). Candidates for such 'passenger' segments are the sensor helices of the voltage-sensitive channels and the amino termini of secreted proteins.

More generally, some sort of concerted mechanism of insertion has to be postulated every time transmembrane segments are proposed that are not individually stable, e.g. β-strands, or charged or very short helices. A concerted mode of folding is where the crucial difference lies between models that incorporate only individually stable helices and more complex, 'holistic' models. How often models of the second type will be encountered is perhaps one of the most interesting questions of membrane biochemistry. What is presently known of the structure of the nAChR and its congeners suggests that even channel-forming proteins, which expose a relatively large transmembrane region to water, can turn out to be made up of canonical hydrophobic helices. This could be taken as an indication that 'holistic' structures may be in the minority.

Import and export of integral membrane proteins

Typical stable folding domains in soluble proteins are 70–150 amino acids long and this is reflected in the size of soluble subunits, which are seldom smaller unless they are stabilized by disulfide bridges or prosthetic groups. If transmembrane hydrophobic α-helices behave as folding domains, one might expect integral membrane protein complexes to comprise much smaller subunits. This is borne out by an examination of known amino acid sequences[24,25]. Interestingly, the distribution of very small subunits, often barely more than one hydrophobic helix, is not uniform. They are mainly found in the inner membranes of mitochondria and chloroplasts. Most organelle proteins with multiple potential transmembrane helices are synthesized locally, imported proteins very rarely having more than three hydrophobic segments. This suggests that there are restrictions to import, perhaps related to the stability of unfolded, import-competent precursors. Such a 'micro-assembly' of organellar inner membrane complexes is made possible because the transmembrane segments of these small subunits are hydrophobic and therefore individually stable.

The same kind of constraint might apply, even more strongly, to bacterial outer membranes. It appears that none of the proteins in these membranes has a long hydrophobic stretch. Indeed, insertion of such sequences into engineered proteins prevents export[26,27]. It appears likely that the β-sheet structure of porins is an answer to this biosynthetic constraint[24,27].

Conclusion

Can we draw from the above considerations strict rules as to how to predict the topography of a membrane protein from its sequence? Obviously not. Can we infer tendencies? Probably. It seems likely that most transmembrane segments are hydrophobic helices long enough to span the bilayer. When exceptions to this rule are certain or highly probable (porins, voltage-gated channels), there may be specific reasons, either related to biosynthesis or to the function of the protein. Being large or forming an aqueous channel does not appear to require a departure from the general model. Given the difficulty of testing models by biochemical, genetic or computational means, initial modelling choices should be based on the strongest probabilities. While it may seem somewhat monotonous to look at the same string of long hydrophobic helices in every membrane protein sequence paper, this simple scheme probably represents the best first guess when approaching an unknown structure. Departures ought to be based on strong rationales.

References

1 McCrea, P. D., Engelman, D. M. and Popot, J-L. (1988) *Trends Biochem. Sci.* 13, 289–290
2 Ferenci, T. (1989) *Trends Biochem. Sci.* 14, 96
3 Lodish, H. F. (1988) *Trends Biochem. Sci.* 13, 332–334
4 Deisenhofer, J. and Michel, H. (1989) *Science* 245, 1463–1473
5 Wallace, B. A. and Teeters, C. L. (1987) *Biochemistry* 26, 65–70
6 Henderson, R. and Unwin, P. N. T. (1975) *Nature* 257, 28–32
7 Henderson, R. (1975) *J. Mol. Biol.* 93, 123–138
8 Blaurock, A. E. (1975) *J. Mol. Biol.* 93, 139–158
9 Henderson, R. (1977) *Annu. Rev. Biophys. Bioeng.* 6, 87–109
10 Vogel, H. and Gartner, W. (1987) *J. Biol. Chem.* 262, 11464–11469
11 Gibson, N. J. and Cassim, J. Y. (1989) *Biochemistry* 28, 2134–2139
12 Jap, B. K., Maestre, M. F., Hayward, S. B. and Glaeser, R. M. (1983) *J. Biophys.* 43, 81–89
13 Tsygannik, I. N. and Baldwin, J. M. (1987) *Eur. Biophys. J.* 14, 263–272
14 Popot, J-L., Engelman, D. M., Gurel, O. and Zaccaï, G. (1989) *J. Mol. Biol.* 210, 829–847
15 Noda, M. *et al.* (1984) *Nature* 312, 121–127
16 Tempel, B. A., Papazian, D. M., Schwartz, T. L., Jan, Y. N. and Jan, L. Y. (1987) *Science* 237, 770–775
17 Tanabe, T., Takeshima, H., Mikami, A., Flockerzi, V., Takahashi, H., Kangawa, K., Kojima, M., Matsuo, H., Hirose, T. and Numa, S. (1987) *Nature* 328, 313–318
18 Popot, J-L., Gerchman, S-E. and Engelman, D. M. (1987) *J. Mol. Biol.* 198, 655–676
19 Popot, J-L. and Engelman, D. M. *Biochemistry* (in press)
20 Popot, J-L. and Changeux, J-P. (1984) *Physiol. Rev.* 64, 1162–1239
21 Hucho, F. (1986) *Eur. J. Biochem.* 158, 211–226
22 von Heijne, G. and Blomberg, C. (1979) *Eur. J. Biochem.* 97, 175–181
23 Engelman, D. M. and Steitz, T. A. (1981) *Cell* 23, 411–422
24 Popot, J-L. and de Vitry, C. (1990) *Annu. Rev. Biophys. Biophys. Chem.* 19, 369–403
25 de Vitry, C. and Popot, J-L. (1989) *C. R. Acad. Sci. Paris Ser. 3* 309, 709–714
26 Davis, N. G. and Model, P. (1985) *Cell* 41, 607–614
27 MacIntyre, S., Freudl, R., Eschbach, M-L. and Henning, U. (1988) *J. Biol. Chem.* 263, 19053–19059

D. M. Engelman is in the Department of Molecular Biophysics and Biochemistry at Yale University, New Haven, CT 06511, USA. J-L. Popot is with the College de France at the Institut de Biologie Physico-Chimique, 75005 Paris, France.

Protein Motifs and Data-base Searching

Janet M. Thornton and Stephen P. Gardner

Protein structure and sequence motifs are now recognized for many different protein families and topologies. To aid identification and use of these motifs in modelling and prediction, it has become necessary to establish consistent data bases of protein structure, including not only coordinates, but also derived data such as secondary structure location and solvent accessibilities. This article describes first attempts to construct such data bases and explains how they can be used.

Although over 15 000 protein sequences are now available, there are only around 400 proteins of known structure in the Brookhaven Databank[1], including coordinates for about 100 non-homologous protein sequences. In principle it should be possible to predict protein structure from sequence, but this remains one of the great unsolved problems of molecular biology today. We need to know three-dimensional structure before we can begin to understand protein function at the molecular level. Enzyme catalysis, antibody–antigen interactions and receptor–ligand interactions are prime examples. Rational design of modified proteins, ligands and even novel molecules must be based at the three-dimensional coordinate level.

Despite the limited scope of available structural data, it has become increasingly clear that proteins fall into distinct structural families, built from an apparently limited set of motifs[2]. Peptide geometry and the restriction to L-amino acids limit the conformational space available to polypeptide chains. Consequently the α-helix, β-strand and composite secondary structure patterns (e.g. βαβ-unit) dominate protein structures. The same structural motifs recur many times in 'unrelated' proteins. A surprising number of new structures which appear in the literature look similar to other known structures, whose sequences may or may not exhibit a visible similarity. The extent of the structural similarity may extend through the whole sequence (e.g. tumour necrosis factor and the icosahedral viral coat proteins[3]), or be confined to a small section of the protein (e.g. DNA-binding motifs[4]). Interleukin 1β was found quite unexpectedly to have the same topology as soybean trypsin inhibitor[5], although the sequences are difficult to align even when this similarity is known! These recurrent motifs have led to the concept of a hierarchy of protein structure with building blocks of gradually increasing complexity (see Fig. 1). It is possible to envisage a library of protein fragments which can be used to 'construct' proteins by piecing together the appropriate fragments. In parallel, comparisons of protein sequences are revealing many sequence motifs, which are usually functionally important. This review will not explicitly consider sequence comparisons nor protein structure prediction, which are described by Doolittle (p. 71) and by Fasman (p. 135), respectively, elsewhere in this book. We will concen-

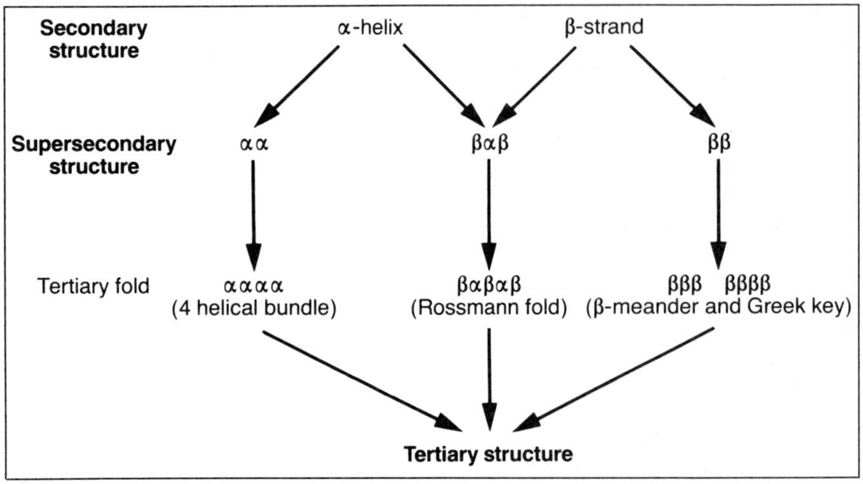

FIGURE 1

A summary of the basic hierarchy of protein structure.

trate on structural motifs and structure-related sequence motifs and describe how these can be extracted from the newly established data bases of protein structure for use in prediction and modelling (see also p. 163).

PROTEIN STRUCTURE DATA BASES

Traditionally protein structures have been examined by eye using sophisticated computer graphics software. Whilst these systems are excellent for studying an individual protein, they are very time-consuming for scanning many structures. Four factors have led to the requirement for protein structure databases:

(1) The increase in the number of known structures.
(2) Structural data, although encapsulated by coordinates, can more easily be described and understood at a variety of structural levels (e.g. secondary structure and hydrogen bonding). Such data must be derived from coordinates and stored.
(3) The data need to be rationalized, so that properties are defined consistently over all structures. In Brookhaven files for example, secondary structure is usually defined by eye by the author, with consequent variation between structures. It is necessary to derive algorithms that can be applied without bias to all structures to obtain a consistent set of data, e.g. the program written by Kabsch & Sander to assign secondary structures[6]. A set of consistent data is a basic requirement before algorithmic searches for higher order motifs can be made.
(4) A requirement for easy access to all structures and their derived properties using a simple query system.

To fulfil these requirements several laboratories around the world have been setting up protein structure data bases. Initially commercial relational data-base management systems, such as Oracle (Oracle Corporation UK Ltd, Bristol, UK), have been used. They provide the shell structure into which the data can be loaded and the means of access, using a simple query language, usually SQL. These are robust, flexible, well-documented and easy to use, but are not ideally suited to the sequential architecture of proteins. Object-oriented[7] and protein-specific data-base systems are alternatives, which should have advantages in speed, cost, storage requirements, graphics linkage, portability and query flexibility. All these data bases are still being developed, but the concept of extracting data is universal and we shall use examples from the data-base systems being developed at Birkbeck to illustrate their use for motif specification and searching.

At Birkbeck we have established a protein structure data base BIPED[8], using the Oracle Relational Database Management System. This has been part of a program to develop an integrated protein sequence and structure data base (ISIS)[9] which has been supported by the Protein Engineering Club in the UK. We are also writing (SPG & JMT) a protein-specific data-base system, 3-D SCAN, specifically to allow extraction of the more complex motifs.

PROTEIN MOTIFS

The term 'motif' has been used to describe a variety of features in proteins (see Table I).

Sequence motifs are derived from homologous sequences, for which there may or may not be any structural information available. The identification of sequence motifs has traditionally been performed by aligning two or more homologous sequences and then constructing a consensus sequence for residue positions which are absolutely conserved, or conservatively varied. Increasingly, this procedure is being automated so that alignments and consensus sequences are derived computationally and the sequence data bases can be searched using this consensus sequence[10].

Structure-related sequence motifs are defined as those sequence patterns which are reliably associated with a specific structure. For secondary structure prediction from sequence the most straightforward pattern matching approach is to search for sequence patterns which always adopt the same secondary structure (with a specified tolerance). It has been shown[11] that there are many examples of identical pentapeptides which adopt different secondary structures, i.e. the local structure is dependent on the context. However there are some short sequence patterns which reliably predict secondary structure independent of context[12,13]. Table I shows some of these patterns which were extracted by rigorous searching and matching algorithms applied over the whole data base.

Structural motifs recur in many proteins, although their sequences may vary enormously. The basic structural motifs are the ubiquitous α-helix and β-strand. Composite patterns of increasing complexity are often observed ranging from simple supersecond-

TABLE I
Examples of protein motifs[a]

Sequence motifs:	
{GPP}	Collagen triple helix (multiple repeats)
FDTGS	Acid proteinase cleavage site
G.G..G	Nucleotide binding motif (simplified)
Structure-derived sequence motifs:	
AA..K	Peptide containing at least two helical residues
GG.L	Structure XXββ
Structural motifs:	
Helix; strand, turn	Secondary structure
β-hairpin, βαβ-unit, Greek key, 4α-bundle	Supersecondary structure
Jelly roll, TIM barrel	Tertiary structure

[a]The one-letter code is used to define the sequence: . indicates no conservation. The code for structure is β=β-strand; X=not α, β or turn.

```
Select all example of the sequence motif GLY GLY ANY LEU
from a list of proteins with moderate or high resolution

    1>SELECT UNIQID,SEQ:4,STRK:4
    2>FROM MOD_RES.LIS
    3>WHERE FUZZY='GGXL'

      UNIQID        SEQ     STRK
      -----------   ------  -------
      2APR.0.0095   (G)GLL  (T)TEE
      5CHA.A.0043   (G)GSL  (E)EEE
      5CHA.A.0198   (G)GPL  (E)EEE
      1TPP.0.0043   (G)GSL  (E)EEE
      2EST.E.0043   (G)GTL  (E)EEE
      2EST.E.0196   (G)GPL  (T)TEE
      2PKA.A.0043   (G)GVL  (E)EEE
      2PKA.B.0196   (G)GPL  (T)TEE
      2SGA.0.0196   (G)GSL  (T)TEE
      3SGB.E.0196   (G)GPL  (T)TEE
      3RP2.A.0043   (G)GFL  (E)EEE
      3RP2.A.0196   (G)GPL  (T)TEE
      3WGA.A.0094   (G)GKL  (G)TB
      2YHX.0.0071   (G)GDL  (T)TEE

       13 hits matched query specifications
```

(a)

```
Select all disulphide bridges occurring between two
residues in alpha helical conformation

1   SELECT BRCODE,SEQNO,SEQNOP,STRK,STRKP,GAP,CADIST
2   FROM DISULPHIDE
3*  WHERE STRK='H' AND STRKP='H'

    BRCO  SEQNO    SEQNOP S S  GAP   CADIST
    ----  ------  -------- - -  ----  ------
    1CRN     16        26 H H   10    5.80
    1HMG    144       148 H H    4    5.19
    1PP2     28        44 H H   16    5.34
    1PP2     43        95 H H   52    5.80
    1PP2     49       122 H H   73    5.35
    1PP2     50        88 H H   38    5.60
    1PP2     57        81 H H   24    6.09
    2ACT     56        98 H H   42    4.66
    2GRS     58        63 H H    5    4.65
    2HFL     30       115 H H   85    6.23
    2LYM     30       115 H H   85    6.12
    2PAD     56        95 H H   39    4.57
    2TGP      5        55 H H   50    5.67

    13 records selected.
```

(b)

```
Select all type I ß-turns from the database.
The first to last residue must form a main chain hydrogen bond
and have a CA - CA distance of less than 7.0 Å. The second
residue must be a PRO with a psi value of less than 0.0°.

    1>SELECT UNIQID,SEQ:3,STRK:3,PHI_1,PSI_1
    2>FROM DB$SCAN:MASTER.LIS
    3>WHERE GAP = 3  AND CADIST_3 < 7.0
    4>AND FUZZY='XP'
    5>AND SSTRC='X( BETS)( BETS)X' AND PSI_1 < 0.0

     UNIQID        SEQ     STRK    PHI_ 1  PSI_ 1
     -----------   ------  ------  ------  ------
     155C.0.0023   (A)PDG  (T)TTT   -57.4    -1.4
     1APR.0.0060   (N)PSD  (T)TTT   -38.6   -33.8
     1BP2.0.0013   (I)PSS  (T)TTH   -61.0   -11.6
     1CAC.0.0041   (D)PSL  (T)TTT   -57.4   -62.0
     1CHG.0.0027   (W)PWQ  (T)TTE   -35.6   -17.4
          .           .       .        .
          .           .       .        .
          .           .       .        .
     8CAT.B.0294   (N)PFD  (T)TTT   -51.8   -26.4
     8CAT.B.0334   (D)PSN  (T)TTT   -45.8   -25.6
     8CAT.B.0366   (G)PNY  (T)TTT   -48.8   -29.8

      268 hits matched query specifications
```

(c)

```
Select all examples where a PRO in beta sheet conformation
has been inserted in the aspartic proteases.

1   SELECT ANUM1 "P1APR",ANUM2 "P2APP",ANUM3 "P4APE",
2   RTYP1,RTYP2,RTYP3,"   ",SSTRC1,SSTRC2,SSTRC3
3   FROM APALIGN
4   WHERE (RTYP1<>'P' OR RTYP2<>'P' OR RTYP3<>'P') AND
5   ((  RTYP1='P' AND SSTRC1='E')
6   OR (RTYP2='P' AND SSTRC2='E')
7*  OR (RTYP3='P' AND SSTRC3='E'))

    P1APR P2APP P4APE R R R      S S S
    ----- ----- ----- - - - ---  - - -
        7     7     3 P T T      E E E
       10    10     6 D P P      E E E
       20    20    17 Q P P      E E E
      187   183   183 P G A      E E E
      227   222   224 P D P      E T T
      251   248   249 S D P      E E
      259   255   256 P D S      E E E
      322   321   325 P P S      E E B

     8 records selected.
```

(d)

FIGURE 2

(a) Query and response using the tabular 3-D SCAN program which searches flat protein data files. This query illustrates the use of Order Based SQL in 3-D SCAN, demonstrating the additional flexibility possible in a protein-dedicated data base. The query selects four sequential residues with specified secondary structure assignments and sequence. The column labels are as follows: UNIQID is a field uniquely identifying each residue from within a multi-protein data bank. SEQ is the amino acid sequence (first Gly bracketed). STRK is the calculated secondary structure assignment (value for first Gly bracketed). **(b)** Query and response using the relational data base BIPED, under ORACLE accessed using Structured Query Language (SQL). The data are stored in a table called DISULPHIDE and the columns are as follows: BRCO is four-letter Brookhaven code for protein. SEQNO is logical number of first Cys residue. SEQNOP is logical number of partner Cys residue. S S are the columns holding the secondary structure assignments for the first and partner Cys, respectively. GAP is the number of residues separating the two Cys residues in the protein. CADIST is the distance (in Å) between the C^{α} atoms of the two Cys residues. **(c)** Query and response using the 3-D SCAN data base, with the following columns: UNIQID is a field uniquely identifying each residue from within a multi-protein data bank. SEQ is the amino acid sequence (first turn residue bracketed). STRK is the calculated secondary structure assignment (value for first turn residue bracketed). $PHI\_{_1}$ and $PSI\_{_1}$ hold the Φ, Ψ values for the residue at $i+1$ (i.e. second residue of turn). **(d)** Query and response using the tables of structurally aligned protein families in the BIPED data base. The table APALIGN holds the structural alignment of the aspartyl protease family: rhizopuspepsin (P1APR), penicillopepsin (P2APP) and endothiapepsin (P4APE) with coordinates aligned from Brookhaven[1]. The columns are labelled as follows: P1APR, P2APP, P4APE columns hold structurally equivalent residue numbers, so that each row represents a residue position in the family alignment. R R R are the respective sequences associated with the three residue numbers. S S S are the respective secondary structural assignments associated with the three residue numbers.

ary structures (e.g. the ββ-hairpin with two secondary units) to complete tertiary structures such as the $(\beta\alpha)_8$ barrel first observed in triose phosphate isomerase and subsequently found in over 14 different proteins (see Ref. 14 and references therein). These structural motifs have been identified by detailed comparison of many protein structures using computer graphics by eye. For example, the families of ββ-hairpins and other supersecondary structures with 'special' loop conformations were extracted by searching manually for all such hairpins and then comparing loop conformation using Φ, ψ values, computer-aided superpositions and visual inspection[15,16]. It soon became apparent that we needed automatic methods to search for and extract such motifs.

Motif searching

Given a consistent data set and a motif definition, we can use the data-base query language to extract all examples of a given motif from proteins of known structure. There are many criteria which can be used to define a motif including sequence, conformation (coordinates, secondary structure, Φ, ψ values), accessibility and tertiary interactions. Here we give a few examples taken from BIPED and 3-D SCAN to illustrate the use of the query language and the types of enquiries that can be made.

The general format of an SQL is:

SELECT properties
FROM table name(s)
WHERE conditions are met (if any)

This format is simple but extremely flexible allowing information to be extracted from more than one table and cross-correlated. Sample queries are shown in Fig. 2.

The most common query is to extract structural information for a given sequence motif. Figure 2(a) shows the query used to find the observed secondary structure for all examples of the sequence Gly-Gly-X-Leu. This is one of the structure-related sequence motifs cited by Rooman & Wodak[12]. We find 12 out of the 13 examples adopt a XXββ conformation.

The second case illustrates how to extract all examples of a disulphide bridge connection between two α-helices. These are relatively rare and only 13 non-homologous examples were found in 300 structures (Fig. 2b).

The β-turns are more complex motifs, which can be explicitly defined and characterized. We can search for all examples of a turn, defined as four residues, where the separation between the end α-carbons is less than 7 Å, and the central two residues should not be helical. We can also add the constraints that the turn should include a proline at residue $i+1$, for which $\psi<0$, and a main chain...main chain hydrogen bond should form between the first and last residues. These constraints are satisfied by 268 turns in the data base (see Fig. 2c).

Having extracted data for a given motif the searcher will usually want to superpose all the examples and display them graphically. This is not possible within the conventional SQL framework and the data base must be linked to other software and graphics facilities. Currently we use 3-D SCAN, a non-relational data base and query system, to superpose all examples in a 'hit list' automatically and then load the coordinates into one of the standard display packages. Two examples of superposed motifs are shown in Fig. 3. These were both extracted using 3-D SCAN on the basis of sequence, dihedral angle and coordinate constraints.

The last important example concerns extracting data from a homologous family of structures and sequences. In order to explore the structural effects of sequence replacements we need to store sequence and structural alignments and be able to query them. There is currently much activity in the area of automatic structure alignment[17] and we have established tables for only a subset of the 17 families in Brookhaven for which more than one structure is available. In the query shown in Fig. 3 we select all instances where a residue in a β-sheet is replaced by a proline in the aspartyl proteinase family of structures. In addition these tables are essential to identify homology between extracted motifs.

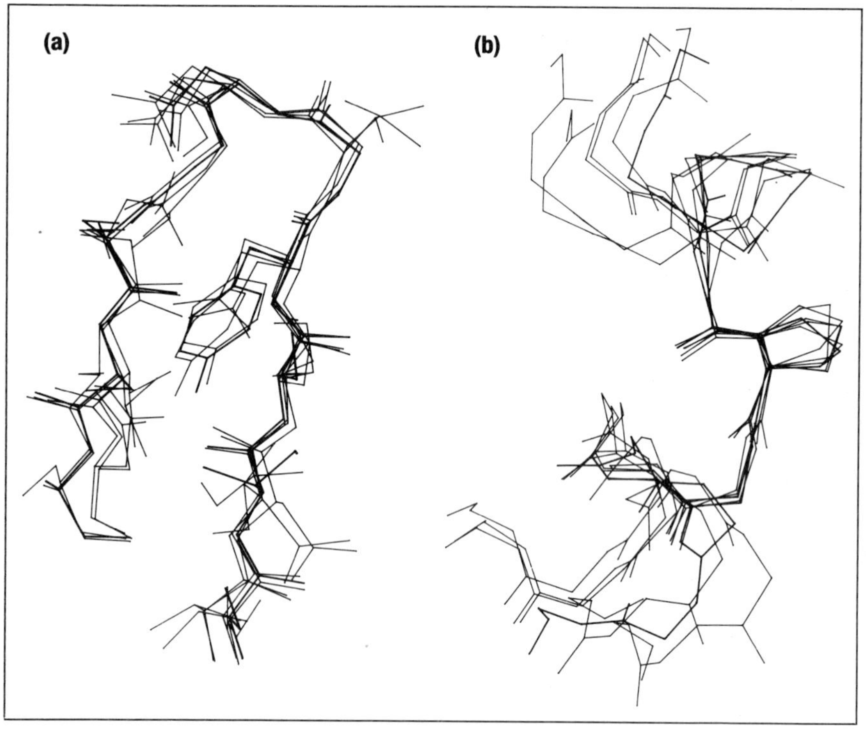

FIGURE 3

(**a**) Six superposed motifs extracted from 3-D SCAN on the basis of geometry constraints matching closely a template structure (shown in bold). (**b**) Six superposed motifs extracted from 3-D SCAN on the basis of sequence and secondary structural similarity to a target structure (PRO linking two α-helices).

APPLICATIONS

Some important applications of protein structure data bases are summarized below:

(1) *Fitting the polypeptide backbone into electron density*

Interpretation of an electron density map can be facilitated by extracting fragments from one protein stored in a 'dedicated' data base to construct another protein, using the electron density to choose the fragments. The elegant method of using the distance matrix of C^α separations to sift the fragments was first developed by Jones & Thirup[18] and has been incorporated into standard electron density fitting software. This method is particularly useful for fitting the loop regions, by specifying the end-point geometries and the length of the loop in residues.

(2) *Modelling on the basis of homology*

When modelling a protein using coordinates from a homologous protein structure, it is the loop regions which differ most in sequence and which are therefore the most difficult to model. Knowledge of preferred loop conformations or motifs derived from the whole data base of structures, and combined with identification of critical residues (especially glycines), can guide the modelling process[19,20].

(3) *Prediction by recognition of motifs*

Since we know that most protein structures will be built from fragments already observed in other proteins, one aim is to predict structure from sequence by recognizing motifs. Early attempts to predict super-secondary structures were based on broad structural principles[21], such as helix and sheet probabilities, and lengths of helices (see review by Fasman on p. 135). With the increasing size of the data base it should be possible to include sequence-specific information as well. The recent prediction of the structure of the zinc-binding finger by Berg[22] is an excellent example of a knowledge-based approach to structure prediction. This was based on the recognition of two motifs:

His-X-X-X-His and Cys-. . .-Cys

involving residues known to be used for zinc binding. Berg used similar motifs taken from proteins of known structure to guide his prediction. Whilst this is a special case, since these residues are involved in binding zinc, it illustrates the principle and if we could extend this to more general motifs it would be very powerful. With an improved motif library and the ability to scan the data base very rapidly progress in this approach should be possible.

(4) *Prediction of side chain interactions*

Most of this review has concentrated on backbone fragments, but there is increasing interest in modelling side chain interactions, and estimating their contribution to stability. Again a knowledge-based approach can be helpful, combined with conventional energy calculations. For example, the observed limited set of disulphide bridge geometries[23] will be important in determining detailed structure, as will aromatic–aromatic interactions[24,25] and salt-bridge geometries. Currently the data base is sufficiently large to extract overall preferences in interaction geometry. However, if backbone constraints are included[26] there are too few examples available to extract statistically significant preferences, except in rare cases such as disulphide bridges or adjacent aromatic groups on helices.

(5) *Protein design*

The *de novo* design and manufacture of proteins is now technically possible (see review by Richardson and Richardson on p. 173). Given our incomplete understanding of the energetics of protein folding, most attempts have been knowledge-based. The designer can interrogate the data base to find if specific aspects of the proposed design are observed in 'real' proteins. For example, we can easily extract all examples of disulphide bridges between helices (see Fig. 3) or strands, or examples of asparagines hydrogen-bonding to the free amide groups at the N-terminus of an α-helix.

(6) *Protein engineering*

In designing site-directed mutagenesis experiments it may well be valuable to know if the proposed mutation has been observed at a similar site in other protein families and if so what was its effect on the structure. The query shown in Fig. 2(d) could be used to assess the effects of changing a residue in a strand to a proline.

FUTURE WORK

This area of protein motifs and data bases is evolving very rapidly. The data-base technology is essential for efficient and easy extraction of sequence and structure data. The developments described in this review are only the beginning. Future data bases will include more information, with improved classification and more sophisticated pattern-matching methods[27,28]. These data bases will hopefully be much used by a wide range of biochemists, as they seek to understand and modify the properties and functions of their 'favourite' proteins. Ultimately, the goal must be to improve our understanding of the relationship between sequence, structure and function, by establishing a consistent set of data accessible to everyone.

REFERENCES

1 Bernstein, F. C., Koetzle, T. F., Williams, G. J. B., Meyer, E. F., Brice, M. D., Rodgers, J. R., Kennard, O., Shimanouchi, T. and Tasumi, M. (1977) *J. Mol. Biol.* 122, 535–542

2 Richardson, J. S. (1981) *Adv. Protein Chem.* 34, 167–339

3 Jones, E. Y., Stuart, D. I. and Walker, N. P. C. (1989) *Nature* 338, 225–228

4 Struhl, K. (1989) *Trends Biochem. Sci.* 14, 137–140; see also p. 259 of this book

5 Priestle, J. P., Schar, H. P. and Grutter, M. G. (1988) *EMBO J.* 7, 339–343

6 Kabsch, W. and Sander, C. (1983) *Biopolymers* 22, 2577–2637

7 Paton, N. W. and Gray, P. M. D. (1988) *PROLOG and Databases: Implementation and New Directions* (Gray and Lucas, eds), Ellis Horwood

8 Islam, S. A. and Sternberg, M. J. E. (1989) *Protein Eng.* 2, 431–442

9 Akrigg, D., Bleasby, A. J., Dix, N. I. M., Findlay, J. B. C., North, A. C. T., Parry-Smith, D., Wootton, J. C., Blundell, T. L., Gardner, S. P., Hayes, F. R. F., Islam, S. A., Sternberg, M. J. E., Thornton, J. M., Tickle, I. J. and Murray-Rust, P. M. (1988) *Nature* 335, 745–746

10 Taylor, W. R. (1988) *Protein Eng.* 2, 77–86

11 Kabsch, W. and Sander, C. (1984) *Proc. Natl Acad. Sci. USA* 81, 1075–1078

12 Rooman, M. J. and Wodak, S. J. (1988) *Nature* 335, 45–49

13 Argos, P. (1987) *J. Mol. Biol.* 197, 331–348

14 Chothia, C. (1988) *Nature* 333, 598–599

15 Thornton, J. M., Sibanda, B. L., Edwards, M. S. and Barlow, D. J. (1988) *Bioessays* 8, 63–69

16 Milner-White, E. J. and Poet, R. (1986) *Trends Biochem. Sci.* 12, 189–192

17 Sutcliffe, M. J., Haneef, I., Carney, D., and Blundell, T. L. (1987) *Protein Eng.* 1, 377–384

18 Jones, T. A. and Thirup, T. (1986) *EMBO J.* 5, 819–822

19 Chothia, C. and Lesk, A. M. (1987) *J. Mol. Biol.* 196, 901–917

20 Blundell, T. L., Carney, D., Gardner, S. P., Hayes, F., Howlin, B., Hubbard, T., Overington, J. P., Singh, D. A., Sibanda, B. L. and Sutcliffe, M. J. (1988) *Eur. J. Biochem.* 172, 513–520

21 Taylor, W. R. and Thornton, J. M. (1984) *J. Mol. Biol.* 173, 487–514

22 Berg, J. M. (1988) *Proc. Natl Acad. Sci. USA* 85, 99–102

23 Pabo, C. O. and Suchanek, E. G. (1986) *Biochemistry* 25, 5987–5991

24 Singh, J. and Thornton, J. M. (1985) *FEBS Lett.* 191, 1–6

25 Burley, S. K. and Petsko, G. A. (1985) *Science* 229, 23–28
26 Reid, L. S. and Thornton, J. M. (1989) *Proteins* 5, 170–182
27 Cohen, F. E., Abarbanel, R. M., Kuntz, I. D. and Fletterick, R. J. (1986) *Biochemistry* 25, 266–275
28 Lathrop, R. H., Webster, T. A. and Smith, T. F. (1987) *Commun. Assoc. Comput. Mach.* 30, 909–921

J. M. Thornton and S. P. Gardner are at the Laboratory of Molecular Biology, Crystallography Department, Birkbeck College, Malet Street, London WC1E 7HX, UK.

FROM COMPARISONS OF PROTEIN SEQUENCES AND STRUCTURES TO PROTEIN MODELLING AND DESIGN

A. ŠALI, J. P. OVERINGTON, M. S. JOHNSON AND T. L. BLUNDELL

A useful approach to modelling proteins exploits knowledge of three-dimensional structures determined by X-ray crystallography together with rules defined by their analysis and comparison.

Let us begin with the assumption that neither quantum mechanical calculations nor molecular dynamics simulations alone will, in the near future, allow us to define which of the multitude of conformers available to a polypeptide will occur in a living organism. The reasons for this assumption are many, but the most important is the immense computational time required to simulate a complex macromolecular system during the folding process.

Many protein sequences, however, adopt the same general fold. Recent structure determinations suggest that the majority of new structures comprise motifs or domains that have been previously identified in other, often functionally different proteins. If we can use this information from related proteins, then we have a feasible route to modelling a protein from its sequence.

The classical approach to this problem was developed for modelling one protein on the three-dimensional structure of an homologue. The first application was the construction of a model for α-lactalbumin on the basis of lysozyme[1]. Other applications included construction of models for relaxins and insulin-like growth factors (see Ref. 2 for a review) and various serine proteinases[3]. Some rules for modelling protein structure[3–5] have been suggested and the advent of computational techniques have made modelling more straightforward.

In this review we shall consider our own approach to protein modelling which can be completely automated, and in which all decisions are rule based (see Fig. 1). We begin by comparing or aligning protein sequences and three-dimensional structures determined by experiment and organized in databases (see, for example, Ref. 6 and p. 153 of this book). We use these comparisons to derive rules about families of protein structures that adopt a common fold. Automation of the approach allows these rules to be tested systematically. The rules are then used to define a template for sequences for each family fold; this can be considered as a projection from three-dimensional structure onto one-dimensional sequence or as a generalized protein sequence summarizing knowledge about the family fold. The sequence of the protein to be modelled is then aligned with the appropriate template, and the alignment is used to extrapolate the three-dimensional features of the known structures to the sequence of interest. A model of the protein is then constructed by making it consistent with the features implied by proteins of known structure.

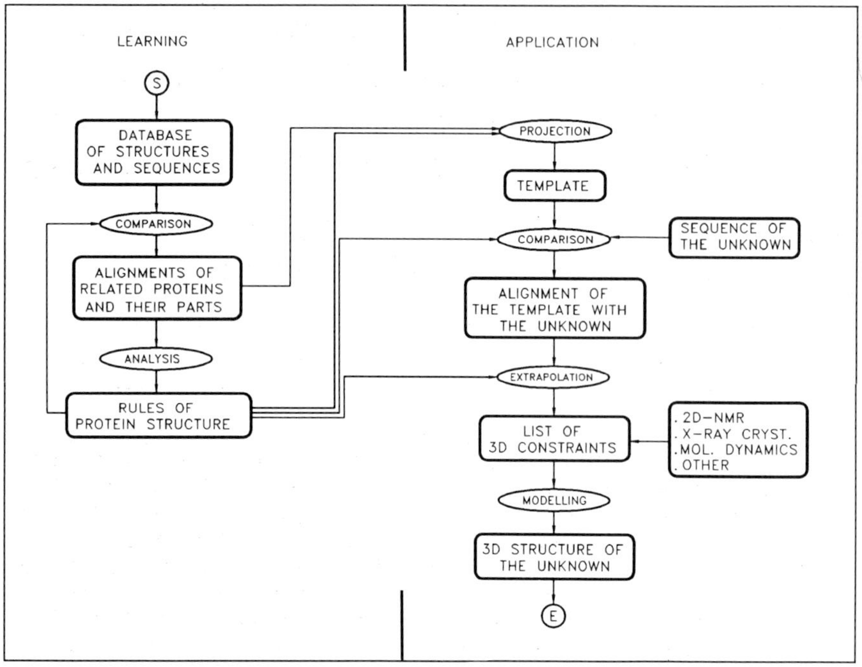

FIGURE 1

A scheme for knowledge-based modelling of proteins. The approach involves the derivation of rules from the comparisons of sequences and three-dimensional structures, and their use in the generation of a template and the construction of the three-dimensional model. Operations involved in comparison, analysis, projection and extrapolation with modelling are described in detail in the text.

Thus we exploit not only the details of protein structures but also the rules developed from their analysis and comparison; these together form the knowledge base.

The most helpful learning set is that of homologous proteins; there are fewer ambiguities in their comparisons, they provide a reliable source for the definition of rules but they remain a challenge for modelling. We also describe new approaches that can be used to model more distantly related protein structures. An alternative rule-based approach, which has been developed for predicting protein structures where no obvious homology or analogy is apparent, has been developed by Cohen and his collaborators[7].

COMPARISON AND CLUSTERING

The first task in knowledge-based modelling is to develop ways of comparing protein structures and of clustering them into related subgroups. The most familiar comparison methods are those for sequences, which use dynamic programming procedures based on the algorithm of Needleman and Wunsch[8] for pairwise or multiple sequence alignments[9]. These methods usually consider the mutation rates of amino acid residues to derive optimal comparison scores and corresponding alignments, but other properties such as physicochemical parameters can also be included[10,11].

Comparison of protein tertiary structures has often involved their superposition in three-dimensional space (see Ref. 12 for a review). If several homologous structures are available[13,14] we usually find that many helices and strands are conserved in the family. This is called the 'framework'. However, in more distantly related proteins differences in the positions and orientations of these strands and helices may preclude their superposition[15–18].

The problem of defining topologically equivalent residues in polypeptides that have little sequence similarity but adopt similar folds was addressed more than a decade ago by Rossmann, Matthews and coworkers. They either included information about main chain direction in the alignment or based their comparisons on superposition of small parts of the whole structure (see Ref. 19 for a review). Others used inter- or intramolecular distances or relationships between secondary structure elements.

In our approach to comparing distantly related proteins[20], we consider features of both sequence and three-dimensional structure simultaneously. Many features depend only on local properties such as hydrophobicity, local conformation and solvent accessibility. These can be aligned using the same approach used for sequence alignment. Furthermore, a similar comparison can be made at each level in the protein hierarchical organization – residue, secondary structure, supersecondary structure, motif or domain. A related approach based principally on intramolecular distances has been described by Taylor and Orengo[21]. Specific relationships such as hydrogen bonding or packing interactions tend to be conserved in protein folds and the patterns of interactions can also be used to align structures[20].

4APEN	t̃ p a q̃ t̃ L n L d F D̃ T G s̃ S D̃ L W̃ V F S̃ s ẽ T̃ t a s e v
2APPN	- - g t t̃ L n L n F d̃ T G s̃ A D̃ L W V F S t̃ ẽ L p a s q q
2APRN	t̃ p G k k̃ F ñ L d F d̃ T g s̃ S D̃ L W̃ I A S t l C t̃ - ñ C g
PEPN	t̃ p a q d F t̃ V i F d̃ T G S̃ S Ñ L W̃ V P S v y C s s l A C
4APEC	k - s t s I d̃ G I A d̃ T G t̃ t l L y L p a t V V s a - - -
2APPC	g - d g - f s G i A d̃ T G t̃ t l L l L d̃ d̃ s V V s q̃ - - -
2APRC	a - s s - f d̃ G i L d̃ T G t̃ t l L i L p ñ n i A a s - - -
PEPC	a C s g g c q̃ A I V d̃ T g t̃ s l L T G p t̃ s a I a n - - -

FIGURE 2

The multiple alignment of eight aspartic proteinase lobes constructed using COMPARER. Protein codes: 4APE, endothiapepsin; 2APP, penicillopepsin; 2APR, rhizopuspepsin; PEP, hexagonal porcine pepsin. The final N or C of the protein code indicates either the N- or C-terminal lobe. The coordinates of the three-dimensional structures were obtained from the Brookhaven data-bank[29], with the exception of the coordinates of porcine pepsin which were the kind gift of Dr Jon Cooper. The amino acid code is the standard one-letter code formatted using the following convention: italic, positive value of the main chain torsion angle Phi; upper case, solvent-inaccessible residues; lower case, solvent-accessible residues; bold type, hydrogen bonds to main chain amide; underline, hydrogen bonds to main chain carbonyl; ~, side chain–side chain hydrogen bonds.

Figure 2 shows part of the alignment of the two domains of pepsins using our program COMPARER. The alignment identifies all those strands and helices that have previously been considered equivalent on a more subjective basis and which have recently been shown to be shared with the retroviral proteinases.

Once protein structures have been aligned, clustering methods using a matrix of similarity scores between all pairs of proteins can be used to construct a tree that describes relationships between them. Whilst there has been much discussion of methods for construction of evolutionary trees from sequences (see Ref. 9 and p. 71 of this book for review), the clustering of protein three-dimensional structures has been less studied. Eventoff and Rossmann[22] constructed dendrograms based on structural features alone to describe distant phylogenetic relationships among the mono- and dinucleotide binding proteins. We have shown that a useful measure of structural difference can be obtained from differences between superposed structures[17,18]. Figure 3 shows an example of a tree constructed for the domains of the aspartic proteinases on the basis of the three-dimensional structures. We have extended this approach by considering additional structural and sequence features in the comparisons[18,20] so that proteins with little sequence similarity can be compared and classified. Such structural

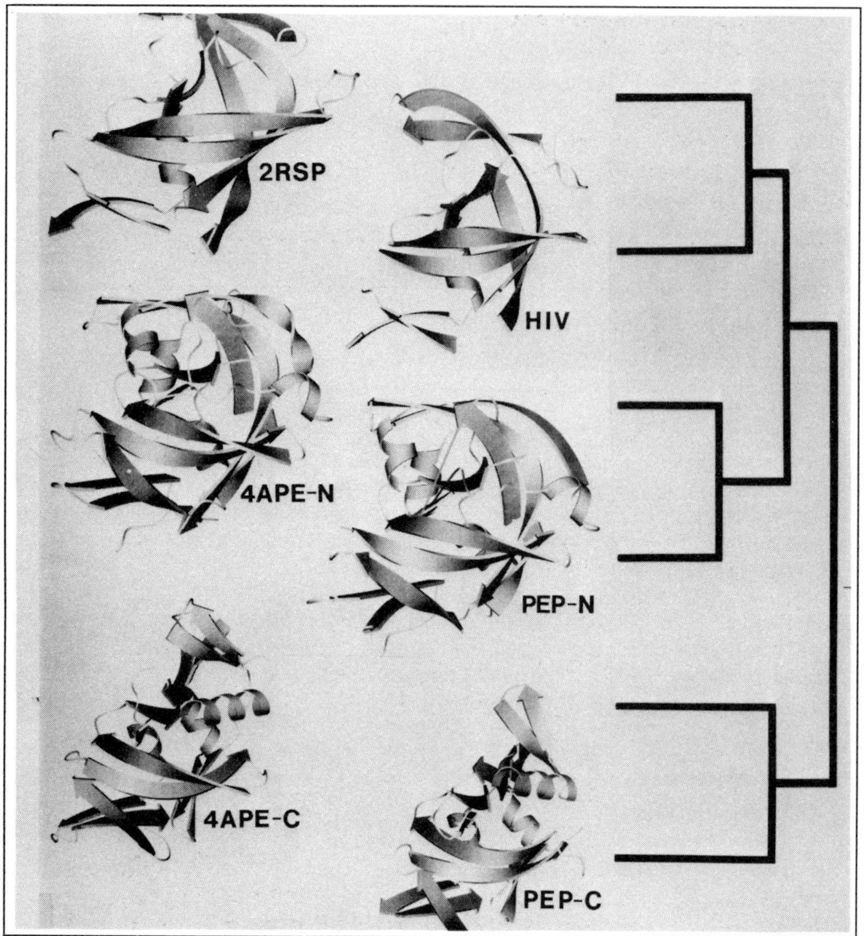

FIGURE 3

The clustering of the N-terminal and C-terminal lobes of the pepsin-like aspartic proteinases and the subunits of the retroviral aspartic proteinases on the basis of structures using methods of Johnson *et al.*[17,18]. HIV, human immunodeficiency virus proteinase; 2RSP, Rous sarcoma virus proteinase; PEP, pepsin; 4APE, endothiapepsin. HIV structure was kindly provided by Lapatto *et al.*[30]

trees can be used in conjunction with the sequence-based trees to provide automatic selection of the best structures and fragments for modelling from sequence.

DERIVATION OF RULES

The alignment of three-dimensional structures allows derivation of rules useful for protein modelling. For example, rules can be obtained that correlate an unknown side chain dihedral angle with the dihedral angles at equivalent positions in related proteins[13,14,23].

Rules for the substitution of amino acids are derived by counting how many times two residue types occur at structurally equivalent positions. We have constructed specific substitution tables in which only a subset of residues that have a certain structural environment are considered. For example, 20 by 20 tables were built separately for solvent-accessible and solvent-inaccessible residues; in this way we can quantify the well-known rule that buried residues tend to be among the more conserved residues in a family. Other structural features included in our analysis were local main chain conformation and side chain hydrogen-bonding to peptide groups or other side chains.

These tables quantify the influence of structural features on the acceptance of amino acid substitutions in evolution. Large differences exist in the substitution patterns of the same residue type in different structural environments. Hydrogen-bonded and inaccessible polar residues such as aspartic acid, serine or threonine are amongst the most highly conserved residues in families of proteins. Their structural roles are relatively specific; as a result it is not easy to vary the amino acid type and also retain the important structural role. For example, as shown in Fig. 2, Thr33 and Thr216 are conserved or conservatively varied to serine in all pepsin-like and retroviral proteinases. These residues play an important role in holding together, through buried hydrogen bonds, the two subunits in retroviral proteinases and the two lobes in pepsins.

PROJECTION FROM STRUCTURE TO SEQUENCE

The substitution tables described above can also be seen as sets of rules that relate the structural environment of an amino acid to the probability for the acceptance of any other of the 20 residue types. This allows us to use three-dimensional structure to project constraints onto the one-dimensional sequence or, in other words, to construct the template representing the sequence of the family of proteins with a particular tertiary fold.

For each topologically equivalent position in each known structure, we use the tables to predict the substitution of amino acid residues. Figure 4 shows both the predicted substitution of four residue positions in a protein on the basis of its three-dimensional structure, and the observed pattern of amino acid substitutions in the equivalent positions of 155 aligned sequences. The example demonstrates how the environment-dependent substitution tables can provide a remarkably good estimate of sequence variation if the three-dimensional structure of at least one protein is known. Likely places for insertions and deletions can also be predicted. This provides a general approach to the generation of templates. An alternative approach has been suggested by Ponder and Richards[24] who have suggested an algorithm for systematically constructing all sequences of amino acid types and their side chain conformations that are consistent with a particular fold.

These templates constructed on the basis of one or more three-dimensional structures are com-

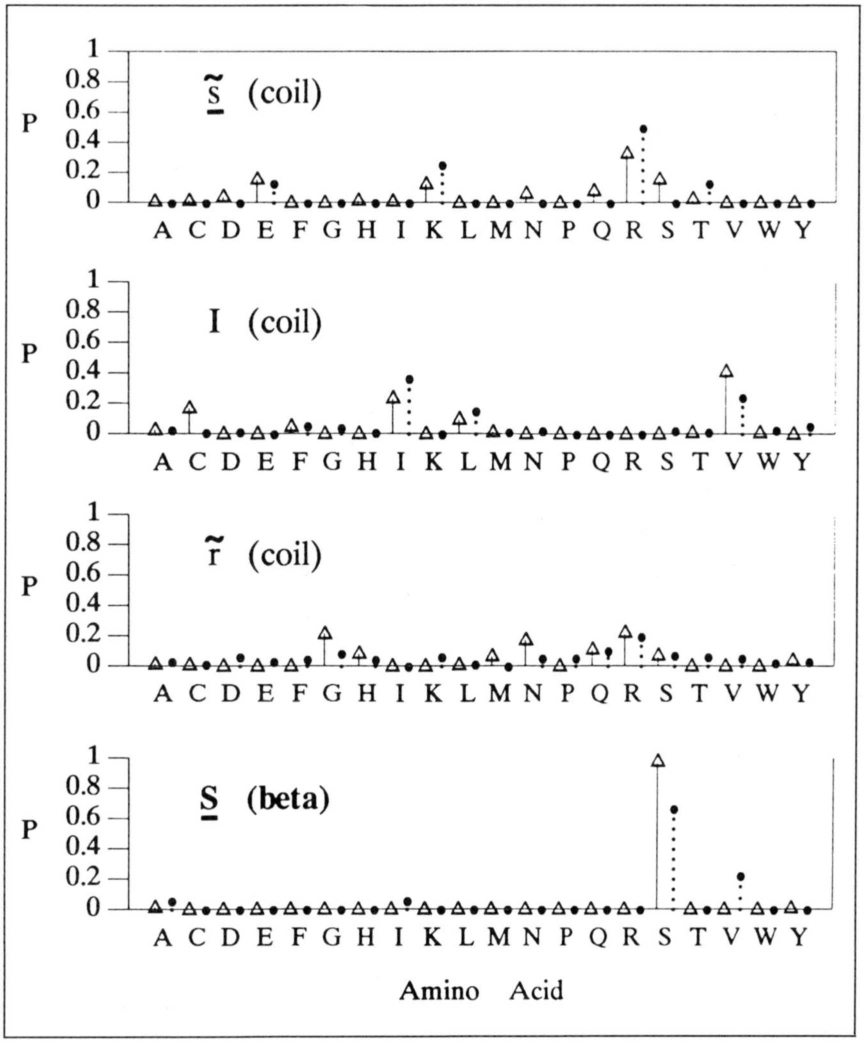

FIGURE 4

Comparison of the predicted pattern of amino acid substitution with the variability observed among real sequences at strand d of motif 3 of γ-crystallins which has the sequence Ser-Ile-Arg-Ser. Residues 2 and 4 are inaccessible to solvent. Residue 4 is serine which is hydrogen-bonded to the amide function of a main chain peptide and so is predicted to be highly conserved.

plementary to those constructed from the alignment of many sequences[9,11]. Both kinds of template can define sequence fingerprints that are essential to structure or function. They can be used in the form of consensus sequences or substitution tables to search out distantly related proteins in the sequence database.

The templates of all known three-dimensional structures or families of structures including loops, motifs, domains and complete globular proteins should be precalculated so that a new sequence can be compared with them rather than with individual proteins. This will result in a better alignment of whole proteins or their parts and thereby in a better extrapolation of spatial features from known structures onto the sequence of the unknown in knowledge-based modelling.

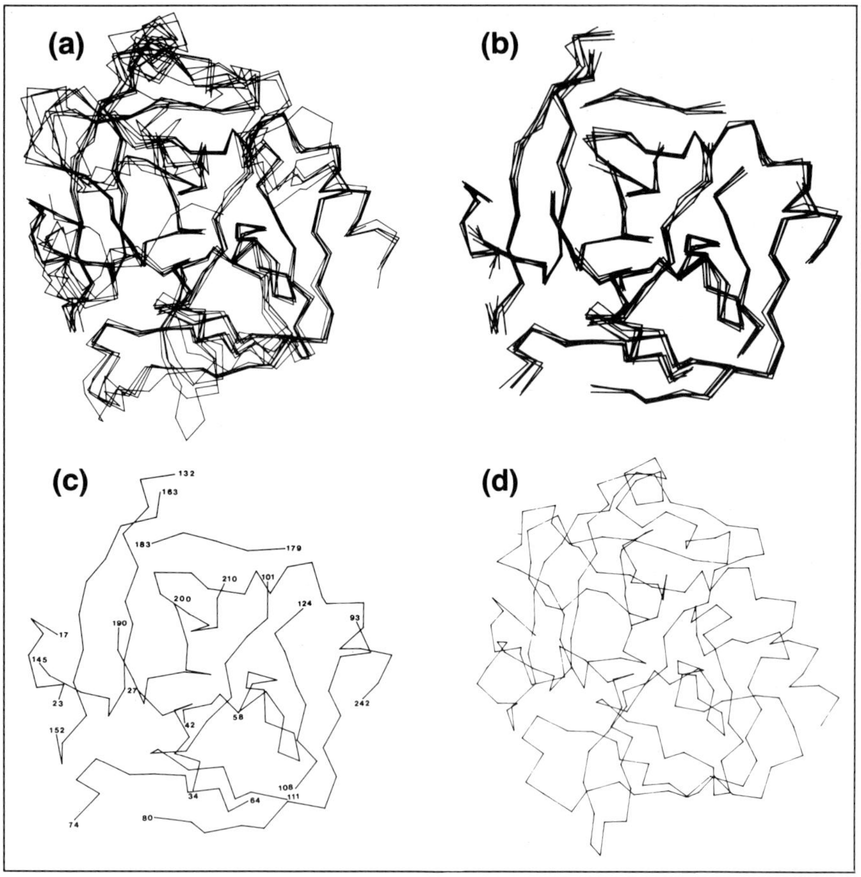

FIGURE 5

Modelling human plasma kallikrein domain from homologous serine proteinases by the program COMPOSER. **(a)** shows the superposition of the structures defined by X-ray analysis. **(b)** indicates the fragments in the structurally conserved regions that contribute towards generation of the framework shown in **(c)**. Fragments, selected using rules from a broader data base of structures, are used to model the structurally variable regions. The C^{α} atom positions of the complete model are shown in **(d)**. Side chain positions (not shown) are also generated by a set of rules derived from comparisons of known structures.

EXTRAPOLATION FROM SEQUENCE TO STRUCTURE

We have shown that analyses of aligned structures of homologous proteins can give rise to simple rules, and shall now consider methods for using these rules to produce a three-dimensional model from a sequence.

Let us first consider the use of such rules to model homologous structures. Most approaches depend on the assembly of fragments of three-dimensional structures[25–27]. In the computer program COMPOSER we select three sets of fragments. The first set is derived from the framework defined by multiple superposition of the chosen structures[14] (Fig. 5). A second set of protein fragments for regions outside the framework is selected from the database of loop sub-structures using a distance filter in a similar way to Jones and Thirup[25]. The third set of fragments, the side chains, is selected using rules derived from the analysis of homologous structures[14]. These 1200 rules include one for each of the 20 by 20 amino acid replacements in each of the three secondary structure types (α-helix, β-strand or irregular). The templates of selected fragments are clustered and ranked using the methods described above, and the top-ranking fragments are annealed together. The model is checked for serious overlaps between fragments; where this occurs the next ranking fragment is used. The final model is energy minimized to remove minor inconsistencies.

This modelling procedure is very successful where the known structures cluster around that to be predicted and where the percentage sequence identity to the unknown is high (>40%). For example, in building a model of porcine trypsin from four other structurally known serine proteinases, the root mean square distance difference between the model and the known structure is 0.64 Å for the 150 residues defined in the framework. Similarly, 80% of side chain conformations are correctly predicted for closely homologous structures. In all cases, the accuracy of the prediction decreases very quickly as the sequence identity between the known and unknown decreases. For these cases, a new modelling technique is required that is not restricted by the idea of assembling rigid fragments of protein structure. In this procedure we construct a model from distances between atoms, similarly to methods for structure analysis using two-dimensional NMR data[28,29].

The sequence to be modelled is first aligned with sequences of known related structures to derive a list of distance constraints. For example, if two equivalent positions in the alignment of known structures are always hydrogen bonded, we can assume that the same hydrogen bond exists in the unknown structure as well. This represents a distance constraint on the atoms involved in the hydrogen bond. In general, we use the tables described above to predict main chain and side chain dihedral angles, C^{α}–C^{α} distances and hydrogen-bonding distances from the known structures aligned with that being modelled.

The predicted distances are expressed as Gaussian probability functions. For non-bonded atoms this often involves taking a mean distance from an homologous protein and a standard deviation proportional to the similarity between the proteins and the magnitude of the distance. For side chain dihedral angles the probability functions are usually trimodal with the relative magnitudes depending on the particular residue type and the values of equivalent dihedral angles in related known structures. In general, every structural feature can be constrained by several knowledge sources. For example, a distance between a particular pair of C^{α} atoms may be constrained by information from several homologous proteins and also by van der Waals criteria. In such cases we combine the individual probability functions.

The goal is now to construct the three-dimensional model that will satisfy these constraints. Obviously, the most probable structure of the molecule as a whole is the one that maximizes the product of all feature probability functions. The optimization is performed in Cartesian coordinate space using a variable target function[28] approach and a combination of conjugate gradients and simulated annealing minimization to make the best use of the speed of the former and large radius of convergence of the latter.

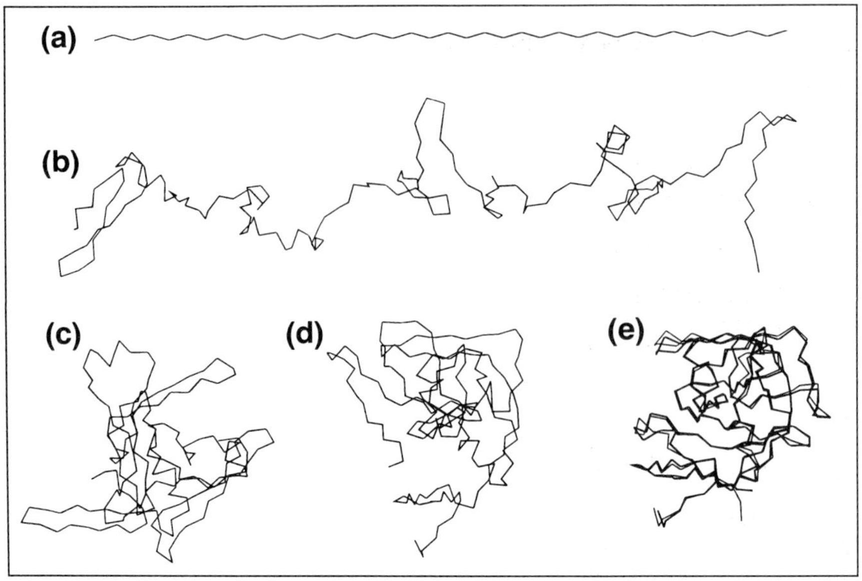

Figure 6

Stepwise generation of a model of N-lobe of endothiapepsin using information from homologous aspartic proteinases and from protein three-dimensional structures in general expressed as distance constraints. **(a)** is the extended chain, **(b)** shows the influence of mainly local constraints, **(c)** and **(d)** are intermediate structures, and **(e)** is the final structure compared with that experimentally defined.

Figure 6 illustrates the generation of a preliminary model for the N-terminal lobe of endothiapepsin; it has a distance root mean square deviation from the crystallographically determined structure of 0.76 Å although only C^{α}–C^{α} constraints were used.

Concluding remarks

Knowledge-based modelling, most often in its simplest form of modelling by homology, is now widely used by biochemists. This reflects the steady advancement in the field including the automation of the algorithms and development of integrated systems synthesizing such diverse tools as databases of sequences and structures, interactive molecular graphics, molecular dynamics and energy minimization, together with methods for pattern recognition, comparison and clustering. It also reflects the steady advance in the numbers of sequences and structures defined experimentally.

Knowledge-based modelling techniques are firmly based on the progress and success of experiment. As a consequence we can expect that the next decade will bring a closer integration of modelling techniques with experimental analyses using crystallography, two-dimensional NMR, image reconstruction in electron microscopy, epitope mapping and crosslinking, all of which have contributed so much to our understanding of complex protein structures and assemblies. The great challenge will be to unify all techniques for determination or prediction of protein structure into a single protocol making the best use of all available information about the structure of a given protein, regardless of whether it is directly based on experiment, on the broader knowledge base, on empirical force potentials or on intuition.

Acknowledgements

AŠ was supported by an ORS Awards Scheme, the Research Council of Slovenia, the J. Stefan Institute and Merck, Sharp and Dohme. MSJ is funded by the American Cancer Society and JPO by Pfizer and SERC. We are grateful to the Imperial Cancer Research Fund, the SERC and the EEC for general financial support. We also thank our colleagues at Birkbeck for many helpful discussions.

References

1 Browne, W. J., North, A. C. T., Phillips, D. C., Brew, K., Vanaman, T. C. and Hill, R. L. (1969) *J. Mol. Biol.* 42, 65–86
2 Blundell, T. L. and Humbel, R. E. (1980) *Nature* 287, 781–787
3 Greer, J. (1981) *J. Mol. Biol.* 153, 1027–1042
4 Sibanda, B. L., Blundell, T. L. and Thornton, J. M. (1989) *J. Mol. Biol.* 206, 759–777
5 Chothia, C., Lesk, A. M., Tramonto, A., Levitt, M., Smith-Gill, S., Air, G., Sheriff, S., Padlan, E. A., Davies, D., Tulip, W. A., Colman, P. M., Spinelli, S., Alzari, P. M. and Poljak, R. J. (1989) *Nature* 342, 877–883
6 Thornton, J. M. and Gardner, S. (1989) *Trends Biochem. Sci.* 14, 300–304; see also p. 153 of this book
7 Presnell, S. R. and Cohen, F. E. (1989) *Proc. Natl Acad. Sci. USA* 86, 6592–6596
8 Needleman, S. B. and Wunsch, C. D. (1970) *J. Mol. Biol.* 48, 443–453
9 Doolittle, R. (1989) *Trends Biochem. Sci.* 14, 244–245; see also p. 71 of this book
10 Argos, P. (1987) *J. Mol. Biol.* 193, 385–396
11 Taylor, W. R. (1986) *J. Mol. Biol.* 188, 233–258
12 KenKnight, C. E. (1984) *Acta Crystallogr.* A40, 708–712
13 Sutcliffe, M. J., Haneef, I., Carney, D. and Blundell, T. L. (1987) *Protein Eng.* 1, 377–384
14 Sutcliffe, M. J., Hayes, F. R. F. and Blundell, T. L. (1987) *Protein Eng.* 1, 385–392
15 Chothia, C. and Lesk, A. M. (1986) *EMBO J.* 5, 823–826
16 Hubbard, T. J. P. and Blundell, T. L. (1987) *Protein Eng.* 1, 159–171
17 Johnson, M. S., Sutcliffe, M. J. and Blundell, T. L. (1990) *J. Mol. Evol.* 30, 43–59
18 Johnson, M. S., Šali, A. and Blundell, T. L. (1989) *Methods Enzymol.* 783, 670–690
19 Matthews, B. W. and Rossmann, M. G. (1985) *Methods Enzymol.* 115, 397–420
20 Šali, A. and Blundell, T. L. (1990) *J. Mol. Biol.* 212, 403–428
21 Taylor, W. R. and Orengo, C. A. (1989) *J. Mol. Biol.* 208, 1–22
22 Eventoff, W. and Rossmann, M. G. (1975) *Crit. Rev. Biochem.* 3, 111–140
23 Summers, N. L., Carson, W. D. and Karplus, M. (1987) *J. Mol. Biol.* 196, 175–198
24 Ponder, J. W. and Richards, F. M. (1987) *Proteins* 193, 775–791
25 Jones, T. H. and Thirup, S. (1986) *EMBO J.* 5, 819–822
26 Blundell, T. L., Carney, D., Gardner, S., Hayes, F., Howlin, B., Hubbard, T., Overington, J., Singh, D. A., Sibanda, B. L. and Sutcliffe, M. J. (1988) *Eur. J. Biochem.* 172, 513–520
27 Claessens, M., Cutsem, E. V., Lasters, I. and Wodak, S. (1989) *Prot. Eng.* 2, 335–345
28 Braun, W. and Go, N. (1985) *J. Mol. Biol.* 186, 611–626
29 Havel, T. F., Kuntz, I. D. and Crippen, G. M. (1983) *Bull. Math. Biol.* 45, 665–720
30 Lapatto, R. *et al.* (1989) *Nature* 342, 299–302

A. Šali, J. P. Overington, M. S. Johnson and T. L. Blundell are at the Laboratory of Molecular Biology and ICRF Unit of Structural Molecular Biology, Department of Crystallography, Birkbeck College, University of London, Malet Street, London WC1E 7HX, UK.

THE *DE NOVO* DESIGN OF PROTEIN STRUCTURES

JANE S. RICHARDSON AND DAVID C. RICHARDSON

The revolution in molecular biology, along with other technical advances, has led to a far more active and meddlesome approach to both the study and the exploitation of proteins. No more need we be content with admiring the elegant and varied proteins produced by living things for doing their chemical, structural, and organizational work – now we can tinker with those proteins however we please, and can even start from scratch and try to invent new ones. Just as by designing and building a house you will learn many things that you would never have found out just from living in, or even redecorating one (such as the real purpose of moldings, and why mortar joints are at least ½″ thick) so also by designing and building our own new proteins we can expect to learn things we would not have learned just by studying natural proteins. Technique is still in advance of understanding, but freewheeling exploration is encouraged by the fact that there have been unexpected successes almost as often as unexpected difficulties.

Protein design has two distinct but interrelated components: the design of activity and the design of structure. Eventually, both will dovetail into a single enterprise, but for the moment they tend to be separate. The design of active sites can best be done on top of some structural framework which will not change much during the process. In practice, this has been achieved either by capitalizing on the resilience of naturally evolved proteins, as in mutating one enzyme into another[1] and in selecting for catalytic antibodies[2], or else by using non-peptide templates with geometries enforced by their covalent structure[3]. Such work is improving our understanding of catalysis and in some cases coming up with truly new functions.

The theme of this review is the second component of protein design, which concentrates on the production of three-dimensional structure. The emphasis of a given study may be either on the understanding of protein folding or on devising practical ways of producing desired structures. For understanding natural proteins one must stay reasonably close to natural polypeptide characteristics, while for ease of human comprehension and production the best strategy may be a more modular and perhaps non-linear one[4]. However, the biological advantages of linear information, self-replication and self-assembly are so great that it is certainly worth the attempt at controlling such systems knowledgeably.

Protein design is an extremely ambitious concept and it still lies mostly in the future. However, the achievements so far are worth reviewing because they have demonstrated the feasibility of the overall approach and have progressed farther and faster than could reasonably have been expected even a few years ago.

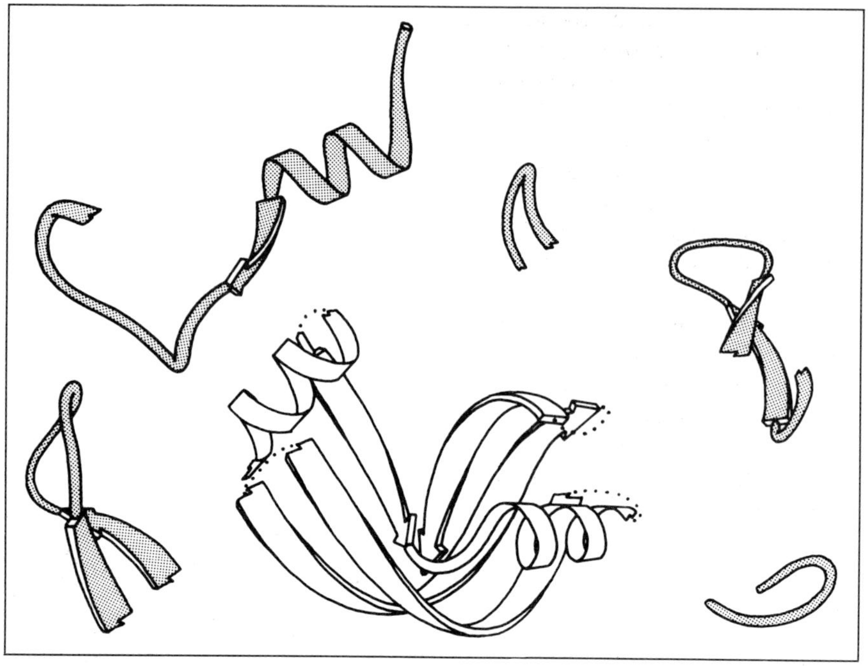

FIGURE 1

A version of bovine ribonuclease[5] trimmed down to just 63 residues (bottom center), with the pieces that were left out shown floating around the edges[6]. Arrows are β-strands and spirals are α-helices.

EARLY HISTORY

The pioneer of protein design was Gutte, first at the Rockefeller University and now at the University of Zurich. His interests are primarily in enzymatic catalysis and binding, but his methods depend on producing novel three-dimensional structures as well. He designed a half-length version of bovine ribonuclease in which the excursions away from the active site were truncated (see Fig. 1). The goal was a synthesizable analog for testing substitutions (in the days before site-specific mutagenesis). It was produced by solid-phase peptide synthesis, and it showed up to 15% of native ribonuclease A activity and similar specificities[6].

Gutte and his colleagues later synthesized a 34-residue polypeptide (FTFTYTDPNCQTGQGQNPNGISEPTAAKVQAHCA) designed to bind the trinucleotide GAA and to have a ββα tertiary structure stabilized by a single disulfide bond[7]. It did indeed bind various polynucleotides including GAA, and unexpectedly cleaved poly-(C). A disulfide-linked 68-residue dimer was much more active than the monomer. Circular dichroism (CD) spectra showed definite denaturable secondary structure, but also a random component.

Then they synthesized the 24-mer (MTFIRPNVGAMSNFYHYPNIIITF) designed to form a four-stranded β-sheet and to bind the insecticide DDT. It did indeed bind DDT two to three orders of magnitude better than controls. Small crystals were grown[8], but not of X-ray quality. The peptide is highly insoluble in water, as is not surprising since it binds a large hydrophobic ligand.

Gutte's work stimulated much of the early interest in protein design, and the functional success of these three molecular designs is very impressive. So far they have not contributed directly to structural design principles since their structures have been difficult to characterize even spectroscopically. Recent *in vivo* expression of the DDT-binding 24-mer[9] may well provide new information.

DESIGN OF SINGLE α-HELICES

Although specifically avoiding tertiary-structure interactions, the design and study of isolated α-helices has been a very important influence on this field. Using the concept of an amphiphilic helix, Kaiser and co-workers at Chicago and Rockefeller successfully built peptide-hormone analogs with minimal homology to the native sequences in the attachment portions except for helical propensity and a segregation of charges and hydrophobics to opposite helix faces[10]. These peptides, like the native ones, are not helical in solution but do form helix at the surface of membranes. The success of such synthetic models tested the binding mechanisms of the individual hormones, and has also clearly demonstrated the power of simple amphiphilic sequences to form helix.

Baldwin and his group at Stanford have extensively tested synthetic variants of the ribonuclease S-peptide helix and other model peptides, in order to dissect the influence of particular factors on formation of an isolated α-helix[11]. They find the interaction of charges with the helix dipole to be of major importance; although it has been known for some time that negatively charged amino acids are commoner near the beginning of helices and positive charges near the end, theirs are the first experimental demonstrations of such charge effects on helix stability. They also find some contribution from salt links and other side chain interactions. For example, the peptide A(EAAAK)$_3$ with *i* to *i*+4 salt links is a better helix-former than (AEAAKA)$_3$ with *i* to *i*+3 (Ref. 12). All of these are position-specific effects, over and above the general helix propensities of particular amino acids. The Stanford group studies helices that do not aggregate and are only weakly amphiphilic; therefore their results nicely complement the work of Kaiser and of DeGrado on strongly amphiphilic, aggregating sequences.

Design of fibrous proteins

Many fibrous proteins are long polypeptides with nearly exact repeats, which coil up into single helices or aggregate into coiled-coils of helices. Much has been learned about these structures by synthesizing peptides that contain several exact repeats of a consensus sequence.

Collagen is a three-strand coiled-coil of repeating Gly-Pro-X sequences which adopt the extended helix of polyproline conformation. The aligned assembly of native collagen is controlled by cleavable domains at the chain ends, so that the interactions of its triple coil are difficult to study. Heidemann and colleagues at Darmstadt have made synthetic, exact-repeat collagen analogs that are covalently connected at one end[13].

Tropomyosin is a coiled-coil of two α-helical chains with an approximate heptapeptide repeat that puts large hydrophobics (typically Leu) at positions 1 and 4, where they provide the packing surface. Hodges *et al.* at the University of Alberta have synthesized a variety of successful peptide models [e.g. (LEALEGK)$_5$] for such α-helical coiled-coils[14].

The elasticity of elastin depends on a decrease of entropy upon stretching. Its sequence has many nearly-exact copies of a five-residue VPGVG repeat, whose structure was demonstrated crystallographically on a cyclic three-copy peptide[15]. The elastin helix, sketched in Fig. 2,

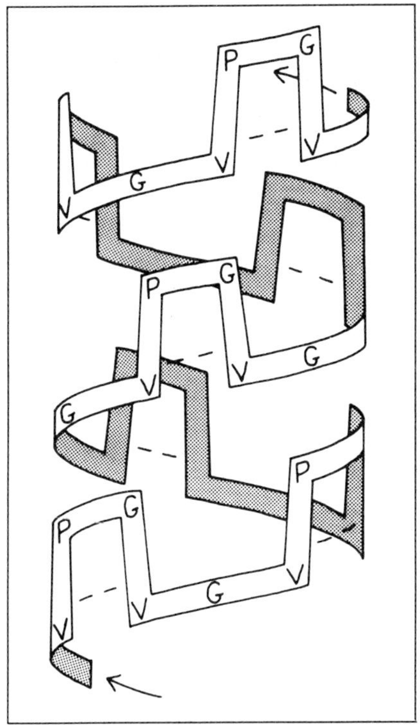

Figure 2

The backbone arrangement and sequence of the elastin helix[15,16], with dashed lines for the hydrogen bonds of the tight turns. Since there are only Van der Waals interactions between adjacent turns of the helix, stretching is fairly easy, but is entropically unfavorable because it restricts rotation of the two peptides in the extended VGV region.

apparently operates by controlling the degree of allowed peptide rotation. Working with a variety of longer repeating peptides, Urry and colleagues at the University of Alabama have made not only models for elastin behavior[16], but also novel elastomers with temperature-transition or pH-transition behavior that can be tailored by adding different numbers of charged residues to the sequence[17]. These elastomers are surely an example of *de novo* protein design.

Design for β-sheet folding

It is frustrating that it is easy to make plausible hypotheses about the determinants of tertiary structure but essentially impossible to test them experimentally. As a way to make such tests, we decided to design from scratch model proteins based on paradigm tertiary-structure types but not homologous to any particular native protein. If they could be successfully synthesized and structurally characterized, then their design principles could be altered and the variants tested. With native proteins one can make small changes and test for stability, but it is probably not feasible to change from one topology to another because all parts of the sequence have been chosen by natural selection to work with the native structure.

In an early collaboration with Ted Hazen and Marty Edelstein at Texas A&M, we helped choose alternating dipeptide sequences such as $(ID)_4$ and $(RA)_4$ that could perhaps be joined by Pro-Asn turns to form a β-barrel. They were made by classical solution-phase peptide synthesis and showed indications of β-structure, but they were difficult to handle and were never linked together.

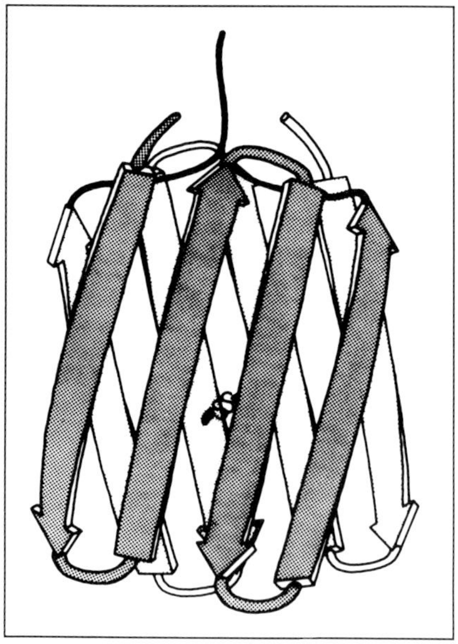

Figure 3

Schematic ribbon drawing of betabellin[19], with the two-armed crosslinker in black. The two β-sheets have identical 32-residue sequences and are joined by a disulfide bond. In recent versions of betabellin, the crosslinker is omitted.

In 1981 we initiated a collaboration with Bruce Erickson (then at the Rockefeller and now at UNC-Chapel Hill) to design and make a β-sheet protein. An antiparallel β-sandwich of two identical, four-stranded sheets (Fig. 3) was chosen as the target structure for 'betabellin' (a *beta*-sheet, *bell*-shaped prote*in*), and Erickson designed a two-armed cross-linker that would allow the two identical chains to be grown simultaneously from a single attachment on the resin[18].

We designed an amino acid sequence for betabellin, using all known or presumed-known criteria, compromising where necessary[19]. The criteria included secondary-structure prediction, placement of hydrophobics and hydrophilics, preferred residue pairings on adjacent β-strands, amino acid composition data for the inside and outside of strands in similar β-barrels, preferred strand twist, position specificity of turn residues, preferred side chain conformations, and patterns of internal packing interactions. Because our aim was to study the determinants of tertiary structure, we avoided on one hand any homology with a natural protein, and on the other hand used a fairly normal amino acid composition in a non-repeating, 'protein-like' sequence. Suitability was also considered for synthesis, NMR assignments, crystallography, and later possibility of adding a binding or catalytic site at one end. Our hope is to overdetermine the well-understood features of such a β-structure in order to make up for the inevitable omissions and mistakes, much as the framework of an immunoglobin variable domain is over-determined so that it can fold correctly no matter what changes in its hypervariable loops.

Betabellin 1 had the sequence STVTARQPNVTYSISPNTATVRLPNΦTLSIG (where Φ is iodo-Phe, as a potential heavy-atom derivative). The synthesized material showed multiple components, missing either a valine or a threonine. The sequence was redesigned to eliminate consecutive branched C^{β} residues, which couple poorly in synthesis. Betabellin 2 did indeed produce a cleaner product, with a CD spectrum showing no α-helix and perhaps 40% β-structure. Version 3 included a cysteine at position 21, to produce a disulfide across the inside of the barrel. The disulfide is indeed formed, and is within the monomer.

As is common for peptides that form β-sheet, these early betabellins were not very water-soluble. The partitioning of hydrophobics inside and hydrophilics outside had been overdesigned to be even better than native proteins, as confirmed in model calculations by Jiti Novotny at Massachusetts General Hospital (now at the Squibb Institute). However, betabellin had an unusually low number of charges per residue for a soluble, globular protein. Therefore, for betabellins 7 and 8, eight additional charges per molecule were added. The solubility improved, but not dramatically. Laser Raman spectra taken by Bob Williams at the Uniformed Services University of the Health Sciences showed very high β-content but still could be done only on the precipitate.

The two-armed crosslinker is the part of the design for which there is the least empirical basis, and it seemed possible that its geometry did not fit well with the rest of betabellin. The internal disulfide had been shown to form successfully, so Erickson and Daniels decided to dispense with the crosslinker altogether for betabellin 9, which was made as the single-chain, 32-residue half sequence. This omission turned out to be the key to producing a truly water-soluble betabellin.

Betabellin 9 under native conditions shows a CD spectrum typical of β-structure, changing to the shape appropriate for 'random coil' in 6 M guanidine HCl. Its laser Raman spectrum looks like that of proteins with 60–80% anti-parallel β-sheet, but can now be measured in aqueous solution at somewhat acid pH. Raman spectra also demonstrate formation of the internal disulfide. Betabellin 9 gives well-resolved two-dimensional NMR spectra (done by Art Pardi at the University of Colorado) that contain crosspeaks which are reasonable candidates for close interactions between adjacent antiparallel β-strands.

The sequence of the successful betabellin 9 is HTLTASIPDLTYSIDPN-TATCKVPDΦTLSIGB (where B stands for β-alanine), which has 12 conservative changes relative to the original betabellin 1. With several spectroscopic probes to assess structural quality, it should certainly be possible to improve the betabellin sequence enough to allow determination of its complete three-dimensional structure by crystallography or by NMR, if it turns out that the current material is not yet suitable. Sequence changes will also be made to test particular structural hypotheses: for example, substitution of DPro,DAsp for the normal

LPro,LAsp at the turns to test the proposal of Sibanda and Thornton[20] that such mirror-image turn conformations should promote formation of tight β-hairpins.

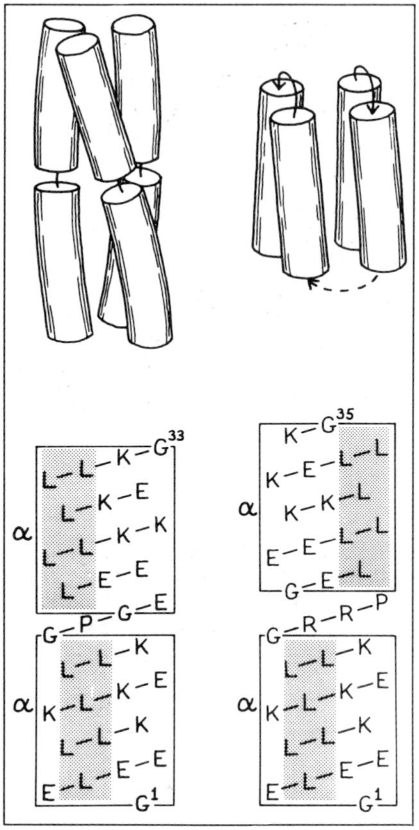

Figure 4

The sequences of two different two-helix intermediates along the way to construction of the 'α-4' protein, one with a single proline in the helix–helix connection and the other with Arg-Arg-Pro[22]. This 'helical net' layout shows the cleanly segregated hydrophobic stripes of leucines designed into the individual α-helices (boxed). With the shorter connection, the two stripes of Leu line up almost perfectly the way they would in a long coiled-coil structure, while the Arg-Arg-Pro connection offsets the two hydrophobic stripes so that a coiled-coil would be disfavored. The sketches (top) show the type of aggregation behavior that was observed for each of these sequences. The successful dimer of helix pairs was then covalently joined with yet a third Arg-Arg-Pro connection (dashed line) to form the final α-4 protein[23].

A MODULAR MINIMALIST HELIX BUNDLE

Eisenberg at UCLA has used his hydrophobic-moment work to design a simple 16-residue sequence that should form an amphiphilic α-helix and associate into a bundle. It was made by DeGrado at Du Pont using peptide synthesis[21]. The strategy is a modular one of optimizing the sequence of the single helix based on CD spectra and association behavior, then linking two of them to form a helical hairpin whose connection and dimer interaction could be optimized, and finally synthesizing a string of four identical helix sequences to form a helix-bundle protein. The group at Du Pont incorporated several sequence changes during this process, some from molecular modeling and one to cure a problem found experimentally at the two-helix stage. The initial connection designed (Gly-Pro-Gly) seemed a very reasonable one, but instead of dimerizing to form a four-helix bundle, the 33-residue piece associated into an elongated trimer. Addition of two arginines to the connection overcame this problem[22]. The change increased both the charge and the flexibility of the connection, and also offset the hydrophobic stripes of leucines along a helical net (see Fig. 4), all of which should help in avoiding the trimerization into a coiled-coil. This illustrates the advantages of the modular construction, and also the interesting point that protein design inherently involves negative as well as positive design, since alternative possible structures must be actively avoided.

The entire 73-residue 'α-4' is now being produced by *in vivo* expression methods, and is the best characterized of any designed protein. It has a highly helical CD spectrum and is remarkably stable, with a cooperative unfolding transition above 6 M guanidine HCl. Disulfides at varying positions are being used to test the uniqueness of the structure and determine more about its arrangement[23].

A 'PROTEIN-LIKE' HELIX BUNDLE

To complement the betabellin project on peptide synthesis of a β-sheet design, we are also working on a four-helix bundle (see Fig. 5) made by gene synthesis and expression methods. 'Felix' was designed in a graduate seminar at Duke in 1986*, using principles for the stabilization of α-helices by patterns of charge or hydrophobicity, helix contact geometries, etc., and collecting additional statistics on helix hairpin connections and individual-position amino acid preferences[24]. Plausible alternative structures were considered and actively avoided. Felix has a non-repetitive sequence of 79 residues that uses 19 of the 20 amino acids (MPEVAENFQQCLERWAKLSVGGELAHMANQAAEAILKGGNEAQLKNA-

* Seminar students: Susan Gardlik, Paul Gardner, Ali Khorrami, Diana Pitterle, and Kim Schandel; instructors: DCR, JSR and Richard Ogden.

QALMHEAMKTRKYSEQLAQEFAHCAYKARASQ) and has no homology with any natural protein, although its architecture is based on four-helix clusters such as hemerythrin and cytochrome b_{562}.

Richard Ogden designed a DNA sequence for felix, and he and Linda Musick at the Agouron Institute synthesized the gene. Both they, and Michael Hecht in our lab at Duke, have worked on a large variety of expression systems, including several types of fusion proteins with dihydrofolate reductase. Recently Hecht was successful using a system[25] in which the phage T7 RNA polymerase is expressed in *E. coli* under inducible control and felix is in turn expressed from a plasmid containing the T7 promoter. A large amount of felix protein is produced in inclusion bodies, and it can be readily solubilized and purified[26]. There is no degradation, and the amino acid composition and N-terminal sequence are correct except that the initiating Met has all been cleaved off *in vivo*. The CD spectrum of felix shows very high α-helix content, and the solubility is excellent. The disulfide designed between the first and fourth helices forms readily by exchange with glutathione, and at reasonable concentrations it is all intramolecular with no detectable dimer linkage. That is pleasing, because the geometry of disulfides between α-helices is very critical and would not work for most of the possible alternative tertiary structures. Felix is probably the nearest thing so far to a native-like sequence designed from scratch that folds up successfully.

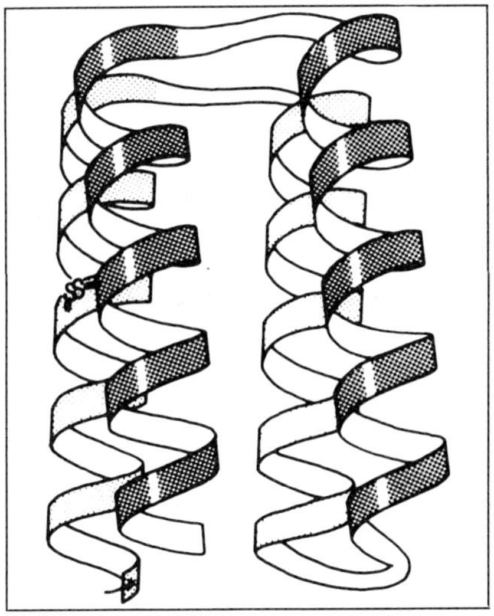

FIGURE 5

Schematic ribbon drawing of felix[26]. The single chain of 79 amino acids is designed to form a four-helix bundle with a disulfide bond linking the first and fourth helices.

TEMPLATE DESIGNS TO AVOID THE FOLDING PROBLEM

Mutter and his colleagues, at Basel, are interested in the *de novo* design of protein framework molecules on which new functions can be built. After early experiences with such things as designed βαβ units[27] they have now devised a strategy for avoiding the difficulties of the protein folding problem altogether by making what they call 'TASPs', or template-assisted synthetic proteins[28]. These have the desired tertiary relationships built in by branching covalent connectivities, rather than using biological-style self-assembly from a linear chain. TASPs must, of course, be synthesized chemically rather than biologically.

The prototype TASP has a short peptide as the central template, with four Lys side chains which form attachment sites for four identical peptides designed to be amphiphilic α-helices (e.g. the sequence EALEKALKEALAKLG). Their C-termini are constrained to be close together, and they should assemble into a parallel four-helix bundle (see Fig. 6). The individual peptides show concentration-dependent but quite weak helicity, but together on the TASP they have a much more helical CD spectrum that is no longer concentration dependent. A 4-β version uses the sequence $(VK)_4G$; it shows a β-type CD but only at pH 12, presumably because of charge repulsion. They have also made TASPs that include both α-forming and β-forming peptide branches in various arrangements; these may quite likely have assembled successfully, but that is inherently harder to demonstrate for the mixed systems. All of the TASPs are quite soluble.

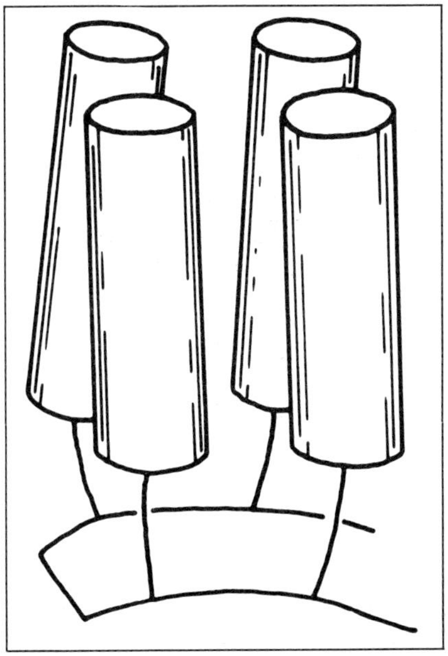

FIGURE 6

Simplified representation of a four-helix TASP (template-assisted synthetic protein), with four identical helical peptides coupled to the Lys side chains of a β-hairpin template peptide[28].

Design of a Membrane Channel

One of the most spectacular successes of *de novo* design so far are the peptides (LSLLLSL)$_3$ and (LSSLLSL)$_3$ designed, synthesized and characterized by the group at Du Pont[29]. These peptides apparently bind to membranes in α-helical conformation and then, when a voltage is applied, show the sort of quantized conductance behavior typical of natural ion channels. Computer modeling suggests that (LSLLLSL)$_3$ could form coiled-coil helix tetramers perpendicular to the membrane and (LSSLLSL)$_3$ could form six-helix coiled-coils with a larger Ser-lined central pore. Indeed, the former molecule is found to conduct only protons while the latter also conducts larger cations. The number and arrangement of the helical peptides will be tested by attaching them to circular template molecules. Although this design was inspired by an inside-out resemblance to the coiled-coil four-helix bundle structure of the ROP protein[30], it was of course done in the absence of any direct knowledge of the tertiary structure of naturally occurring ion channels.

Other Current Design Work

A number of groups are working on other *de novo* designs that are not as far along as the ones described above. For instance, in 1986 Chris Sander organized a design workshop at the EMBL in Heidelberg at which groups collaborated on seven different protein designs: two based on the four-helix coil of ROP protein[31], three on a nucleotide-binding protein similar to flavodoxin[32] and two on the parallel β-barrels typified by triose-phosphate isomerase[33,34]. The strengths and limitations of existing computer-modeling and evaluation tools for this rather demanding application were also explored at the workshop, which motivated several subsequent developments (e.g. Ref. 35). Several ROP designs and redesigns have been cloned and overexpressed by Emery and Bloecker in Braunschweig, in collaboration with Cesareni, Tsernoglou and Sander at EMBL. One of the parallel β-barrels (Fig. 7) is being expressed from a synthetic gene of the quarter-length sequence by Fritz in Göttingen and Kirschner in Basel.

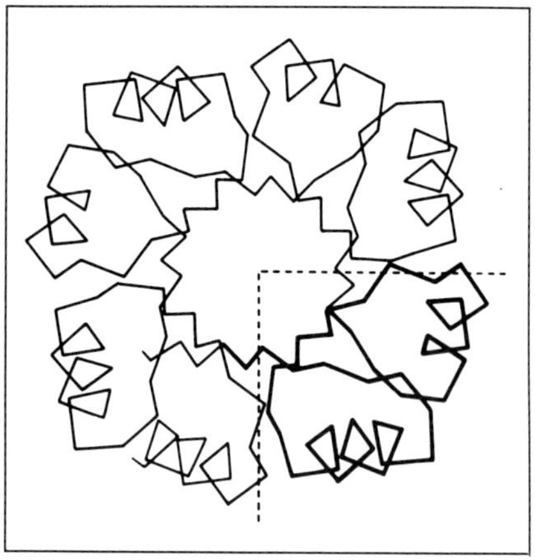

Figure 7

An α-carbon plot of βαβαrellin, a parallel β-barrel protein designed to have fourfold symmetry in both sequence and structure[34]. The computer model was built using Biosym's INSIGHT program for the graphics manipulations and VFF for energy minimizations.

In addition to their separate involvements in α-4 and in betabellin, Eisenberg and Erickson are also collaborating on the design and peptide synthesis of a series of 'betacylindrin' molecules. They are designed to form six-strand cylinders of β-sheet, and are synthesized on a three-armed crosslinker.

Ponder and Fox at Yale are making a *de novo* design of a curved, four-strand antiparallel β-sheet with an α-helix packed against the concave side. So far they have successfully produced the α- and β-components of the sequence separately but are having some problems with hydrophobic behavior of the β-peptide, as often seen for other *de novo* β-structures[8,19,28] and even for some natural β-proteins[36].

At Birkbeck College, London, Travers is working on the biological expression of 'crystallanova', which was designed by Hubbard as an analog of one domain of γ-crystallin, but with an almost exact internal twofold symmetry in both sequence and structure. Four histidines are added in the top loops of the β-barrel to provide a copper-binding site.

Discussion

The most exciting overall conclusion is that it actually is possible to design new proteins from scratch that successfully fold up into something like the intended structure (although not always on the first try). Every one of the *de novo* design projects has indeed worked, at least at the primitive level of initial characterization. That first step is the risky one, whereas any needed further improvements from here should be relatively straightforward. These multiple successes, with such a wide variety of protein models, synthetic methods, and design philosophies, surely imply that some relatively large fraction of amino acid sequences produce approximate tertiary structures. Many of us started out with the hope that reverse prediction (i.e. design) would be easier than prediction of protein structures, and it seems that may be the case.

There is some reason to conclude that membrane proteins may be simpler to design than soluble proteins, and that α-helices are easier to design than β-sheets. Betabellin has required four major changes in sequence before achieving the degree of success that felix had on the first try; the helical TASPs have been much more convincingly characterized than the β ones; and 'α-4' is the stablest design made so far. Helices also, of course, have the important advantage that they produce strong and characteristic CD spectra.

The great diversity of approaches within this field has been extremely valuable. Peptide synthesis is better for short sequences and for producing material quickly and reliably (although not always pure) in an untried system, and it is essential for incorporating non-biological amino acids or topologies. Synthesis *in vivo* is much easier once a successful expression system is found, and is essential for long sequences. Exploring the whole range of types of sequence arrangement (from native-like, to highly simplified but linear, to those with branched connectivity) is very important both because each can contribute in different ways to our understanding, and because probably some elements of each will be incorporated into designs that are eventually of practical use.

Acknowledgements

We thank the researchers mentioned here for their comments on this manuscript, and especially our collaborators Bruce Erickson, Richard Ogden and Michael Hecht, without whom betabellin and felix could not have become realities. Our design work was supported by the NIH Institute of General Medicinal Sciences, and also by the Office of Naval Research, the Life Sciences Foundation, and the MacArthur Foundation. The copyright for the figures is held by JSR.

References

1 Clarke, A. R., Atkinson, T. and Holbrook, J. J. (1989) *Trends Biochem. Sci.* 14, 145–148; see also pp. 31 and 39 of this book

2 Lerner, R. A. and Tramontano, A. (1987) *Trends Biochem. Sci.* 12, 427–430
3 Cram, D. J. and Lam, P. Y-S. (1986) *Tetrahedron* 42, 1607–1615
4 Drexler, K. E. (1986) *Engines of Creation* Anchor Press/Doubleday
5 Wyckoff, H. W., Tsernoglou, D., Hanson, A. W., Knox, J. R., Lee, B. and Richards, F. M. (1970) *J. Biol. Chem.* 245, 305–328
6 Gutte, B. (1975) *J. Biol. Chem.* 250, 889–904; and (1977) *J. Biol. Chem.* 252, 663–670
7 Gutte, B., Däumigen, M. and Wittschieber, E. (1979) *Nature* 281, 650–655
8 Moser, R., Thomas, R. M. and Gutte, B. (1983) *FEBS Lett.* 157, 247–251
9 Moser, R., Frey, S., Münger, K., Hehlgans, T., Klauser, S., Langen, H., Winnacker, E., Mertz, R. and Gutte, B. (1987) *Protein Eng.* 1, 339–343
10 Kaiser, E. T. and Kézdy, F. J. (1983) *Proc. Natl Acad. Sci. USA* 80, 1137–1143
11 Shoemaker, K. R., Kim, P. S., York, E. J., Stewart, J. M. and Baldwin, R. L. (1985) *Proc. Natl Acad. Sci. USA* 82, 2349–2353
12 Marqusee, S. and Baldwin, R. L. (1987) *Proc. Natl Acad. Sci. USA* 84, 8898–8902
13 Germann, H-P. and Heidemann, E. (1988) *Biopolymers* 27, 157–163
14 Talbot, J. A. and Hodges, R. S. (1982) *Acc. Chem. Res.* 15, 224–230
15 Cook, W. J., Einspahr, H., Trapane, T. L., Urry, D. W. and Bugg, C. E. (1980) *J. Am. Chem. Soc.* 102, 5502–5505
16 Urry, D. W. (1988) *J. Protein Chem.* 7, 1–34
17 Urry, D. W. (1988) *Int. J. Quantum Chem., Quantum Biol. Symp.* 15, 235–245
18 Unson, C. G., Erickson, B. W., Richardson, D. C. and Richardson, J. S. (1984) *Fed. Proc.* 43, 1837
19 Richardson, J. S. and Richardson, D. C. (1987) in *Protein Engineering* (D. Oxender and C. Fox, eds), pp. 149–163, Alan R. Liss
20 Sibanda, B. L. and Thornton, J. M. (1985) *Nature* 316, 170–174
21 Eisenberg, D., Wilcox, W., Eshita, S. M., Pryciak, P. M., Ho, S. P. and DeGrado, W. F. (1986) *Proteins* 1, 16–22
22 Ho, S. P. and DeGrado, W. F. (1987) *J. Am. Chem. Soc.* 109, 6751–6758
23 DeGrado, W. F., Wasserman, Z. R. and Lear, J. D. (1989) *Science* 243, 622–628
24 Richardson, J. S. and Richardson, D. C. (1988) *Science* 240, 1648–1652
25 Studier, F. W. and Moffatt, B. A. (1986) *J. Mol. Biol.* 189, 13–130
26 Hecht, M. H., Richardson, D. C., Richardson, J. S. and Ogden, R. (1989) *J. Cell. Biochem.* 13A, 86
27 Mutter, M., Altmann, K-H. and Vorherr, T. (1986) *Z. Naturforschung* 41b, 1315–1322
28 Mutter, M. (1988) *Trends Biochem. Sci.* 13, 260–265
29 DeGrado, W. F., Lear, J. D. and Wasserman, Z. R. (1988) *Science* 240, 1177–1181
30 Banner, D. W., Kokkinidis, M. and Tsernoglou, D. (1987) *J. Mol. Biol.* 196, 657–675
31 Moult, J., Frömmel, C., Postma, J., Skerra, A. and Valencia, A. (1986) *EMBL BIOcomputing Technical Document* 1, 71–86; and DeGrado, W. F., Hubbard, T., Reichelt, J. and Woodward. C. (1986) op. cit. 87–94
32 Lesk, A., Herzberg, O., Schaefer, M. and Wasserman, Z. R., (1986) op. cit.[31] 40–50; and Richardson, D. C., Colonna, F., van Cutsem, E. and Mortensen, K. K. (1986) op. cit.[31] 51–70
33 Jones, T. A., de Vlieg, J., Eliasson, M., Sander, C. and Hirshberg, M. (1986) op. cit.[31] 13–20
34 Richardson, J. S., Argos, P., Kneller, D., Osguthorpe, D. and Scharf, M. (1986) op. cit.[31] 21–39
35 Baumann, G., Frömmel, C. and Sander, C. (1989) *Protein Eng.* 2, 329–334
36 Haase-Pettingell, C. and King, J. (1988) *J. Biol. Chem.* 263, 4977–4983

J. S. and D. C. Richardson are at the Department of Biochemistry, Duke University Medical Center, Durham, NC 27710, USA.

Protein Translocation and Turnover in Eukaryotic Cells

Ralph A. Bradshaw

Although most eukaryotic proteins are synthesized in the cytoplasm from mRNAs originating in the nucleus, many function in specialized compartments and must be specifically translocated co- or post-translationally. A variety of signals contained within the amino acid sequence guide these processes as well as direct turnover by different proteolytic schemes. Several properties of the mature protein are determined as a result of translocation and can serve as predictors of cellular localization.

The synthesis of most eukaryotic proteins occurs in the cytoplasm directed by mRNAs prepared in the nucleus. Contained within the nucleic acid and amino acid sequences is the information for proper folding and for the co- and post-translational modifications essential to achieve activity, regulation and, ultimately, degradation. Importantly, these directions also include trafficking or translocation, the processes whereby proteins are either retained in the cytoplasm or are transferred to various locations, inside or outside the cell, so that the activity that they possess can function in the appropriate environment. It is clearly not sufficient to the physiology of a cell simply to elaborate a protein of a given function; it must, for example, assure that hormones are exported and that tricarboxylic acid cycle enzymes are efficiently transferred to the mitochondrial matrix in order for the organism as a whole to coordinate its myriad of vital cellular functions in a balanced way.

Membrane-bound compartments

With a few notable exceptions (e.g. mitochondria), eukaryotic cells are basically divided into two membrane-bound domains. The dividing barrier is the endoplasmic reticulum (ER) membrane and the various membrane compartments that are a continuum of it including the plasma membrane. Protein translocation across this membrane is a unidirectional process that basically insures that proteins that are extruded through the ER membrane at the time of biosynthesis will not return to the cytoplasmic compartment. Thus, these proteins may remain in the ER or be transferred to the Golgi or any vesicle that can arise from or fuse with either of these organelles, including lysosomes, vacuoles and secretory vesicles[1]. This pathway also includes proteins that are exocytosed into the extracellular milieu and the intrinsic membrane proteins that are found in the membranes of any of these organelles including the plasma membrane[2]. There is, however, not necessarily free transfer between the various compartments of the ER continuum and the routes that are in fact taken are clearly controlled, usually by signals composed of sequences that are an inherent part of the structure of the transported protein[1,3].

The vectoral nature of the synthesis of proteins transferred across the ER membrane has many important consequences for the physiology of eukaryotic cells, not the least of which is the demands that are placed on subsequent communication between these two compartments. With the exception of a few small molecules, which appear to be able to cross these membranes without specific assistance from membrane-bound protein components, information generated on the outside of the cells must be 'communicated' through membrane-bound entities (usually receptors) to the machinery contained in the intracellular compartment. With few if any exceptions, it seems unlikely that protein messengers of any type can function by crossing the membranes of the ER continuum and interacting directly with structures in the intracellular compartments.

The mechanism for the transfer of proteins across the ER membrane is now reasonably well understood[4,5]. In general, an N-terminal signal peptide is required that is recognized cotranslationally by a signal recognition particle, temporarily arresting elongation. This complex directs the ribosome to a docking protein in the ER membrane where protein synthesis is then allowed to continue resulting in the extrusion of the protein into the lumen of the ER. Two other articles in this book [by Bill Wickner (p. 227) and Hans Neurath (p. 201)] address details of this process.

Eukaryotic cells also contain membrane-bound domains that are unconnected to the ER system. These include the mitochondrion and the interior of the nucleus as well as more specialized organelles such as chloroplasts and glyoxysomes. Access to these compartments also generally requires special signals that are a part of the protein sequence[6–9]. Proteins that are translocated to the mitochondrial inner membrane or matrix have N-terminal signal peptides, analogous to the signal peptides required for ER translocation but different in chemistry (ER signal peptides are basically highly hydrophobic in nature, whereas those signalling mitochondrial import are amphipathic helices). The import of proteins into the mitochondrion (and chloroplast) is largely if not completely a post-translational event but is similar to the ER translocation process in that it requires energy, an unfolded state of the transported protein and the ultimate excision of the signal peptide by a peptidase (located in the lumen of the ER or the matrix of the mitochondrion). In addition, the transfer also appears to be unidirectional.

The transfer of proteins into the nucleus is conceptually and functionally different from the transfer of proteins across the ER membrane or into the mitochondrion[10,11]. The nucleus is also surrounded by a double membrane system, the outer membrane of which is actually part of the ER continuum. Thus proteins contained in the outer membrane of the nucleus (or in the intermembrane space) are derived from ER translocations. Proteins located in the nucleoplasm, however, are introduced by transfer through nuclear pores, which span the two membranes. These are elaborate channels of varying diameter (usually about 100 Å) which apparently have the capacity to recognize most if not all proteins synthesized in the cytoplasm that are meant to function in the nucleus. This is particularly important for larger proteins which, after specifically associating with the nuclear pores, cause the expansion of the pore channel to allow transfer into the interior of the organelle. The nuclear recognition signals that have been identified are generally short, contiguous sequences that can be located anywhere throughout the protein. Unlike the ER and mitochondrial systems, these peptides are not processed but remain as part of the protein structure after the transfer is complete. This is important because it allows nuclear proteins to reassemble (after the nucleus dissolves and reforms during mitosis) without the requirement for new protein synthesis. Recent studies have shown that the binding and translocation reactions can be divided into separate processes, with the latter being energy dependent[12,13].

CO-TRANSLATIONAL PROCESSING AND CYTOPLASMIC LOCALIZATION

With the exception of the events surrounding the recognition of signal peptides from either the secretory or organellar import pathways, the principal co-translational event associated with eukaryotic protein synthesis occurs at the N-terminus. This may involve the removal of the initiator methionine residue and the acylation of the α-NH_2 group (either of the initiator methionine or the penultimate amino acid)[14]. The latter modification most commonly involves the transfer of an acetyl moiety from acetyl CoA by the enzyme N^{α}-acetyltransferase (NAT), and to a considerably lesser degree, but perhaps no less importantly, the addition of myristyl groups derived from myristyl CoA[15,16]. The specificities of the methionine aminopeptidase (MAP) and the two transferases appear to be largely governed by the sequences immediately following the initiator methionine. Sherman and colleagues[17] suggested that the specificity of MAP was governed almost completely by the size of the side chain of the penultimate residue based on an analysis of isocytochrome *c* mutants in yeast and a comparison of the known N-terminal structures found in the literature. A systematic study using site-directed mutagenesis of a synthetic gene expressed in yeast confirmed this proposal[18], as did a similar study using globin expression in a cell-free rabbit reticulocyte system[19]. Acetylation is also strongly influenced by the N-terminal residue but is clearly affected by neighboring residues as well (R. Yamada and R. A. Bradshaw, unpublished); thus, its specificity is more difficult to define and a greater variation in acetylated N-termini are observed[19,20]. In the main, serines and alanines, and to a much lesser degree, threonines and glycines, that are exposed by the removal of the methionine become acetylated (see Fig. 1). In yeast, there is no detectable acetylation of valine, cysteine, and proline when exposed as the N-terminal residue by MAP[18]. The remaining 13 amino acids in the penultimate position, are generally not substrates for MAP and therefore remain protected by the initiator methionine. Proteins containing aspartic acid, glutamic acid, and asparagine at this position can subsequently be acetylated on the α-NH_2 of the methionine. Significant variations from this profile have been observed with the isocytochrome *c* mutants[21] as well as with other proteins in yeast[20] and in the

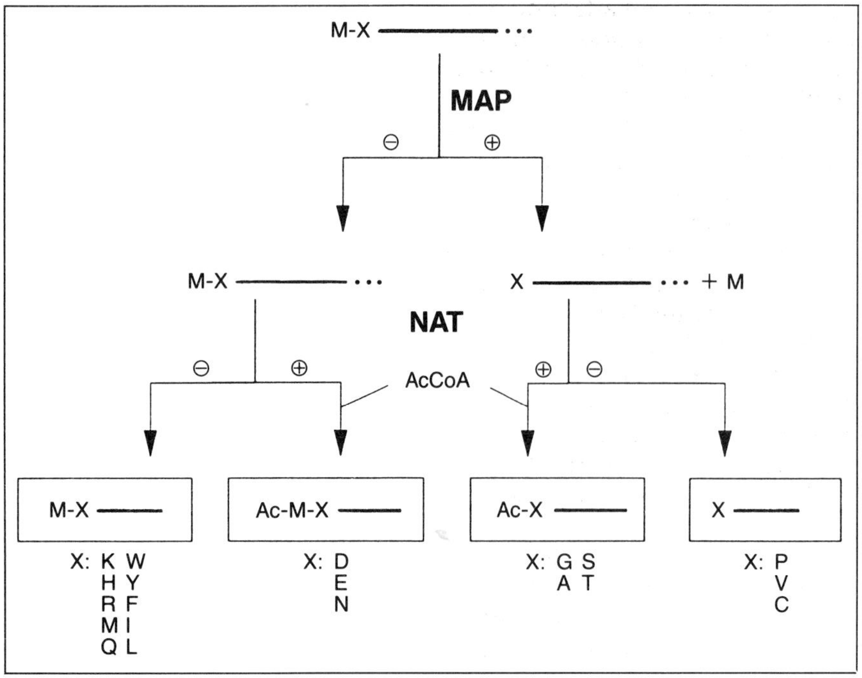

FIGURE 1

Proposed pathway for the co-translational modification of eukaryotic proteins by methionine aminopeptidase (MAP) and N^{α}-acetyl transferase (NAT). X represents the penultimate residue; dots indicate nascent polypeptide chains; (+) and (−) indicate positive or negative actions of the enzymes. *Reprinted in modified form from Ref. 14 with permission of The American Chemical Society.*

modified globin molecules expressed in rabbit reticulocyte and wheat germ systems[19].

The addition of myristyl moieties is clearly more specific, occurring only on selected proteins with N-terminal glycine (after removal of initiator methionine)[15,16]. Residues in the second and fifth position also affect the specificity of this transfer[22]. Although these requirements greatly proscribe the number of substrates for the myristyl CoA transferase, the importance of this enzyme, at least in yeast, has been rigorously established[23]. Importantly it has also been shown that these modifications occur co-translationally[24] and therefore must be competitive with NAT. The manner in which this selection occurs at the ribosomal level is not yet known.

The net effect of these modifications is clearly manifested in proteins that are found in the cytoplasm and within the nucleus as well as certain specialized locales such as the intermembrane compartment of the mitochondria which do not have signal peptides. On the other hand, proteins which contain signal peptides that are subsequently removed by germane peptidases, do not retain any characteristics of this N-terminal co-translational processing.

PROTEIN TURNOVER

The overall concentration of any given protein is governed by the rates of its synthesis and degradation. Considerably more is known about biosynthetic processes, including post-translational modifications required to produce a mature protein [see also the articles by Yan *et al.* (p. 191) and Neurath (p. 201) elsewhere in this book] than the way in which proteins are degraded[25]. However, as with translocation, there is accumulating evidence that the signals that ultimately dictate turnover are inherent elements of the structure. It appears that at some point during the lifetime of the protein, these signals can be 'read', leading to modifications that ultimately direct the proteolytic destruction of the molecule.

Several schemes of varying complexity operate for intracellular protein degradation in eukaryotic cells[25]. In the ubiquitin-dependent pathway[26,27], proteins first become 'marked' by the covalent attachment of ubiquitin molecules through their C-terminal glycine residues, initially to either α- or ε-NH_2 groups of the protein. The tagged protein is then degraded by a multi-subunit protease with the ubiquitin released to be reutilized for further degradation cycles. The calcium-dependent proteases (calpains) have also been implicated in turnover and many intracellular proteins are modified or fully degraded by these proteases[28], which may be particularly important in modifying cytoskeletal structures. A third group of proteases, found in the lysosomes, are clearly very important for protein degradation providing the principle route for the turnover of extracellular and membrane bound proteins that enter the cell by receptor-mediated endocytosis and lysosomal fusion as well as for long-lived cytoplasmic proteins[25,29]. They, of course, also function following cell lysis. In addition, there are proposed pathways that are energy-dependent but do not utilize ubiquitin and pathways that are independent of both ATP and ubiquitin. Protein degradation may also occur in specialized organelles such as mitochondria and the ER–Golgi complex.

Certain signals have now been recognized that allow access to at least some of these degradation pathways. The presence of amino acid sequences rich in proline, glutamic acid, serine, and threonine (PEST sequences) clearly can direct the degradation of short-lived proteins in a number of cases[30]. It is also important that the PEST sequences are flanked by the basic residues arginine, lysine and histidine. However, not all PEST sequences act as degradation signals since several long-lived proteins have been identified to contain them; presumably these are unavailable because of the active conformation of the molecule. A particularly good example of PEST-directed degradation is the recent studies with ornithine decarboxylase which showed that truncated versions of this enzyme have a greatly increased stability and that the C-terminal sequence, absent in the derivatized molecule, contains a PEST sequence[31]. Interestingly, a second PEST sequence also occurs in this enzyme in the non-deleted portion which presumably does not effect degradation. At present it is unclear what proteolytic enzymes are actually involved. A second sequence directing protein degradation was recognized to occur in the N-terminal segment of ribonuclease A and can target proteins to lysosomes at an enhanced rate during serum withdrawal[32]. Similar five-residue motifs have been found in other proteins and Dice and colleagues[25] have identified a binding protein which may assist in transferring the targeted proteins to the lysosome.

The contribution of N-terminal sequences to protein degradation has been emphasized by the observations of Varshavsky and colleagues[33] using a ubiquitin–β-galactosidase fusion protein expressed in yeast. Following the cleavage of the ubiquitin molecule, the longevity of the resulting β-galactosidase depended on the identity of the newly exposed N-terminal residue. On the basis of these results, they proposed the N-end rule that suggested that the presence of certain residues at the N-terminus of a protein was a destabilizing influence because it led directly to ubiquitin-mediated proteolytic destruction. The destabilizing residues correlated closely with those amino acids that prevented the removal of initiator methionine by MAP[14], suggesting that one function of this enzyme is to protect proteins from premature degradation by the ubiquitin-directed proteolysis.

Acylation of penultimate residues from which methionine had been removed as well as the methionine residues preceding aspartic and glutamic acids (and possible asparagine) provides a further mechanism for protein degradation. In one case, Johnson and colleagues[34] have shown that the products of two alleles encoding hypoxanthine phosphoribosyl-transferase (HPRT) vary in their stability and in their N-terminal residue. The more stable form contains an acetylated alanine in the N-terminal position while the less stable form contains an unmodified proline. Apparently, the acetyl group increases the stability of this enzyme. A different signal can be generated by the removal of acetylated methionines that precede aspartic and glutamic acid by a specific acyl amino hydrolase which renders them excellent substrates for the arginine-tRNA protein transferase[14]. These proteins, now bearing N-terminal arginine, are rapidly turned over, consistent with the findings of Bachmair *et al.*[33] that N-terminal arginine residues are the most destabilizing of all amino acids. Thus N-terminal modifications, particularly those introduced by MAP and NAT, help to direct both short-term and long-term protein turnover, particularly by the ubiquitin system. Even when the turnover mechanism will ultimately employ other enzymes such as calpains or the cathepsins of the lysosome, the N-terminal modifications still serve to protect against premature degradation by the ubiquitin system (or potentially other proteolytic enzymes of the cytoplasm).

Structural consequences of protein translocation

From the foregoing remarks, it is clear that as proteins are directed to their various compartments by the mechanisms described, there are certain general features of their chemistry that are predictable indicators of these processes. In particular, the nature of the N-terminus, the presence or absence of different carbohydrate groups, and the oxidation state of sulfhydryl groups are determined or at least affected by the translocation process. These properties are summarized in Table I. Clearly proteins containing N-terminal signal peptides that are subsequently removed will have un-blocked α-NH_2 groups (unless they are subsequently modified by other post-translational alterations, a relatively rare event). In contrast, pro-

Table I

Properties of proteins related to intra- and extracellular translocation

	Location			
Property	Cytoplasm	ER/Golgi/secretory	Mitochondrial matrix	Nucleus
N-terminus	Acylated[a]/open	Open	Open	Acylated/open
1/2-Cystine	Reduced	Oxidized	Reduced	Reduced
Complex glycosylation[b]	No[c]	Yes	No	No[c]
Subunit structure	Oligomeric	Mono- /oligomeric[d]	Oligomeric	Mono- /oligomeric

[a]Majority are acylated.
[b]See also article by Paulson on p. 209 of this book.
[c]Selected cytoplasmic and nuclear proteins contain *O*-linked *N*-acetylglucosamine[37].
[d]Majority are monomeric; oligomeric proteins usually contain inter-chain disulfide bonds.

teins of the cytoplasm and nucleoplasm may contain blocked or unblocked residues depending on their N-terminal sequences. For cytoplasmic proteins, surveys[35,36] have indicated that a high percentage are in fact blocked by the N^{α}-acetyl moiety. This probably reflects the stable population of proteins and enzymes that contribute to the general make-up of the cytoplasm; unblocked proteins, protected by initiator methionine are much more likely to be short-lived and transient in nature. Similarly, the addition of complex *N*- and *O*-linked carbohydrate and the insertion of disulfide bonds occurs primarily in the ER/Golgi and therefore they are found almost exclusively on secreted proteins or proteins contained in the ER continuum. In contrast, cytoplasmic proteins are generally devoid of carbohydrate, although certain nuclear and cytoplasmic structures can be modified by *O*-linked *N*-acetylglucosamine (Ref. 37; see also pp. 209 and 219 of this book) and contain thiol groups as opposed to disulfide bonds. These latter properties also apply to proteins incorporated into the mitochondria which are basically cytoplasmic-like except for the unblocked N-termini resulting from the excision of the signal peptide. Interestingly, proteins that span the ER membrane (and membranes derived from it) show the expected properties for each compartment; extracellular domains contain *N*- and *O*-linked complex carbohydrates and disulfide bonds, while intracellular domains appear to be free of carbohydrate and are reduced. An additional common characteristic of proteins of the cytoplasm and the mitochondrion is their oligomeric structures; as opposed to their secreted counterparts, intracellular proteins tend to contain multiple subunits, associated by non-covalent forces. Excreted proteins and proteins of the ER continuum are much more rarely found in multimeric structures and, when they are, they are generally linked by disulfide bonds.

These features are useful predictors of the location of a newly isolated protein or conversely can be used to anticipate the properties of a protein isolated from a known compartment. It should be emphasized, however, that they are general in nature and distinct exceptions are, of course, well known.

Acknowledgement

Studies described emanating from the author's laboratory were supported by US PHS research grant DK 32465. The expert assistance of Kisma Stepanich in preparing the manuscript and Owen Bates in producing the figure is greatly appreciated. The author would also like to thank Dr Stuart Arfin for his thoughtful reading of the manuscript.

References

1 Pfeffer, S. R. and Rothman, J. E. (1987) *Annu. Rev. Biochem.* 56, 829–852
2 Wickner, W. T. and Lodish, H. F. (1985) *Science* 230, 400–407
3 Kelly, R. B. (1985) *Science* 230, 25–32
4 Walter, P., Gilmore, R. and Blobel, G. (1984) *Cell* 38, 5–8
5 Walter, P. and Lingappa, V. R. (1986) *Annu. Rev. Cell. Biol.* 2, 499–516
6 Douglas, M. G., McCammon, M. T. and Vassarotti, A. (1986) *Microbiol. Rev.* 50, 166–178

7 Roise, D. and Schatz, G. (1988) *J. Biol. Chem.* 263, 4509–4511
8 Nicholson, D. W. and Neupert, W. (1988) in *Protein Transfer and Organelle Biogenesis* (Das, R. C. and Robbins, P. W., eds), pp. 677–746, Academic Press
9 Opperdoes, F. R. (1988) *Trends Biochem. Sci.* 13, 255–260
10 Dingwall, C. and Laskey, R. A. (1986) *Annu. Rev. Cell. Biol.* 2, 367–390
11 Newport, J. W. and Forbes, D. J. (1987) *Annu. Rev. Biochem.* 56, 535–565
12 Newmeyer, D. D. and Forbes, D. J. (1988) *Cell* 52, 641–653
13 Richardson, W. D., Mills, A. D., Dilworth, S. M., Laskey, R. A. and Dingwall, C. (1988) *Cell* 52, 655–664
14 Arfin, S. M. and Bradshaw, R. A. (1988) *Biochemistry* 27, 7979–7984
15 Sefton, B. M. and Buss, J. E. (1987) *J. Cell. Biol.* 104, 1449–1453
16 Towler, D. A., Gordon, J. I., Adams, S. P. and Glaser, L. (1988) *Annu. Rev. Biochem.* 57, 69–99
17 Sherman, F., Stewart, J. W. and Tsunasawa, S. (1986) *BioEssays* 3, 27–31
18 Huang, S., Elliott, R. C., Lui, P. S., Koduri, R. K., Weickmann, J. L., Lee, J. H., Blair, L. C., Ghosh-Dastidar, P., Bradshaw, R. A., Bryan, K. M., Einarson, B., Kendall, R. L., Kolacz, K. H. and Saito, K. (1987) *Biochemistry* 26, 8242–8246
19 Boissel, J-P., Kaspar, T. J. and Bunn, H. F. (1988) *J. Biol. Chem.* 263, 8443–8449
20 Lee, F-J. S., Lin, L-W. and Smith, J. A. (1988) *J. Biol. Chem.* 263, 14948–14955
21 Tsunasawa, S., Stewart, J. W. and Sherman, F. (1985) *J. Biol. Chem.* 260, 5382–5391
22 Towler, D. A., Adams, S. P., Eubanks, S. R., Towery, D. S., Jackson-Machelski, E., Glaser, L. and Gordon, J. I. (1988) *J. Biol. Chem.* 263, 1784–1790
23 Duronio, R. J., Towler, D. A., Heuckeroth, R. O. and Gordon, J. I. (1989) *Science* 243, 796–800
24 Wilcox, C., Hu, J-S. and Olson, E. N. (1987) *Science* 238, 1275–1278
25 Dice, J. F. (1987) *FASEB J.* 1, 349–357
26 Hershko, A. and Ciechanover, A. (1986) *Prog. Nucleic Acid Res. Mol. Biol.* 33, 19–56
27 Rechsteiner, M. (1987) *Annu. Rev. Cell. Biol.* 3, 1–30
28 Pontremoli, S. and Melloni, E. (1986) *Annu. Rev. Biochem.* 55, 455–482
29 Amenta. J. and Brocher, S. (1981) *Life Sci.* 28, 1195–1208
30 Rogers, S., Wells, R. and Rechsteiner, M. (1986) *Science* 234, 364–368
31 Ghoda, L., Van Daalen Wetters, T., Macrae, M., Ascherman, D. and Coffino, P. (1989) *Science* 243, 1493–1495
32 Backer, J., Bourret, I. and Dice, J. (1983) *Proc. Natl Acad. Sci. USA* 80, 2166–2170
33 Bachmair, A., Finley, D. and Varshavsky, A. (1986) *Science* 243, 179–186
34 Johnson, G. G., Kronert, W. A., Bernstein, S. I., Chapman, V. M. and Smith, K. D. (1988) *J. Biol. Chem.* 263, 9079–9082
35 Jornvall, H. (1975) *J. Theor. Biol.* 55, 1–12
36 Driessen, H. P. C., de Jong, W. W., Tesser, G. I. and Bloemendal, H. (1985) *CRC Crit. Rev. Biochem.* 18, 281–306
37 Hart, G. W., Holt, G. D. and Haltiwanger, R. S. (1988) *Trends Biochem. Sci.* 13, 380–384; see also p. 219 of this book

R. A. Bradshaw is at the Department of Biological Chemistry, California College of Medicine, University of California, Irvine, CA 92717, USA.

Post-translational Modifications of Proteins: Some Problems Left to Solve

Sou Chi B. Yan, Brian W. Grinnell and Finn Wold

Three major questions regarding the post-translational modification of amino acid side chains in proteins are briefly considered: (1) what are the biological functions of the reactions, (2) what is the specificity of the processing reactions in selecting only a few or sometimes even only one residue for modification, and (3) how do we solve the uniqueness of the processing steps in the production of recombinant proteins? The answers to these questions are not obvious at this time.

Defining the products of protein biosynthesis as the active proteins in their proper compartments, protein synthesis consists of several processes: the polymerization of the 20 encoded, 'primary' amino acids according to the sequence information specified by the DNA and transcribed to mRNA, the transport from the site of synthesis to the compartment of action, the folding of the linear polymer into transient and permanent three-dimensional structures and the covalent modification of peptide bonds and amino acid side chains in co- and post-translational reactions. While these individual steps can be listed and discussed separately, they are obviously tightly coordinated into the integrated process by which biologically active proteins are made. The purpose of this brief overview is to summarize some aspects of post-translational modification reactions involving the amino acid side chains. The overview will be mostly in the form of questions without any answers, but some of these questions may well represent areas of research activities in years to come. Other articles in this book consider specific post-translational reactions such as proteolytic processing (Neurath p. 201) and protein glycosylations (Paulson p. 209).

There are lists of all the hundreds of derivatized amino acids known to be present in proteins[1–3], although these are probably never quite up-to-date and accurate. The main message conveyed by the multitude of modified amino acids is that since nature has established such an extensive machinery for post-translational modification reactions, the products presumably must be important. It is remarkable, then, how little we really know about the biological function of these modified residues.

Functional aspects

In the broadest sense there are two kinds of post-translational modification reactions: those catalysed by a single enzyme, or set of enzymes, in a unidirectional manner to yield stable new amino acid derivatives or altered polypeptide chains; and those that involve two sets of enzymes, one for the derivatization and the other for the reverse reaction, the return to the original primary amino acid. The reversible modifications can be assigned the broad general role of on–off switches for biological activities. The main difficulty with this functional generalization is that our basic information may be incomplete or that we may treat the generalization too literally. ADP-ribosylation of protein has, for example, been known for a long time and enzymes both for the forward and reverse reactions have been well established in animal tissues[4] without permitting the assignment of a specific regulatory role to the reaction. Several bacterial toxins kill host cells by catalysing unidirectional ADP-ribosylation, suggesting that the reverse reaction is not active, but in one case, a clear cut regulatory on–off mechanism has been documented. The activity of *Rodospirillum rubrum* dinitrogenase–reductase is turned off by ADP-ribosylation of an arginine residue, and reactivated by the removal of the ADP-ribosyl group[5]. On the other hand, while it is not difficult to document well-defined regulatory roles to the reversible phosphorylation of Ser-P catalysed by protein kinases and phosphatases, it seems clear that some Ser-P residues may not satisfy the criterion of reversibility and as permanent, unidirectional modification products will have to be treated with the second group of reactions. The biological functions of the products of unidirectional reactions derivatives should be expected to be much more diverse and it seems that we have a long way to go to gain a real understanding of their full significance.

Generalizing again, it is possible to consider subsets of reactions in terms of broad functional consequences. Thus, it appears safe to accept the recently emerging group of glycosyl phosphatidyl inositol derivatives attached to the C-terminal carboxyl group of a variety of proteins as one of nature's strategies toward anchoring a protein to a membrane[6], and also perhaps to consider the principal role of crosslinks as one of stabilizing certain protein structures. It is also interesting to speculate about specific advantages of permanently attached cofactors (such as, for example, biotin, lipoate and flavin) over the corresponding noncovalent equivalents[7]. However, these general considerations are not very satisfactory; there are too many exceptions to the generalities, and too many individual reactions that do not fit in any broad category. Proteins presumably evolved in part to place particular primary amino acid residues in proper juxtaposition to form unique reaction centers for binding and catalysis, and nature has been extremely successful in generating high specificity and efficient catalysis with that strategy; why should it be necessary to generate unique side chains for some reactions? Why make γ-carboxyglutamate when other protein structures can chelate Ca^{2+} equally well or better, why make a single residue of diphthamide {2-[3-carboxyamido-3-(trimethylammonio) propyl]histidine} in a single protein (eEF2) in all eukaryotes[8], and of hypusine (N^{ε}[4-amino-2 hydroxybutyl]lysine), in eIF-D[9]? And, why is it often so difficult to establish the presumed importance of the derivatives? In the case of diphthamide, the yeast mutants unable to make the derivative appeared to do well when grown under laboratory conditions[10]. The eventual answers will probably require the broadest possible definition of biological function. The early attempts to correlate the total structure of an enzyme to its most obvious functional parameters (substrate binding and catalysis), left a lot of apparently useless structure. A consideration of the complete functional purpose built into a polypeptide chain – including regulatory properties, subunit interaction sites and effector binding sites, compartment-specifying properties, biosynthetic trafficking, and probably many more that have not yet been recognized – will be required to understand fully the role of a given post-translational modification. The covalently attached heme in cytochrome *c* may be an interesting example of this type of broad functional model. Most of the heme proteins can function without the thioether linkage between protein and heme, and so the covalent bond has no rationale in terms of electron transfer, catalysis or ligand binding. Recent model studies suggest, however, that the apoprotein can pass freely through the mitochondrial membrane in a diffusion-controlled reversible fashion, but that the mature heme protein cannot[11]. Thus, covalent attachment of the heme in the proper compartment may render the transmembrane transport an irreversible, highly specific process, and compartmental delivery rather than electron transport may dictate the need for the covalent heme linkage.

This area of compartmentalization and intracellular traffic-direction is emerging as a highly sophisticated and complex one. An exciting example of an unexpected mechanism of confining several apparently unrelated protein processing enzymes in the ER is the use of a common subunit with the 'ER-directing' C-terminal sequence Lys-Asp-Glu-Leu. The subunit is itself an active protein-processing enzyme, disulfide isomerase; it can be isolated as such as an active β_4-tetramer or it can be isolated as the β-subunit in active ($\alpha_2\beta_2$) prolyl 4-hydroxylase[12] and oligosaccharyl transferase[13]. Does this subunit association represent only a compartment-anchoring, or is it possible that disulfide isomerization is obligatorily linked to both proline hydroxylation and protein glycosylation as well?

SPECIFICITY

Another area that presents innumerable puzzles concerns the specificity determinants involved in selecting one or a few amino acid residues for modification. Specificity of biochemical reactions, whether they involve protein–ligand binding, or binding and catalysis, has been an active area of investigation for a long time, moving from studies of simple enzyme–substrate specificity through the more complex interaction of antigens (and haptens) with immunoglobulins, to the broad spectrum of receptor–effector signaling and macromolecular interactions. The progression also has undergone conceptual evolution from the rigid lock-and-key models to the consideration of conformational modulations, volume constraints, and the detailed topographical description of binding surfaces. Post-translational modifications could involve all these types of considerations: the substrate for the co- and post-translational reactions is a protein, the site to be altered is generally a small, covalently well-defined moiety, whether an amino acid side chain, a peptide bond or a C- or N-terminus, but the site to be modified is displayed in the unique environment of its particular protein molecule, either in a transient mode during synthesis (co-translational reactions) or in a more static mode in the finished product (post-translational). Traditionally the reductionist approach to assessing chemical reactivity of amino acid side chains and peptide bonds has been to think about average properties and acknowledge that the range averaged may be broad because of 'exposure', microenvironment, or other factors. It would seem more appropriate to start with the opposite extreme and assume that no two residues of a given kind, or no two peptide bonds linking a fixed pair of amino acids, are identical. Some properties may be very similar, but as more and more criteria are examined with increasing sensitivity, subtle differences might appear to make each individual reaction center a unique structure for protein processing enzymes. This picture would be consistent with what is observed. 'Low resolution' specificity may, for example, permit certain protein kinases to use a large number of serine residues (but not all!) as acceptors for P_i, while the most discriminating environmental effects would permit, for example, one unique histidine to exist in eEF-2 or one unique lysine in eIF-D to permit the formation of diphthamide and hypusine respectively in these proteins, but apparently in no other ones. This concept of unique protein matrix-induced structures certainly also applies to glycan processing in glycoproteins; it is well established that the common *N*-linked precursor structure is processed very differently at different glycosylation sites within the same protein, presumably reflecting how they are displayed in different protein environments[14]. This concept of unique amino residues in proteins is also consistent with the frequent lack of success encountered in our search for consensus peptide sequences required for certain reactions. Often they cannot be discerned at all, and when consensus does appear to exist, the exceptions where the proper structure is inactive inevitably lead to the designation 'required but insufficient for the reaction to take place'. The unique environment of each residue clearly reflects more than just the primary sequence.

A key specificity determinant that appears to remain quite elusive is the cellular location of the processing enzymes in relation to the processing pathway. A given enzyme may be found only in certain species, a given reaction may be specific to certain organs or cell types, and even within a given cell with a fixed location for the processing enzyme, different potential substrates may or may not encounter the enzyme at all. Even if they do encounter the enzyme, their current conformational state may be all wrong for the reaction to take place. Since it is the substrate structure in the compartment of modification that determines the unique properties of the group modified, a complete understanding of specificity may be a very elusive object indeed.

Recombinant proteins

A third area that is, and will continue to be, actively pursued for post-translational modification reactions is the practical problem of how to handle all these reactions in recombinant proteins. The problem has really been stated above in the discussion of both specificity and function. There are dramatic differences in the processing capabilities of different organisms as well as more subtle differences between different organs and cells in a single organism. Prokaryotes cannot handle most of the eukaryotic modifications, and eukaryotes with similar overall capabilities may still make different end products. Glycosylation illustrates this well; when glycoprotein genes are expressed in different eukaryotic cells, the *N*-glycosylation of specific asparagine residues appears to proceed normally across the board; however, the subsequent processing of the starting glycan yields very different structures, polymannose in yeast, complex and hybrid glycans in vertebrates, and xylose-containing glycans in plants and some invertebrates[14]. There is no obvious, rational solution to this problem at this time, it can only be tackled by trial and error. In the following, a brief consideration of the production of recombinant protein C will be reviewed to illustrate some of the pertinent issues.

An important example

Protein C is one of the vitamin K-dependent proteins involved in the hemostasis cascade. Along with coagulation Factor II (prothrombin), Factor VII, Factor IX, Factor X and Protein S it constitutes a group of proteins with very similar and extensive post-translational modification[15]. Because of the potential importance of Factors VII and IX in the treatment of hemophilic patients, and of proteins C and S in the treatment of various thrombotic diseases, a good deal of exploration of the production of the recombinant proteins has been carried out. Protein C is the zymogen form of a serine protease. Once activated (by removing residues 158–169) the enzyme is a potent anti-coagulant and antithrombotic agent.

Human Protein C has five types of post-translational modifications: (1) γ-Carboxylation on the first nine glutamic acid residues in the protein sequence. This post-translational modification event is performed on all the above listed proteins by a vitamin K-dependent microsomal carboxylase[16], although the number of γ-carboxyglutamates (Gla) in the different proteins varies from 9 to 12. The full complement of Gla on each protein is required to give full activity of all the proteins[17–19]. (2) β-Hydroxylation of Asp71 in one of the two EGF-like domains to give erythro-L-β-hydroxy-aspartate (bHA). The β-hydroxylation can be on either aspartates or asparagines that satisfy a certain stringent primary sequence structural requirement[20]. The modification is required for functional activity[21]; as was demonstrated by mutating Asp71 to Glu. The enzyme responsible for this modification has not yet been definitively identified; it appears to be coupled to α-ketoglutarate consumption in the same way other protein (lysine and proline) hydroxylases are coupled[20]. (3) *N*-linked glycosylation at three possible glycosylation sites. Plasma human Protein C has been reported to be 23% carbohydrate by weight[22]. (4) Disulfide formation. (5) Multiple proteolytic cleavages of the polypeptide backbone to remove an 18 amino acid signal peptide, a 24 amino acid propeptide and then cleavages at 155–156 and 157–158 to yield the two-chain structure of the circulating zymogen. The signal peptide is for targeting the nascent polypeptide into the lumen of endoplasmic reticulum. The propeptide is a signal for the vitamin K-dependent carboxylase[23,24]. A final cleavage at 169–170, converts the zymogen to the active enzyme.

To express the recombinant gene of protein C and achieve a fully processed, active product, it is clearly necessary to use a eukaryotic cell line that can carry out all the post-translational modifications. Ideally the cells should be easy to grow and manipulate genetically, and be able to secrete high levels [$> 1\ \mu g\ (10^6\ \text{cells})^{-1}\ (24\ h)^{-1}$] of the protein. Liver cell lines were prime candidates since liver is the natural biosynthetic site for all of the vitamin K-dependent proteins in the hemostasis cascade. When recombinant human Protein C gene was expressed in a rat hepatoma cell line, FAZA (Ref. 25) and recombinant Factor IX in rat hepatoma line, H4-11-E-C3 (Ref. 26) and in a human hepatoma line, HepG2 (Ref. 27), the results were disappointing. The expression levels were low and the recombinant proteins were not fully active. Since the recombinant proteins were not characterized in detail, it is not known exactly which one or more types of post-translational modifications were not properly performed.

Over a dozen other mammalian cell lines have been tried for the expression of vitamin K-dependent proteins, but most of these cell lines cannot perform the post-translational modifications properly. Chinese hamster ovary cell (CHO) was the most widely used. Recombinant genes of Factor II[28], Factor IX[29,30] and Protein C[25] were expressed in this cell line. CHO cells could only γ-carboxylate Factor II properly if the expression level was $0.1\ \mu g\ (10^6\ \text{cells})^{-1}\ (24\ h)^{-1}$, or below. For Factor IX, even at very low level of expression, the protein was not fully active. CHO cells were not able to perform polypeptide backbone processing efficiently in the case of recombinant human Protein C; 90% of the protein was not processed into the two-chain active form, and 40% of the protein secreted by the CHO cells still had the propeptide attached. Similar results were also observed for recombinant Protein C from mouse fibroblast cell line C127 cells[25]. Two other mouse fibroblast cell lines, LtK^- (Ref. 26) and $LM\text{-}TK^-$ (Ref. 27) have been used to express recombinant Factor IX. At expression level of only $0.1\ \mu g\ (10^6\ \text{cells})^{-1}\ (24\ h)^{-1}$, the functional activity of the recombinant Factor IX were only 5–10% of that of plasma Factor IX.

In addition to liver, both kidney and bone cells are known to synthesize Gla-containing proteins[27]. Although bone cell lines have not been rigorously examined as a host cell for expression, our experience with the rat osteosarcoma ROS-11 cell line would suggest that they are difficult to transfect and that the resulting transformants are unstable. Kidney cell lines may have the machinery for both making the Gla and the β-hydroxyaspartate residues in proteins, since one of the proteins, uromodulin, synthesized normally in kidney, has the unusual amino acid, erythro-L-β-hydroxyasparagine[31]. Human kidney cell line 293 cells are capable of producing and secreting high levels of recombinant human Protein C [1.5–20 μg $(10^6 \text{ cells})^{-1}$ $(24 \text{ h})^{-1}$; (Refs 19, 32]. Regardless of the expression levels, the recombinant human Protein C was fully functional. The 293 cell line is also capable of properly performing all five types of post-translational modifications. Two other kidney cell lines, MK_2 and SS78 did not produce fully functional human Protein C even at low levels of expression[32]. The 293 cell line is an adenovirus-transformed cell which constitutively produces the E1a tumor antigen. In the construction of the human protein C expression vector, the enhancer of the human polyomavirus BK was incorporated. It was found[32] that the BK enhancer was 'driven' (*trans*-activated) by the E1a protein. The production of 25–100-fold higher levels of recombinant human Protein C in these cells is thus likely to be a direct consequence of the higher transcriptional rate due to the presence of the BK enhancer.

Two other cell lines, AV12-664 and BHK-Ad cells were tested for their abilities to produce recombinant human Protein C[19,33]. The AV12 cell line is an adenovirus 12-induced hamster tumor line of unknown cell type. The BHK-Ad cell line is a baby hamster kidney (BHK-21) line transformed with adenovirus type 12. The AV12-664 cell line was capable of producing recombinant human Protein C that was proteolytically processed to the correct mature form of human Protein C and had the correct β-hydroxyaspartate content at levels up to 17 μg $(10^6 \text{ cells})^{-1}$ $(24 \text{ h})^{-1}$. However, the production of completely carboxylated fully functional recombinant human Protein C was limited to production below 1 μg $(10^6 \text{ cells})^{-1}$ $(24 \text{ h})^{-1}$. Thus, vitamin K-dependent γ-carboxylation was the only rate limiting post-translational modification. For the BHK-Ad cell line, the specific activity of the recombinant human Protein C was 60–80% of the native protein. In this case, the protein has the proper number of Gla, β-hydroxyaspartate, and the correct N-terminal protein sequences. The reason for the lower activity is not clear. The glycosyl content of recombinant Protein C is quite different among three cell lines (293 cells, AV12-664 cells and BHK-Ad cells), and from plasma human Protein C. Whether glycosylation plays a role in the functional activity of Protein C is currently being investigated. There is precedent for the effect of specific host cell glycosylation on the biological activity of a recombinant protein[34].

An interesting corollary to these observations on Protein C-production comes from the current work on another group of extensively processed proteins, the plasma membrane receptors. As discussed by Olson and Lane, in their recent review of what appears to be a common mechanism for the post-translational activation of several receptors[35], introduction of *N*-linked glycans has been established to be required for the production of active receptors. By also demonstrating that activated receptors retained their activity after removal of the *N*-linked glycans, it is proposed that the glycosylation is essential for the activation process itself. With disulfide bond interchange as the major activation steps in the receptors, this may represent an excellent example of the role of one post-translational reaction on the proper expression of another; such an interrelationship of two processing steps could obviously explain the variation between cell lines discussed above. The mere ability to carry out various reactions may not be sufficient; the relative rate and the precise end product of each step may have to be finely tuned for active products to be formed.

References

1 Uy, R. and Wold, F. (1977) *Science* 198, 890–896
2 Wold, F. (1981) *Annu. Rev. Biochem.* 50, 783–814
3 Wold, F. (1983) in *Post-Translational Covalent Modification of Proteins for Function* (Johnson, C. B., ed.), pp. 1–17, Academic Press
4 Ueda, K. and Hayaishi, O. (1985) *Annu. Rev. Biochem.* 54, 73–100
5 Pope, M. R., Saari, L. L. and Ludden, P. W. (1986) *J. Biol. Chem.* 261, 10104–10111
6 Ferguson, M. A. J. and Williams, A. F. (1988) *Annu. Rev. Biochem.* 57, 285–320
7 Rucker, R. B. and Wold, F. (1988) *FASEB J.* 2, 2252–2261
8 Van Ness, B. G., Howard, J. B. and Bodley, J. W. (1980) *J. Biol. Chem.* 255, 10710–10716
9 Park, M. H. and Folk, J. E. (1986) *J. Biol. Chem.* 261, 14108–14111
10 Chen, J-Y., Bodley, J. W. and Livingston, D. M. (1985) *Mol. Cell Biol.* 5, 3357–3360
11 Zhou, L-X., Jordi, W. and DeKruijff, B. (1988) *Biochim. Biophys. Acta* 942, 115–124
12 Kivirikko, K. I., Myllylä, R. and Pihlajaniemi, T. (1989) *FASEB J.* 3, 1609–1617
13 Geetha-Habib, M., Noiva, R., Kaplan, H. A. and Lennarz, W. J. (1988) *Cell* 54, 1053–1060
14 Yet, M-G., Shao, M-C. and Wold, F. (1988) *FASEB J.* 2, 22–31
15 Davie, E. W. (1987) in *Hemostasis and Thrombosis: Basic Principles and Clinical Practice* (Coleman, R. W. *et al.*, eds), pp. 242–267, Lippincott
16 Suttie, J. W. (1985) *Annu. Rev. Biochem.* 54, 459–477
17 Borowski, M., Furie, B. C., Goldsmith, G. and Furie, B. (1985) *J. Biol. Chem.* 260, 9258–9264
18 Jones, M., Griffith, M., Monroe, D., Roberts, H. and Lentz, B. (1985) *Biochemistry* 24, 8064–8069
19 Grinnell, B., Walls, J., Berg, D., Boston, J., McClure, D. and Yan, S. B. in *Genetics and Molecular Biology of Industrial Microorganisms* (Hersberger, C., Hegman, G. and Queener, S. W., eds), Am. Soc. Microbiol. (in press)
20 Stenflo, J., Holme, E., Lindstedt, S., Chandramouli, N., Huang, L., Tam, J. and Merrifield, R. (1989) *Proc. Natl Acad. Sci. USA* 86, 444–447
21 Ohlin, A., Landes, G., Bourdon, P., Oppenheimer, G., Wydro, R. and Stenflo, J. (1988) *J. Biol. Chem.* 263, 19240–19248
22 Kisiel, W. and Davie, E. W. (1981) *Methods Enzymol.* 80, 320–332
23 Foster, D., Rudinski, M., Schach, B., Berkner, K., Kumar, A., Hagen, F., Sprecher, C., Insley, M. and Davie, E. W. (1987) *Biochemistry* 26, 7003–7011
24 Suttie, J., Hoskins, J., Engelke, J., Hopfgartner, A., Ehrlich, H., Bang, N., Belagaje, R., Schoner, B. and Long, G. (1987) *Proc. Natl Acad. Sci. USA* 84, 634–637
25 Suttie, J. W. (1986) *Thromb. Res.* 44, 129–134
26 Anson, D., Austen, D. and Brownlee, G. (1985) *Nature* 325, 683–685
27 de la Salle, H., Altenburger, W., Elkain, R., Dott, K., Dieterle, A., Drillien, R., Cazenave, J., Tolstoshev, P. and Lococq, J. (1985) *Nature* 316, 268–270
28 Jorgensen, M., Cantor, A., Furie, B. C. and Furie, B. (1987) *J. Biol. Chem.* 262, 6729–6734
29 Rabiet, M., Horgensen, M., Furie, B. and Furie, B. (1987) *J. Biol. Chem.* 262, 14895–14898
30 Kaufman, R., Wasley, L., Furie, B. C., Furie, B. and Shoemaker, C. (1986) *J. Biol. Chem.* 261, 9622–9628
31 Przysiecki, C., Staggers, J., Ramjit, H., Musson, D., Stern, A., Bennett, C. and Friedman, P. (1987) *Proc. Natl Acad. Sci. USA* 84, 7856–7860
32 Grinnell, B., Berg, D., Walls, J. and Yan, S. B. (1987) *Bio/Technology* 5, 1189–1192
33 Yan, S. B., Razzano, P., Chao, B., McClure, D. and Grinnell, B. (1989) *J. Cell Biochem.* 13A (Abstr. for the 1989 UCLA Symposium on the Molecular Biology of the Cardiovascular Diseases)

34 Goto, M., Akai, K., Murakami, A., Hashimoto, C., Tsuda, E., Ueda, M., Kawanishi, G., Takahashi, N., Ishimoto, A., Chiba, H. and Sakai, R. (1988) *Bio/Technology* 6, 67–71
35 Olson, T. and Lane, M. (1989) *FASEB J.* 3, 1618–1624

S. C. B. Yan and B. W. Grinnell are at the Lilly Research Laboratories Eli Lilly and Company, Indianapolis, IN 46285, USA. F. Wold is at the Department of Biochemistry and Molecular Biology, University of Texas Medical School, Houston, TX 77225, USA.

Proteolytic processing is the final step in the biosynthesis of a great variety of proteins. Processing occurs in many different ways and is directed to the cleavage of specific peptide bonds in the precursor of the target protein. Such *limited proteolysis* depends on the specificity of the processing enzyme and on the accessibility of the scissile peptide bonds to the attacking protease[1]. The repertoire of physiological processes regulated by proteolytic processing is large and varied[2] (see Table I for examples). In this overview, attention is focused on recent developments in this field of research, with special emphasis on the molecular mechanisms and physiological consequences in representative systems.

PROTEOLYTIC PROCESSING AND PHYSIOLOGICAL REGULATION

HANS NEURATH

Proteolytic processing is a common and effective mechanism of physiological regulation. The basic principle is a conformational change induced in the protein precursor by the post-translational proteolytic cleavage of a specific peptide bond. The extension of earlier studies of model zymogens to more complex systems of physiological regulation, using methods of both protein chemistry and molecular biology, has enormously extended knowledge of the repertoire of proteolytic processing reactions and has contributed significantly to current studies of the structure, domain organization and evolution of proteins.

LIMITED PROTEOLYSIS AND PROTEOLYTIC PROCESSING

The physiological significance of limited proteolysis was first recognized in studies of the activation of zymogens of pancreatic proteases, notably trypsin and chymotrypsin[3], two proteases which still serve as models for more complex systems. The chemical principle generally involves the proteolytic cleavage of a specific peptide bond in the N-terminal region of the precursor protein. Full understanding of the conformational and functional consequences requires a detailed knowledge of the structure and mechanism of action of the active protein. For example, in the case of trypsin, proteolytic activation induces the insertion of the newly formed N-terminal residue into a cleft where it forms an ion pair with an aspartic acid residue adjacent to the active-site serine[4]. In the case of the aspartate protease penicillopeptidase, the initiating event is the disruption at low pH of salt bridges that position the N-terminal activation segment across the active site. This is followed by the proteolytic release of the activation peptide as the catalytic and substrate-binding sites are exposed[5].

In general, proteolytic processing releases a conformational restraint on the inactive precursor to generate a biologically active protein. The sites of proteolytic processing are generally in relatively flexible interdomain segments or surface loops, in contrast to the less accessible intradomain peptide bonds. These principles underlie all known proteolytic processing events exemplified in Table I. They include: zymogens generally, precursors of protein and peptide hormones,

TABLE I

Examples of zymogens that are converted to active proteins in response to various stimuli[16]

Normal physiological response	Response to foreign stimulus	Programmed response in development or repair
Vasoactive products	Digestion	Development
Angiotensinogen	Pepsinogen	Prochitin synthetase
Prekallikrein[a]	Prochymosin	Prococoonase
Kininogen	Trypsinogen[a]	Procollagenase
Hormones	Chymotrypsinogen	Proacrosin
Proinsulin	Procarboxypeptidases	Fibrinolysis
Proglucagon	Prophospholipase	Plasminogen proactivator[a]
Proparathyroid hormone	Proelastase	Plasminogen
Enzymes	Blood coagulation	Toxin
Protyrosinase	Factors VII[a], IX[a], X[a]	Promellitin
Prephenoloxidase	Prothrombin[a]	Self-assembly
Prorenin	Factor XIII	Procollagen
	Fibrinogen	Phage head proteins
	Complement	Processing of zymogen
	Properdin precursor	precursors
	Factors C_3, C_5[a]	

[a]An intermediate component in a series of consecutive zymogen activations.

growth factors, precursors of multifunctional proteins that participate in consecutive activation reactions (e.g. the cascades of blood coagulation), the macromolecular assembly of collagen, phages, picorna- and retroviruses, and the programmed interactions of proteins involved in developmental processes. The identification of the processing proteases and the elucidation of the cleavage-induced conformational transformations are subjects of active research by methods of both protein chemistry and molecular biology (see below).

PREPROTEINS – SIGNAL PEPTIDES

Proteolytic processing can occur co-translationally, for instance in the release by signal peptidase of the N-terminal signal peptides of secretory proteins during their transport across membranes of eukaryotic and prokaryotic cells. Since several reviews of protein translocation across membranes have been recently published[6,7] (see also Bill Wickner's article on p. 227 of this book), only a few salient aspects will be mentioned here. These membrane-bound proteases recognize the peptide bond at the junction of the signal peptide and the protein or its precursor. The exact substrate specificity of the enzyme is not fully understood. Signal peptide sequences are degenerate but have a common residue distribution in both eukaryotic and prokaryotic cells suggestive of common conformational traits. The isolation and characterization of signal peptidases from eukaryotic membranes has proven a difficult task[8,9]. A prokaryotic signal peptidase from *E. coli* has been cloned, sequenced and expressed[10]. Perhaps of greatest interest to the protein chemist is the mechanism of protein transport across membranes that are impermeable to much smaller molecules. Proteins that normally do not cross the mitochondrial membrane will do so post-translationally when fused to certain natural as well as synthetic leader peptides of the required characteristics[11]. This observation raises fundamental questions of the unfolding and refolding of polypeptide chains as they traverse the membrane and the kinetics and energetics of this conformational transition. The possibility of the participation of structure-determining cofactors or even enzymes has been raised and awaits experimental evidence.

PRO-PROTEINS

Whereas the processing of signal peptides regulates a transport mechanism that directs proteins to their site of action, the processing of pro-proteins is regulated by different enzymes and in different cell compartments[12].

In many cases, pro-protein cleavage is catalysed by serine proteases; in others, e.g. pro-insulin[13], processing involves a carboxypeptidase B-like enzyme that removes the C-terminal residue generated by the prior action of a trypsin-like protease. In yet other cases, processing appears to involve dipeptidyl aminopeptidases, cysteine or aspartate proteases.

Cascades of consecutive zymogen activation (e.g. in the blood coagulation[14], fibrinolysis[14] and complement[15] systems) owe their effectiveness to signal amplification, a high degree of specificity and fine-tuning by protease inhibitors[1]. Rapid progress in understanding these systems has been achieved in recent years by the application of methods of molecular biology, such as cDNA cloning and sequencing, and site-specific mutagenesis. These methods have yielded unexpected insights into the domain organization, homology and evolution of the proteins of these complex systems[16].

Regulatory proteases

Regulation by these proteases and their precursors can be largely attributed to their organization as a covalent assembly of structural and functional domains that have no counterpart in the simpler digestive enzymes. The term '*domain*' is used herein to denote compact, autonomous folding units with a minimum surface : volume ratio[16]. Some domains mediate interaction with cofactors and thereby program the consecutive reaction sequence of the cascade. For example, the vitamin K-dependent blood coagulation proteins prothrombin, factors IX, X and VII, as well as protein C, contain homologous domains that bind calcium ions due to the presence of γ-carboxyl glutamic acid (Gla) residues. Some of these proteins contain kringle domains (i.e. triple disulfide-bonded structures) that bind fibrin or other interacting proteins, and other domains show striking homology to epidermal growth factor[14]. In these systems, processing is a complex series of reactions. For instance, in the case of proteins V and VIII, which are cofactors in the activation of prothrombin and factor X, respectively, processing involves the excision of a large connecting region between the N-terminal and C-terminal domains, resulting in a disulfide-bonded two-chain structure[17]. Activation of the precursor of protein C, which results in specific inactivation of factors V_a and VIII_a, involves the successive removal of an 18 amino acid residue pre-peptide, followed by a 24 residue pro-peptide, a dipeptide and an activation peptide, in that order[18]. Analogous domain organizations and interrelationships can be seen among the proteins of the fibrinolysis cascade (e.g. factor XII, tissue plasminogen activator, urokinase and plasminogen). These, too, are assemblies of structural and/or functional domains that undergo complex processing reactions. Multifunctional protease precursors are also found in the complement system[15], which in the initial phase is triggered by an autocatalytic activation of a trypsinogen-like zymogen (C1r), a homodimer which together with two other proteins forms a pentameric, calcium-dependent aggregation product. Another example is the group of calcium-activated proteases[19] (calpains) that are regulated by a calmodulin-like calcium-binding domain and contain a cysteine protease as the catalytic domain. Proteolytic processing *in vitro* has proven a useful and powerful tool to identify and isolate domains of multifunctional proteins generally, using pre-selected proteases of defined specificity as dissecting instruments. In this connection, the recent introduction of monoclonal antibodies capable of hydrolysing specific peptide bonds[20] constitutes an important adjunct to the proteolytic repertoire.

CELLULAR LOCALIZATION: THE GRANULOCYTE PROTEASES

The repertoire of proteolytic enzymes that are integral components of cells is enormous and largely unexplored. It includes the entire class of

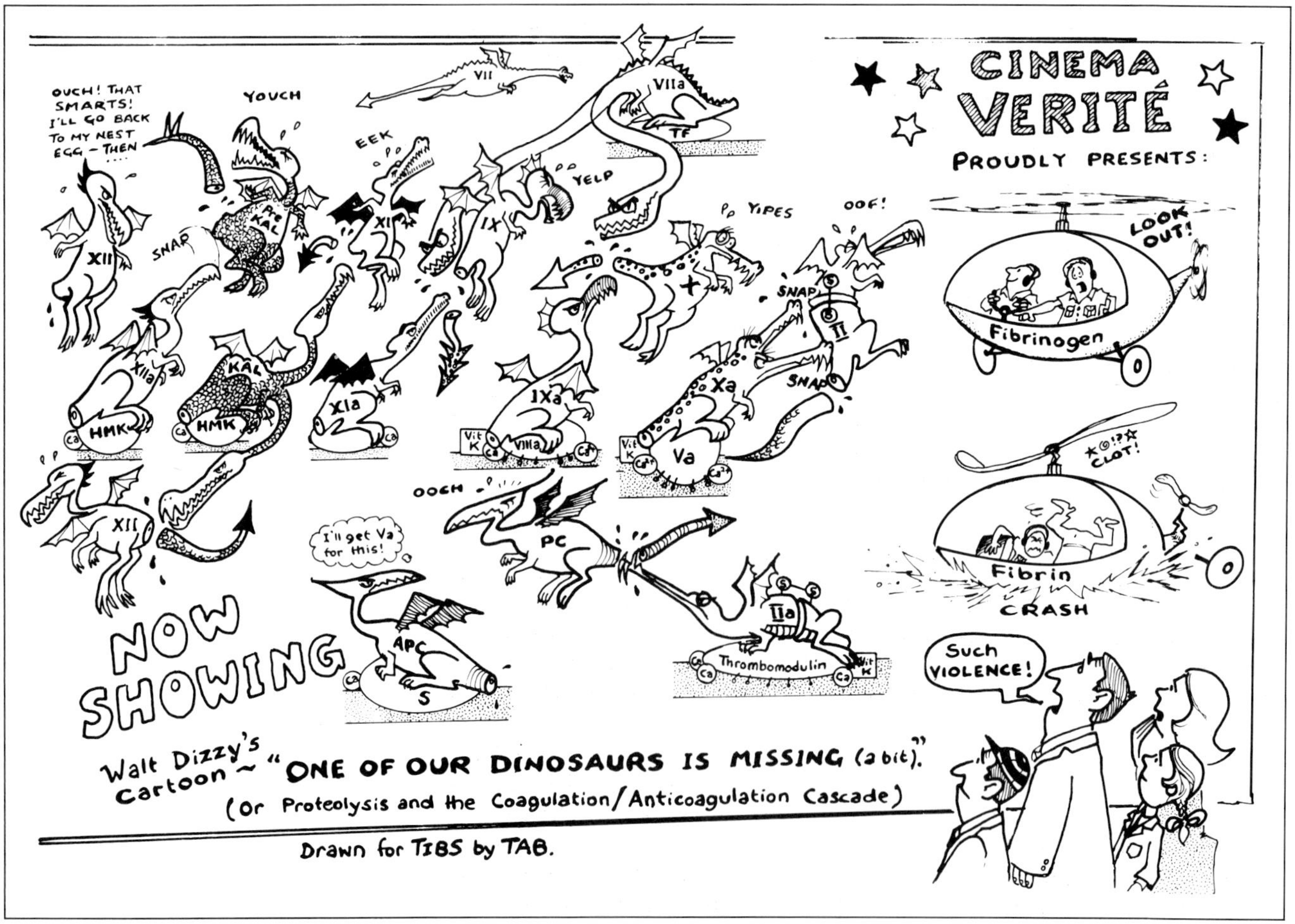

lysosomal proteases, proteases of specialized tissues, the ATP-dependent proteases and the granulocyte proteases associated in particular with leukocytes and mast cells[21].

Mast cell proteases are stored in granules and released in response to inflammatory or allergic reactions. The physiological roles of mast cell proteases and the mechanism whereby their action is controlled and regulated are still open issues[22]. Two quite different observations are of particular interest in this connection. (1) The chymotrypsin-like serine protease of rat serosal (i.e. connective tissue) mast cells is associated with heparin in granules which are inaccessible to proteins larger than about 15–20 kDa, which excludes all known plasma protease inhibitors from playing a regulatory role[23]. (2) No precursor forms have yet been found for any of the mast cell or other granulocyte proteases but cDNAs encoding human neutrophil cathepsin G, mucosal rat mast cell protease RMCP II and several cytotoxic T-cell proteases (granzymes), respectively, indicate the presence of N-terminal signal peptides and both N-terminal and C-terminal activation peptides[24]. The N-terminal activation dipeptide of the rat mast cell protease RMCP II has the sequence Glu-Glu, suggesting the action of a dipeptidyl aminotransferase or of a Glu-X specific protease on the cleavage site. A C-terminal tripeptide (Thr-Ser-Ser), removed during maturation might conceivably direct targeting of rat mast cell protease II into granules. However, there is at present no experimental evidence for the existence or location of the putative processing proteases or for their cellular location.

The chymotrypsin-like enzymes of both mast cells and cytotoxic T cells show sequence similarities to each other and to mammalian chymotrypsin, indicative of homology[25]. However, they differ in the absence of a disulfide bond near the active site serine, present in all other known mammalian serine proteases, and in that respect constitute a separate class of the superfamily of serine proteases. Comparative model building of two serine proteases from cytotoxic T cells, using the known structures of trypsin, chymotrypsin and rat mast cell protease II as models, has defined the similarities and differences in both structure and substrate specificities among these proteases[26].

Polyprotein-encoded processing proteases

Proteases have been discovered in recent years in the replication cycle of certain viruses. They are of particular interest since the processing protease is encoded in the large polyprotein precursors and is released autocatalytically, without the intervention of external enzymes. Moreover, the processing protease is produced in stoichiometric rather than catalytic quantities relative to its substrate. Since the rapidly growing literature on viral-encoded protease has been recently reviewed[27], the present discussion will be limited to certain aspects of particular relevance to the general theme of the present overview. Two types of viral-encoded proteases have been studied in particular detail. One of these is related to cellular aspartate proteases and contains the Asp-Ser/Thr-Gly sequence characteristic of the active

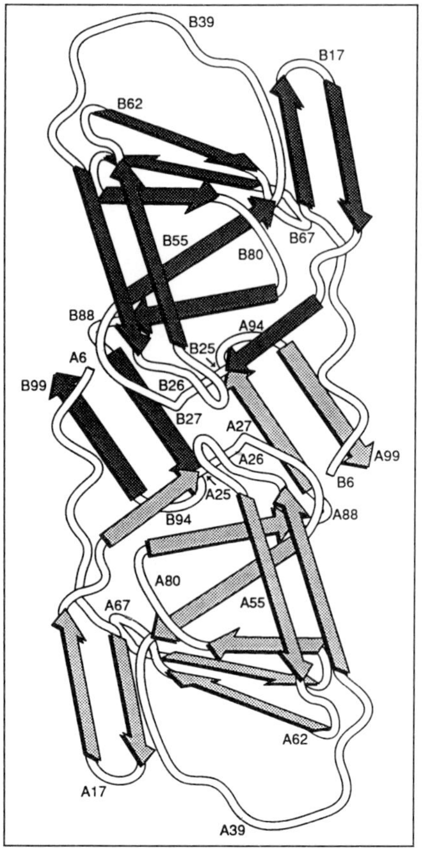

Figure 1

Ribbon drawing of the structure of HIV-1 protease. Residues in one monomer (stippled in light grey) are labeled A, residues in the second monomer (stippled in darker grey) are labeled B. *Reproduced, with permission, from Ref. 30. © 1989 Macmillan Magazines Ltd.*

site of the cellular aspartate proteases, e.g. penicillopepsin, pepsin and renin.

The molecular weight of these viral aspartate proteases is only about one half of that of pepsin but the molecule forms symmetrical dimers that simulate the monomeric cellular pepsins[28]. This has been confirmed by X-ray analysis of the aspartate protease of the Rous sarcoma virus (RSV) as well as the homologous HIV-1 viral protease[29,30] (Fig. 1). Infectivity of the HIV-1 virus is dependent on the viral protease since an Asp→Asn mutation inactivates both the protease and the virus. Another type of virus-encoded protease is a cysteine protease found in human rhinoviruses and other picorna viruses, as well as in certain plant viruses[31]. When cloned and expressed in *E. coli* membranes, it cleaves Gln-Gly peptide bonds in both natural and synthetic substrates. Although the catalytic role of its Cys residue has not been proven directly, it has been inferred from the high degree of conservation in other viral proteases and by inactivation when Cys is replaced by site-directed mutagenesis. Molecular modeling and secondary structure analysis suggest that this family of viral proteases is homologous to subclasses of trypsin-like serine proteases rather than to the cysteine proteases of the papain family[32].

Conclusions and Future Outlook

Proteolytic processing is an effective post-translational mechanism to regulate physiological processes. It responds rapidly to a signal by creating an irreversible change that directs the destination (translocation) or the timing of appearance of activity of the target protein. Both of these characteristics (speed and irreversibility) manifest themselves at the post-translational level rather than at the transcriptional level. The extension of the basic principles of zymogen activation to more complex systems, together with the application of methods of molecular biology, has enormously enlarged our knowledge of the repertoire of proteolytic processing reactions and has advanced our ideas on fundamental issues of protein structure and evolution, e.g. folding, domain structure, homology, gene organization and signal transduction. It is sometimes easier to detect and characterize precursor proteins by cloning and sequencing the corresponding cDNA rather than by methods of protein chemistry. Studies of the gene organization of mammalian serine proteases have contributed significantly to ongoing discussions of their evolutionary relationship and of the structural correlation, if any, between domains and exons. Just as site-directed mutagenesis can be used to probe for the functional role of pre-selected regions or domains of multifunctional proteins, so is it feasible to design recombinant DNA constructs that encode combinations of protein domains with novel functions and physiological properties. By the same token, the opportunity exists to alter the stability or substrate specificity of well defined enzymes and to enhance their practical and industrial use.

ACKNOWLEDGEMENTS

My thanks are due to Dr Kenneth A. Walsh for valuable suggestions in the preparation of this manuscript which was aided by the National Institutes of Health (GM-15731).

REFERENCES

1 Neurath, H. and Walsh, K. A. (1976) *Proc. Natl Acad. Sci. USA* 73, 3825–3832
2 Neurath, H. (1986) *J. Cell. Biochem.* 32, 35–49
3 Neurath, H. (1957) *Adv. Protein Chem.* 12, 319–386
4 Huber, R. and Bode, W. (1978) *Acc. Chem. Res.* 11, 114–122
5 James, M. N. G. and Sielecki, A. R. (1986) *Nature* 319, 33–38
6 Wickner, W. T. and Lodish, H. F. (1985) *Science* 230, 400–407
7 Briggs, M. S. and Gierasch, L. M. (1986) *Adv. Protein Chem.* 38, 109–180
8 Baker, R. K. and Lively, M. O. (1987) *Biochemistry* 26, 8561–8567
9 YaDeau, J. T. and Blobel, G. (1989) *J. Biol. Chem.* 264, 2928–2934
10 Wolfe, P. B., Wickner, W. T. and Goodman, J. M. (1983) *J. Biol. Chem.* 258, 12073–12080
11 Roise, D. and Schatz, G. (1988) *J. Biol. Chem.* 263, 4509–4511
12 Chretien, M., Boileau, G., Lazare, C. and Seidah, N. G. (1984) in *Cell Biology of the Secretory Proteins* (Cantin, M., ed.), pp. 214–246, Karger
13 Steiner, D. F. (1984) in *The Harvey Lectures Series 78*, pp. 191–228, Academic Press
14 Davie, E. W. (1987) in *Hemostasis and Thrombosis, 2nd Edn* (Colman, R. W., Hirsh, J., Marder, V. J. and Salzman, E. W., eds), pp. 242–267, B. Lippincott
15 Porter, R. R., Lachman, P. J. and Reid, K. B. M. (1984) in *Proc. R. Soc. Discussion Meeting*, pp. 279–340, Royal Society of London
16 Neurath, H. (1986) *Chemica Scripta* 27B, 221–229
17 Foster, D. C., Rudinski, N. S., Schach, B. G., Berkner, K. L., Kumar, A. A., Hagen, F. S., Sprecher, C. A., Insley, M. Y. and Davie, E. W. (1987) *Biochemistry* 26, 7003–7011
18 Kane, W. H. and Davie, E. W. (1988) *Blood* 71, 539–555
19 Suzuki, K., Emori, Y., Imajoh, S., Ohno, S., Kawaski, H., Minami, Y. and Miyake, S. (1987) in *Methods in Protein Sequence Analysis 1986* (Walsh, K. A., ed.), pp. 537–549, Humana Press
20 Iverson, B. L. and Lerner, R. A. (1989) *Science* 243, 1184–1188
21 Bond, J. S. and Butler, P. E. (1987) *Annu. Rev. Biochem.* 56, 333–364
22 Woodbury, R. G., Le Trong, H. and Neurath, H. (1987) *Acta Histochem. Cytochem.* 20, 261–269
23 Le Trong, H., Neurath, H. and Woodbury, R. G. (1987) *Proc. Natl Acad. Sci. USA* 84, 364–367
24 Salvesen, G., Farley, D., Shuman, J., Przylyba, A., Reilly, C. and Travis, J. (1987) *Biochemistry* 26, 2289–2293
25 Jenne, D., Rey, C., Haefliger, J-A., Qiao, B-Y., Groscurth, P. and Tschopp, J. (1988) *Proc. Natl Acad. Sci. USA* 85, 4814–4818
26 Murphy, M. E. P., Moult, J., Bleackley, R. C., Gershenfeld, H., Weissman, I. L. and James, M. N. G. (1988) *Proteins: Structure, Function and Genetics* 4, 190–204
27 Kraeusslich, H-G. and Wimmer, E. (1988) *Annu. Rev. Biochem.* 57, 701–754
28 Pearl, L. H. and Taylor, W. R. (1987) *Nature* 329, 351–354
29 Miller, M., Jaskolski, M., Rao, M. J. K., Leis, J. and Wlodawer, A. (1989) *Nature* 337, 576–579
30 Navia, M. A., Fitzgerald P. M. D., McKeever, B. M., Leu, C-T., Heimbach, J. C., Herber, W. K., Sigal, I. S., Darke, P. L. and Springer, J. P. (1989) *Nature* 337, 615–620
31 Libby, R. T., Cosman, D., Cooney, M. K., Merriam, J. E., March, C. J. and Hopp, T. P. (1988) *Biochemistry* 27, 6262–6268
32 Bazan, J. F. and Fletterick, R. J. (1988) *Proc. Natl Acad. Sci. USA* 85, 7872–7876

H. Neurath is at the Department of Biochemistry, University of Washington, Seattle, WA 98195, USA.

Interest in the biological roles of glycoprotein carbohydrate groups has intensified because of increasing examples of their importance as recognition determinants in host–pathogen interactions, in protein targeting, and in cell–cell interactions. The physical properties attributed to carbohydrate groups have also received renewed attention, stemming in large part from the desire of the biotechnology sector to understand the influence of sugar chains on recombinant glycoproteins produced in heterologous cell types. This brief review will address these divergent but overlapping interests in the functional roles of the glycoprotein sugar chains, particularly as they relate to emerging concepts in the control of the cellular glycosylation machinery, giving rise to cell type- and polypeptide-specific glycosylation sequences. Although this review will be concerned primarily with the sugar chains found on glycoproteins of the secretory pathway, some of the concepts discussed apply equally to sugar chains of glycolipids[1], proteoglycans[2] and the newly described GlcNAc-*O*-Thr/Ser sequence on cytoplasmic and nuclear proteins (Ref. 3; see also p. 219 of this book).

Glycoproteins: What are the Sugar Chains For?

James C. Paulson

For many glycoproteins, the carbohydrate groups confer important physical properties such as conformational stability, protease resistance, charge and water-binding capacity. Equally important, however, are the roles of carbohydrate groups in biological recognition, where sequence diversity provides signals for protein targeting and cell–cell interactions.

Structural diversity of sugar chains

From an examination of the sequence elements that produce structural diversity in *N*- and *O*-linked sugar chains, it can be seen that they differ primarily in their core structures, while peripheral sequences exhibit similarities, especially in the larger sugar chains.

Core structures. Three representative *N*-linked oligosaccharides are shown in Fig. 1a. The high mannose type oligosaccharide (structure I) is a biosynthetic precursor of the two complex type sugar chains (structures II and III)[4,5]. Note that they all share the common $Man_3GlcNAc_2$ core sequence (shown in bold type in Fig. 1). By contrast, the only common structural element in *O*-linked sugar chains is the GalNAc attached to the hydroxyl group of threonine or serine (Fig. 1b). The *O*-linked structures can be further divided into groups based on five core structures, defined by the sugars directly attached to the initial GalNAc residue[6]. Structures IV–VIII show representative sugar chains containing the five core structures indicated in bold type in Fig. 1.

Sequence diversity. For both types of sugar chains, extension of the oligosaccharide chains commonly results from the use of a $(Gal\beta1,4)_3$-GlcNAcβ1,3 sequence repeat (shown in plain type), which is 'decorated' with various terminal sugars (shown in italics). Note the similarity in the extension and terminal sequences of the *N*-linked and *O*-linked sugar

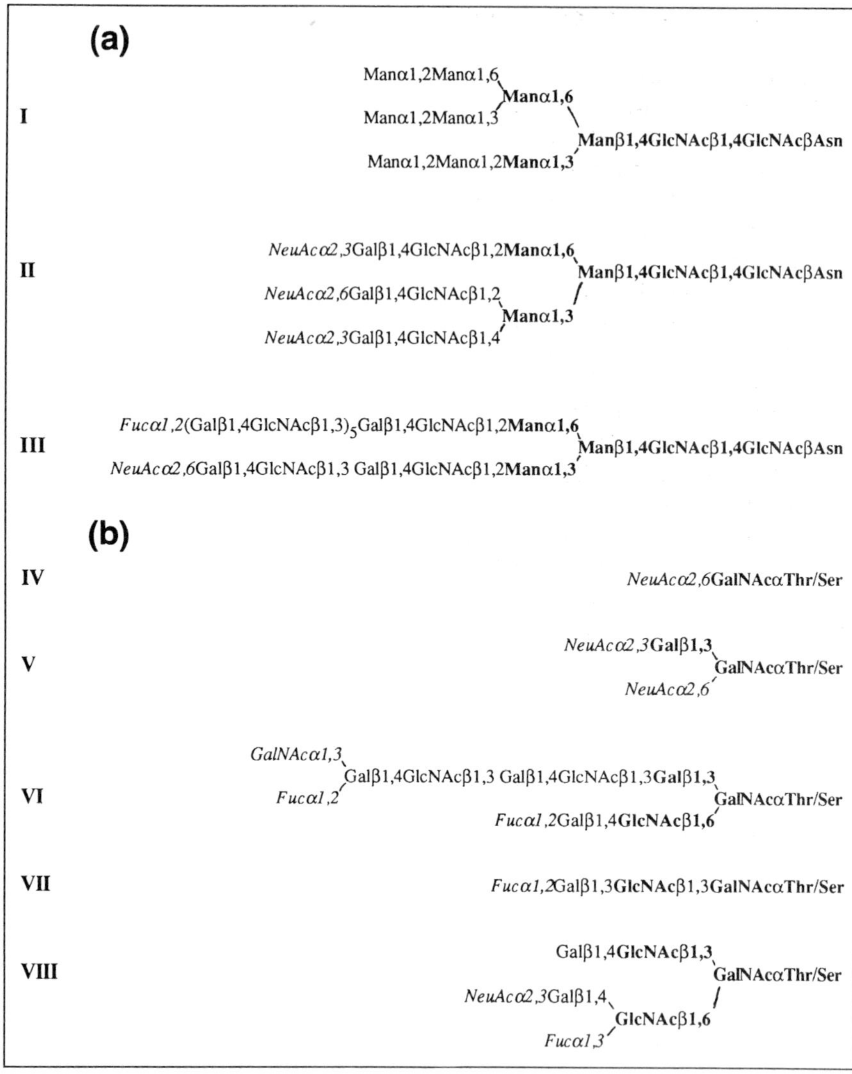

FIGURE 1

Examples of *N*-linked (**a**) and *O*-linked (**b**) sugar chains of glycoproteins. Structures were selected to contrast the two classes of sugar chains, which show differences in the core sequences, and similarities in the peripheral sequences[5–7,37–39]. For a fuller description, see text.

chains represented by structures II, III and VI–VIII. While the extension sequences are often short, *N*- and *O*-linked sugars can reach sizes of 15–40 residues or more. Branched chains result from the attachment of extension sequences to galactose residues at both the 3 and 6 positions (not shown)[5–7].

The major source of structural diversity for both the *N*- and *O*-linked oligosaccharide chains comes from variation in the attachment of the terminal sugars[5–7]. Typically the terminal sugars are α-linked sialic acid, fucose, galactose, *N*-acetylgalactosamine or *N*-acetylglucosamine residues, attached either singly or in combination to one or more hydroxyl groups of the extension or core structures. A survey of known *N*- and *O*-linked sugar chain structures reveals over 20 different terminal glycosylation sequences (e.g. NeuAcα2,3Gal,Galα1,4Gal), and when used in combination (structures V–VII), the true structural diversity is enormous.

Cell type-specific, and developmentally regulated variation

It is well established that terminal glycosylation sequences differ from tissue to tissue, from cell types within each tissue, and that these sequences are differentially expressed during early development[5,8]. To cite one example of the role of cell type in determining the structures of glycoprotein sugar chains, the recombinant human interferon-β1 produced in CHO cells yields *N*-linked sugar chains with the terminal sequence NeuAcα2,3Galβ1,4GlcNAc-R (where R=$Man_3GlcNAc_2$), while the same glycoprotein produced in C127 cells gives a mixture of two terminal sequences (NeuAcα2,6Galβ1,4GlcNAc-R and Galα1,3Galβ1,4GlcNAc-R)[9]. Such cell type-specific variation in structure may prove to be an important consideration in the choice of eukaryotic cell lines used to produce recombinant glycoproteins.

Because the synthesis of glycoprotein sugar chains represents a post-translational modification, their ultimate structures are determined largely by the glycosyltransferases that synthesize them. Thus, the cell type-specific and developmentally regulated variation in sugar chain structures is believed to result from the differential expression of the many glycosyltransferase genes[5,6,8]. This idea is supported by several lines of evidence. Examination of the levels of the mRNA for a sialyltransferase which forms a common terminal glycosylation sequence (NeuAcα2,6Gal) on *N*- and *O*-linked carbohydrate groups has revealed a 50–100-fold variation in various rat tissues, which correlates with the levels of the enzyme activity in the same tissues[10]. Transfection of CHO cells and Cos-1 cells with expression vectors containing the cloned cDNAs for specific sialyl-, fucosyl- or galactosyltransferases, results in the *de novo* appearance of the corresponding NeuAcα2,6Gal,-Fucα1,2Gal and Galα1,4Gal terminal glycosylation sequences not normally expressed by the cells (E. Lee and J. Paulson, unpublished; J. Lowe, Univ. Michigan, pers. commun.). Recent reports have also provided evidence for changes in glycosyltransferase expression during differentiation of cells. Indeed, differentiation of teratocarcinoma cells[11] and embryocarcinoma cells[12] in response to retinoic acid results in the altered expression of terminal glycosylation sequences corresponding to changes in the expression of the glycosyltransferases responsible for their synthesis. Together, these observations imply that expression of glycosyltransferases determines the carbohydrate structures produced by a cell, and that the expression of glycosyltransferase genes is responsive to control of the differentiation state of the cells.

Polypeptide-specific glycosylation sequences

There are now several examples where the polypeptide appears to direct the attachment of a specific glycosylation sequence unique to a single protein or class of proteins (Table I). Best known of these is the mannose 6-phosphate (Man-6-P) substitution on the high-mannose type *N*-linked sugar chains (structure I in Fig. 1) of lysosomal enzymes, which serves as the targeting signal for localization of the enzymes to lysosomes[13]. The initial transfer of phosphate to mannose residues is catalysed by an enzyme that recognizes a specific polypeptide determinant on lysosomal enzymes, which is not present on other glycoproteins with the same high-mannose type *N*-linked sugar chain[13]. Similarly, acquisition of the terminal sequence SO_4-4GalNAcβ1,4GlcNAc-R on the

TABLE I

Polypeptide-specific glycosylation sequences

Glycoprotein	Sequence[a]
(A) Lysosomal enzymes	PO_4^{2-}-6-Man-R
(B) Glycopeptide hormones (LH, TSH)	SO_4^{2-}-4-GalNAcβ1,4GlcNAcβ1,2Man-R
(C) Neural cell adhesion molecule (N-CAM)	NeuAcα2,8(NeuAcα2,8)$_n$NeuAcα2,8NeuAcα2,3Gal-R (where n = 20–200)

[a]R represents the remainder of the carbohydrate group which is in each case an *N*-linked sugar chain. Sequence A is attached to structure I (see Fig. 1), while sequences B and C represent alternative terminal sequences of structure II.

N-linked oligosaccharides of pituitary hormones (LH and TSH) appears to result from the recognition of a peptide determinant by the GalNAc transferase[14]. A third polypeptide-specific glycosylation appears to be developmentally regulated: the polysialic acid chain of the sequence NeuAcα2,8(NeuAcα2,8)$_n$NeuAcα2,3Gal-R (where n = 20–100), on the *N*-linked sugar chains of the neural cell adhesion molecule, N-CAM[15]. To date, N-CAM is the only glycoprotein that has been demonstrated to receive the polysialic acid substituent. Biosynthesis of the polysialic acid moiety is developmentally regulated as evidenced by its presence on embryonic N-CAM, its absence in adult N-CAM, and its reexpression on the highly metastatic Wilm's tumor[15,16].

In addition to these specific glycosylation sequences, the polypeptide chain can also influence the structure of its sugar chains in more subtle ways. Indeed, *N*-linked sugar chains at different glycosylation sites of the same protein can exhibit structures that represent different degrees of processing and terminal glycosylation[8]. Such variation most likely reflects the effect of the polypeptide structure surrounding the glycosylation sites, sterically preventing access of one or more processing glycosidases or terminal glycosyltransferases.

Biological roles – physical properties

From an evolutionary standpoint, once the glycosylation of proteins was introduced, it is reasonable to expect that these rather large hydrophilic substituents were selected for by virtue of certain favorable physical properties they provided. Thus, it is not surprising that many glycoproteins require their carbohydrate groups for their activities *in vivo*. Some generalizations can be made.

Facilitation of the proper folding and stabilization of the conformation of glycoproteins is perhaps the single most important function of *N*-linked oligosaccharides. The potential involvement of *N*-linked sugar chains in folding stems from the fact that the initial *en bloc* transfer of the $Glc_3Man_9GlcNAc_2$ structure from the lipid-linked precursor to Asn occurs as the nascent polypeptide is extruded into the ER, before polypeptide folding is complete[4]. Yet, the relative requirement for *N*-linked oligosaccharides varies between glycoproteins. Following exposure of cells to inhibitors of *N*-linked glycosylation such as tunicamycin, some glycoproteins are found to aggregate in the ER and are then degraded intracellularly. Others are less affected and are secreted, but have compromised biological activities, while some appear to be totally unaffected[4,8]. Using site-directed mutagenesis and expression experiments, more systematic analyses of individual *N*-linked sugar chains on glycoproteins with multiple *N*-linked groups reveal that some sugar chains can be completely eliminated, while others are critically important for normal biosynthesis[17,18].

For recombinant glycoproteins produced commercially for administration to humans and other animals by the blood (e.g. tissue plasminogen activator or erythropoietin), there is an additional requirement for having *N*-linked carbohydrate groups with terminal sialic acid. In this case, the sialic acid provides a masking function to prevent rapid clearance from circulation by carbohydrate-specific receptors on hepatocytes and macrophages[19]. Indeed, while glycoproteins produced in yeast or insect cells may be properly folded and exhibit bioactivity *in vivo*, the fact that both types of cells produce *N*-linked carbohydrate groups with terminal mannose residues means that they would be rapidly degraded *in vivo*. Such considerations dictate the use of mammalian cell lines for the production of many commercially important recombinant glycoproteins.

To a lesser extent, different terminal glycosylation sequences can influence the bioactivity of glycoproteins[8]. However, for most glycoproteins, the *N*-linked carbohydrate groups elaborated by different tissues appear to have no effect on bioactivity[5,8]. Indeed, mutant mammalian cell lines deficient in *N*-linked sugar chain synthesis grow normally in cell culture despite having cell surface glycoproteins with truncated carbohydrate groups[20]. Thus the role of terminal glycosylation sequences must be considered on a case-by-case basis.

The importance of *O*-linked sugar chains on the general properties of glycoproteins have been less well studied. Certainly, they provide charge and water-binding properties important for the function of highly glycosylated mucins. Kreiger and co-workers have described a CHO cell line deficient in the enzyme glucose 4-epimerase, allowing control over *O*-linked glycosylation by omitting or including *N*-acetylgalactosamine in the media[21]. Examination of the expression of recombinant glycoproteins such as the LDL receptor in these cells has revealed that they are expressed and transported to the plasma membrane, but are then rapidly degraded. These results clearly demonstrate that *O*-linked glycosylation is required for the stability of these proteins, and suggest that similar approaches will be important for elucidating the properties of *O*-linked carbohydrate groups for specific proteins.

Roles in Biological Recognition

The challenge in understanding the roles of carbohydrate groups in biological recognition has been to identify when the potential information coded by terminal glycosylation sequences serves some purpose, and when it is merely an indirect consequence of the cellular glycosylation machinery. Several general concepts are evident. If a glycosylation sequence serves as a glycoprotein targeting signal, it must be on every molecule. If a glycosylation sequence is a ligand on the surface of a cell, it can potentially serve as a receptor regardless of which cell surface glycoproteins (or glycolipids) carry it. Finally, a specific glycosylation sequence may mediate cell–cell interactions or other biological events when expressed in one biological context, but have no function when expressed in another.

Cell surface receptors

Sugar chains of glycoproteins and glycolipids are a predominate feature of the surface of cells[22]. Thus, it is not surprising that they serve as cell surface ligands for blood group and tumor-associated antibodies[1,22,23], toxins[22], the cell attachment proteins of some bacteria[1,22], mycoplasma[1], various animal viruses[24], and a variety of plant and animal lectins[22]. The specificities of many of these carbohydrate binding proteins have been studied in detail, and represent a vast literature, covered in part by the comprehensive reviews listed above. Of current medical importance are the monoclonal antibodies to tumor-associated antigens, both for diagnosis, and promising approaches to therapy[23,25]. Accumulating knowledge on the receptor specificity of human pathogens (bacteria, viruses, mycoplasma) holds the promise of developing inhibitors of cellular adsorption for prophylactic approaches to disease. In this regard, it is of interest that Weiss *et al.*[26] have recently described the three-dimensional crystal structure of the influenza virus hemagglutinin complexed with its receptor sialic acid, providing detailed information to aid the design of receptor analog inhibitors of adsorption for this virus.

Signals for protein targeting

Historically important is the observation by Ashwell and Morell that asialo- (Gal terminated) and asialoagalacto- (GlcNAc terminated) glycoproteins are cleared from the blood and targeted to liver hepatocytes and macrophages, respectively[19]. In each case the clearance was found to be mediated by the well known asialo- and asialoagalacto-glycoprotein receptor proteins on these cells. Although the biochemistry and molecular biology of these receptors have been extensively studied, their precise biological roles are incompletely understood. Nonetheless, the intensive research into the structural basis of their carbohydrate-binding properties has revealed homologous carbohydrate-binding domains that are being found in other receptor proteins (described below)[27].

The best documented example of a glycosylation sequence serving as a signal for protein targeting is the Man-6-P sequence on the *N*-linked sugar chains of lysosomal enzymes. This sequence mediates the binding of lysosomal enzymes to the Man-6-P receptor in the Golgi apparatus, which then initiates the sorting events that target the enzymes to the lysosomes[13]. Another polypeptide-specific glycosylation sequence with potential for a targeting function is the SO_4-GalNAc substitution on the *N*-linked oligosaccharides of certain pituitary hormones. Indeed, the differential sulfation of lutinizing hormone (LH) and follicle stimulating hormone (FSH) has been proposed to aid in the sorting of these glycopeptides into separate storage granules in pituitary gonadotrophs prior to secretion[14].

Cell–cell interactions

There is now excellent evidence that sperm adhesion to the egg is mediated by a protein–carbohydrate interaction[28]. Although there are several candidates for the sperm-derived carbohydrate binding protein, the murine egg receptor has been identified as a glycoprotein ZP3. The receptor binding activity has been identified as an *O*-linked carbohydrate group of ZP3, as evidenced by its ability to inhibit sperm–egg binding[28]. The precise sequence that mediates receptor binding activity has yet to be identified.

Sugar groups have long been suspected as mediating recognition functions during early development. In embryogenesis of the mouse, a series of stage-specific embryonic antigens (SSEA) have been shown to be carbohydrate epitopes representing terminal glycosylation structures[1]. One of these, SSEA-1 [Galβ1,4(Fucα1,3)GlcNAc-R] appears at the 8–16 cell stage, just as the embryo compacts from a group of loose cells to a smooth ball. Fenderson *et al.*[29] have demonstrated that a soluble multivalent analog of SSEA-1 inhibits the compaction process and disrupts embryogenesis, while closely related isomers have no effect. Thus, SSEA-1 appears to play a direct role in the process of compaction.

Much later in development, the polypeptide-specific polysialic acid substitution appears on the *N*-linked oligosaccharides of the neural cell adhesion molecule, N-CAM, expressed primarily but not exclusively on neuronal cells (Table I). The function of the large polysialic acid moiety is believed to reduce the affinity of the homotypic N-CAM binding interactions that mediate cell–cell adhesion. Thus, it has been proposed that the polysialic acid prevents tight adhesions during periods of active cell migration[15]. Later in development, the polysialylated N-CAM is replaced with the adult form of N-CAM which has typical *N*-linked sugar chains.

Several recent reports have implicated the involvement of specific glycolipid sequences in mediating various aspects of early development and in the differentiation of cells[30,31]. To the extent that glycoproteins and glycolipids have similar terminal glycosylation sequences[1,23], it can be expected that additional roles of glycoprotein sugar chains in these areas will be forthcoming.

The roles of cell surface glycosylation sequences in mediating cell–cell recognition in fully differentiated cells is also emerging as an active area. In a seminal observation, Drickamer discovered invariant residues in the carbohydrate binding domains of several cloned calcium-dependent lectins, which were known to function as cell surface receptors in the clearance of galactose and *N*-acetylglucosamine terminated glycoproteins from the blood[27]. Homologous carbohydrate domains were also found in other animal Ca^{2+} dependent lectins, defining a C-type lectin family. By searching the known protein sequences for the carbohydrate binding motif, additional members of the C-type lectin family have been found, including pulmonary surfactant proteins, proteoglycan core proteins, and lymphocyte receptor for the F_c portion of IgE. Most of these have now been shown to exhibit carbohydrate binding properties[27]. Although these findings have implications for carbohydrate-mediated recognition events at the cell surface, in most cases the precise physiological functions of carbohydrate binding domains have yet to be identified.

Recently, cloned receptor proteins mediating lymphocyte adhesion (Mel-14 antigens) at sites of recirculation on high endothelial venules

(HEV)[32,33], leukocyte adhesion to wounded epithelium (ELAM-1)[34], and platelet aggregation (GMP-140)[35] have all been shown to have the lectin-binding domains of the C-type family. Since previous studies have shown lymphocyte adhesion to HEVs to be mediated by carbohydrate groups on cell surface sugar chains, it has been proposed that lymphocyte homing and other cellular targeting and adhesion processes might be mediated in part by specific carbohydrate sequences, 'addressins', elaborated on the surface of the target cells[33,36]. As evidenced by these provocative findings, it is likely that additional roles for proteins with carbohydrate-binding motifs will be identified in the foreseeable future.

SUMMARY AND PERSPECTIVE

In the last decade, progress in the understanding of the biological roles of the sugar chains of glycoproteins has been rapid, particularly for the importance of glycosylation sequences in biological recognition. Multidisciplinary approaches involving biologists, chemists, molecular biologists and geneticists will ultimately be required to unravel the emerging roles of carbohydrate-recognition signals in complex biological systems governing the 'social' interactions of cells.

ACKNOWLEDGEMENT

I wish to thank Jasminder Weinstein and Sørge Kelm for critical reading of this manuscript and for offering helpful suggestions.

REFERENCES

1 Feizi, T. (1985) *Nature* 314, 53–57
2 White, T. N. and Mecham, R. P. (eds) (1987) *Biology of Proteoglycans*, pp. 1–400, Academic Press
3 Hart, G. W., Holt, G. D. and Haltiwanger, R. S. (1988) *Trends Biochem. Sci.* 13, 380–384
4 Kornfeld, R. and Kornfeld, S. (1985) *Annu. Rev. Biochem.* 54, 631–664
5 Kobata, A. (1984) in *Biology of Carbohydrates, Vol. 2, 2nd Edn* (Ginsburg, V. and Robbins, P. W., eds), pp. 87–161, John Wiley & Sons
6 Sadler, J. E. (1984) in *Biology of Carbohydrates, Vol. 2, 2nd Edn* (Ginsburg, V. and Robbins, P. W., eds), pp. 199–288, John Wiley & Sons
7 Fukuda, M., Dell, A., Oates, J. E. and Fukuda, M. N. (1984) *J. Biol. Chem.* 259, 8260–8273
8 Rademacher, R. W., Parekh, R. B. and Dwek, R. A. (1988) *Annu. Rev. Biochem.* 57, 785–838
9 Kawaga, Y., Takasaki, S., Utsumi, J., Hosoi, K., Shimizu, H., Kochibe, N. and Kobata, A. (1988) *J. Biol. Chem.* 263, 17508–17515
10 Paulson, J. C., Weinstein, J. and Schauer, A. (1989) *J. Biol. Chem.* 264, 10931–10934
11 Cummings, R. D. and Mattox, S. A. (1988) *J. Biol. Chem.* 263, 511–519
12 Chen, C., Fenderson, B. A., Andrews, P. W. and Hakomori, S. (1989) *Biochemistry* 28, 2229–2238
13 Kornfeld, S. (1987) *FASEB J.* 1, 462–468
14 Smith, P. L. and Baenziger, J. U. (1988) *Science* 242, 930–932
15 Rutishauser, U., Acheson, A., Hall, A. K., Mann, D. M. and Sunshine, J. (1988) *Science* 240, 53–57
16 Roth, J., Zuber, C., Wagner, P., Taatjes, D. J., Weisgerber, C., Heitz, P. U., Goridis, C. and Bitter-Suermann, D. (1988) *Proc. Natl Acad. Sci. USA* 85, 2999–3003
17 Gallager, P., Henneberry, J., Wilson, I., Sambrook, J. and Gething, M-J. (1988) *J. Cell Biol.* 107, 2059–2073
18 Dubé, S., Fisher, J. W. and Powell, J. S. (1988) *J. Biol. Chem.* 263, 17516–17521
19 Ashwell, G. and Hartford, J. (1982) *Annu. Rev. Biochem.* 51, 531–554
20 Stanley, P. (1987) *Trends Genet.* 3, 77–81
21 Kingsley, D. L. and Krieger, M. (1986) *Cell* 44, 749–756

22 Sharon, N. and Lis, H. (1982) in *The Proteins Vol. 5, 3rd Edn* (Neurath, H. and Hill, R. L., eds), pp. 1–144, Academic Press
23 Hakomori, S. (1984) *Annu. Rev. Immunol.* 2, 103–126
24 Paulson, J. C. (1985) in *The Receptors, Vol. 2.* (Conn, P. M., ed.), pp. 131–219, Academic Press
25 Livingston, P. O., Natoli, E. J., Calves, M. J., Stockert, E., Oettgen, H. F. and Old, L. J. (1987) *Proc. Natl Acad. Sci. USA* 83, 2911–2915
26 Weis, W., Brown, J. H., Cusack, S., Paulson, J. C., Skehel, J. J. and Wiley, D. C. (1988) *Nature* 333, 426–431
27 Drickamer, K. (1988) *J. Biol. Chem.* 263, 9557–9560
28 Wasserman, P. M. (1987) *Annu. Rev. Cell Biol.* 3, 109–142
29 Fenderson, B. A., Zehavi, U. and Hakomori, S. (1985) *J. Exp. Med.* 160, 1591–1596
30 Norjiri, H., Takaku, F., Terui, Y., Miura, Y. and Saito, M. (1988) *J. Biol. Chem.* 263, 7443–7449
31 Sariola, H., Aufderheide, E., Bernhard, H., Henke-Fahle, S., Dippold, W. and Ekblom, P. (1988) *Cell* 54, 235–245
32 Siegelman, M. H., van de Rijn, M. and Weissman, I. L. (1989) *Science* 243, 1165–1172
33 Lasky, L. A., Singer, M. S., Yednock, T. A., Dowbenko, D., Fennie, C., Rodriguez, H., Nguyen, T., Stachel, S. and Rosen, S. (1989) *Cell* 56, 1045–1055
34 Bevilacqua, M. P., Stengelin, S., Gimbrone, M. A. and Seed, B. (1989) *Science* 243, 1160–1164
35 Johnston, G. I., Cook, R. G. and McEver, R. P. (1989) *Cell* 56, 1033–1044
36 Stoolman, L. M. (1989) *Cell* 56, 907–910
37 Chandrasekekaran, E. V., Rana, S. S., Davila, M. and Mendicino, J. (1984) *J. Biol. Chem.* 259, 12908–12914
38 Van Halbeek, H., Dorland, L., Vliegenthart, J. F. G., Hull, W. E., Lamblin, G., Lhermitte, M., Boersma, A. and Roussel, P. (1982) *Eur. J. Biochem.* 127, 7–20
39 Lamblin, G., Boersma, A., Klein, A., Roussel, P., Van Halbeek, H. and Vliegenthart, J. F. G. (1984) *J. Biol. Chem.* 259, 9051–9058

J. C. Paulson is at the Department of Biological Chemistry, UCLA School of Medicine, Los Angeles, CA 90024, USA.

Nuclear and Cytoplasmic Glycosylation: Novel Saccharide Linkages in Unexpected Places

Gerald W. Hart, Gordon D. Holt and Robert S. Haltiwanger

In contrast to all other known forms of protein glycosylation, certain recently described protein–saccharide structures are highly localized to the nucleoplasmic and cytoplasmic compartments of the cell. The presence of these novel saccharides on proteins of the nuclear pore, chromatin and cytoskeleton poses fascinating questions concerning their possible functions, modes of biosynthesis, and potential roles in intracellular transport.

Prevailing evidence over the past two decades has led to the belief that glycosylated polypeptides are largely restricted to the cell surface and lumenal compartments of the cell. While a few early studies have suggested that glycoproteins are present within the cytoplasm and nucleus, the preponderance of rigorous biochemical data have not supported these findings. Analyses from many laboratories[1] established that the covalent attachment of most types of saccharides occurs within the lumenal compartments of either the endoplasmic reticulum or Golgi apparatus. Thus, cytoplasmic or nucleoplasmic glycoproteins were not predicted by current models of glycoprotein biosynthesis or transport.

Recently, a novel type of glycosylation in which *N*-acetylglucosamine (GlcNAc) is *O*-glycosidically attached to protein (*O*-GlcNAc) was described[2]. Surprisingly, *O*-GlcNAc was found to be enriched on many proteins localized to the cytoplasmic and nucleoplasmic compartments of the cell[3]. In light of *O*-GlcNAc's unusual localization[4], and with regard to its possible role in the transport of proteins across the nuclear membrane[5–9], interest in this novel form of post-translational modification is rapidly intensifying. In this short review, we will summarize some of the findings that are kindling excitement in this new field of glycoprotein research. Due to space constraints, we will not discuss related findings by several laboratories which demonstrate the presence of unusual glycosaminoglycans and proteoglycans in the nucleus and cytoplasm (described in Ref. 10).

Early evidence for the presence of nuclear glycoconjugates has been reviewed[11] and can be categorized as follows: (1) studies where the incorporation of radioactive sugar precursors into, or the periodic-acid Schiff (PAS) staining of electrophoretically fractionated nuclear proteins were determined; (2) experiments where the specific binding of plant lectins to nuclear components was demonstrated; (3) investigations where carbohydrate compositional analyses on isolated nuclear fractions were performed; and (4) research where various extraction procedures were used to investigate the associations of glycoconjugates with chromatin. However, in all of these studies neither the nature of the carbohydrate–protein linkages nor the structures of the saccharide moieties themselves were elucidated.

LOCALIZATION STUDIES USING LECTINS

Some of the more recent examples of studies which used plant lectins to localize glycoconjugates in subcellular compartments are described below. Chromatin has been shown to contain one concanavalin A (Con A, a mannose-specific lectin) binding site for every 1400 bp of DNA ($K_d = 3 \times 10^{-7}$M) (Ref. 12). Electroblotting of the non-histone chromosomal protein fractions with ^{125}I-Con A labels predominantly three proteins (135 kDa, 125 kDa, and 69 kDa), which were suggested to serve as points of attachment of chromatin to nuclear membranes. In addition, fluorescently-labeled wheat germ agglutinin (WGA, a lectin specific for sialic acid and GlcNAc), but not Con A, intensely stains both mitotic chromosomes and interphase nuclear substructures[13]. Wheat germ agglutinin binds internally at trypsin-sensitive sites on isolated 'nuclear scaffolds' prepared by treating nuclei with high salt and DNase I. In similar studies, fluorescently tagged Con A was found to bind specifically to transcriptionally active regions, called 'puffs', on *Chironomus thummi* polytene chromosomes at different developmental stages[14].

Lectin-blotting studies of electrophoretically separated proteins identified a major WGA-binding protein of 62 kDa and a major Con A-binding protein of 200 kDa in rat liver and hepatoma nuclei[15], which likely correspond to subsequently identified nuclear pore glycoproteins[16,17]. Using fluorescent- or ferritin-conjugated lectins, the nucleoli, chromatin, and cytoplasm of lizard ovarian follicles were found to have specific receptors for both WGA and Con A[18]. Staining by WGA was most intense over the nucleoli and cytoplasm. Lectins labeled with horseradish peroxidase or colloidal gold particles, together with freeze-fracture techniques, have also localized glycoproteins to euchromatic regions of chromosomes from duodenal columnar and exocrine pancreatic cells[19]. Both mannosyl and fucosyl residues appeared to be enriched in areas of euchromatin.

These studies are not conclusive, but suggest that both the nucleoplasmic and cytoplasmic compartments of cells contain glycosylated proteins. However, to date there exists no compelling structural evidence that either of these cellular compartments contain proteins bearing classical asparagine-linked oligosaccharides or serine (threonine)-linked 'mucin-type' oligosaccharides, which are known to be added to proteins in lumenal compartments of the cell[1].

HIGH MOBILITY GROUP CHROMOSOMAL PROTEINS

High mobility group (HMG) proteins are fairly abundant types of chromosomal proteins whose classification is based upon their relative electrophoretic mobilities[20]. These proteins appear to be associated preferentially with regions of active chromatin, and undergo an extraordinarily large number of different post-translational modifications, such as methylation, acetylation, phosphorylation, poly(ADP)-

ribosylation, as well as glycosylation[20]. Purified HMG14 and HMG17 were found by compositional analyses to contain *N*-acetylglucosamine, mannose, galactose, glucose and fucose. They also bind a fucose-specific lectin from *Ulex europeus*, and can be radiolabeled by sugar precursors. The radiolabeled HMG-associated saccharide moieties are insensitive to cleavage by dilute alkali, suggesting a *N*-linkage to protein (reviewed in Ref. 20). Surprisingly, removal of saccharides by mixed glycosidases abrogates the binding of HMG proteins to nuclear matrix[21]. Unfortunately, virtually nothing is known about the structures or biosynthesis of these interesting saccharide moieties on HMG proteins.

A NOVEL PROTEIN-SACCHARIDE LINKAGE (*O*-GlcNAc)

When bovine milk galactosyltransferase was used to probe the cell surface topography of terminal GlcNAc residues on living lymphocytes, a high proportion of these terminal moieties were found to consist of a novel *O*-GlcNAc monosaccharide[2]. Most of these *O*-GlcNAc residues appeared to occur at clustered sites on integral membrane proteins. *O*-GlcNAc accounts for 5–12% of the total macromolecular ^{3}H-glucosamine metabolically incorporated by murine lymphocytes. In fact, B lymphocytes contain over 150 million copies of these structures per cell. Most surprisingly, latency studies indicated that the vast majority of *O*-GlcNAc is localized on intracellular proteins[2].

Shortly after this report, an abstract appeared suggesting that rat liver nuclei contain proteins bearing single GlcNAc residues (see Ref. 22). Subsequently, in a study which used highly purified subcellular fractions from rat liver and galactosyltransferase as a probe for terminal GlcNAc residues, *O*-GlcNAc modified proteins were found to be particularly enriched in the nuclear envelope, and the nucleus as a whole, as well as in the cytosol[3]. In fact, virtually all of the protein-bound terminal GlcNAc residues present in either the nuclear or cytosolic compartments consist of *O*-GlcNAc. Even though serum proteins contain a myriad of different *N*-linked and 'mucin-type' *O*-linked [GalNAc-Ser(Thr)] type oligosaccharide structures, *O*-GlcNAc was found to be completely absent from this

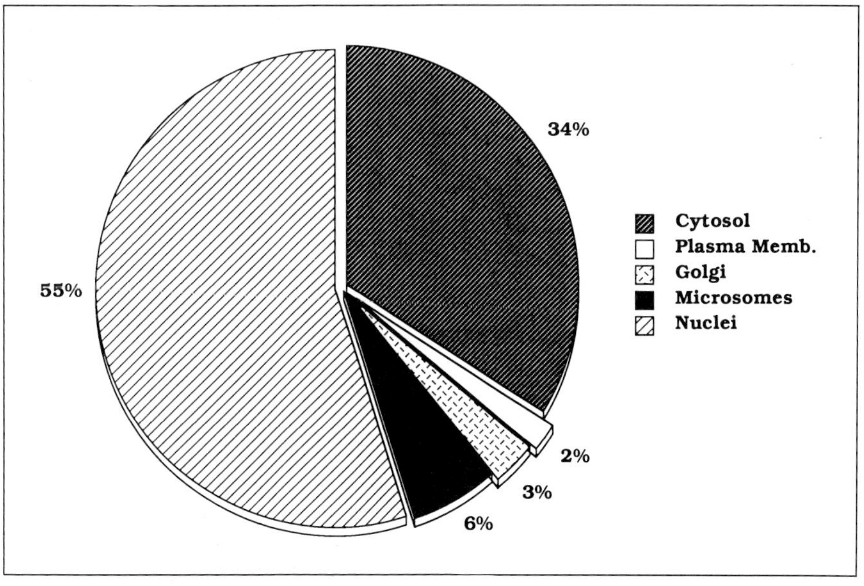

FIGURE 1

Distribution of *O*-GlcNAc modified proteins in the organelles of rat liver cells. The amounts of protein-bound *O*-GlcNAc were determined in highly purified rat liver organelles that were detergent solubilized with Nonidet P-40. To obtain a rough estimate of cellular distribution, the pmol mg^{-1} protein of *O*-GlcNAc for each organelle were multiplied by the percentage of the total protein that the organelles represent in a whole liver homogenate[3]. Mitochondria do not contain detectable *O*-GlcNAc. Inclusion of other organelles would not substantially alter these approximations.

fraction. It was estimated that about 55% of the cellular *O*-GlcNAc in the average liver cell is located in the nucleus, 34% in the cytosol and much less in other cellular compartments (see Fig. 1).

Nuclear pore glycoproteins – cytoplasmically oriented glycosylation

A monoclonal antibody was recently used to identify a 62 kDa (p62) nuclear pore protein, which is first synthesized in the cytoplasm as a 61 kDa precursor[17]. On the basis of the earlier description of a major WGA-binding nuclear protein of the same apparent size[15], and the enrichment of *O*-GlcNAc in nuclear envelope[3], it was suggested that the apparent maturation of this protein might involve the addition of *O*-GlcNAc residues. The p62 nuclear pore protein was indeed shown to bind WGA, but not Con A, on nitrocellulose blots[17]. Subsequently, several distinct monoclonal antibodies recognizing different subsets of the major proteins associated with the nuclear pore complex were described[23]. Concomitantly, the majority of the *O*-GlcNAc-proteins identified in the nuclear envelope[3] were discovered to be the *same* proteins recognized by these nuclear-pore specific antibodies[4] (Fig. 2). *O*-GlcNAc was shown to be a major part of the epitopes recognized by all of these pore-specific antibodies.

Several lines of evidence, however, suggest that the *organization* of the *O*-GlcNAc on these proteins is both protein dependent and nuclear pore specific: (1) Most of the antibodies don't appear to bind *O*-GlcNAc-proteins found in other parts of the cell. (2) Each antibody recognizes a different subset of the pore complex glycoproteins. (3) Free GlcNAc or GlcNAc-containing neo-glycoproteins do not block antibody binding. (4) The antibodies do not bind limit pronase glycopeptide fragments of the nuclear pore glycoproteins. Each nuclear pore

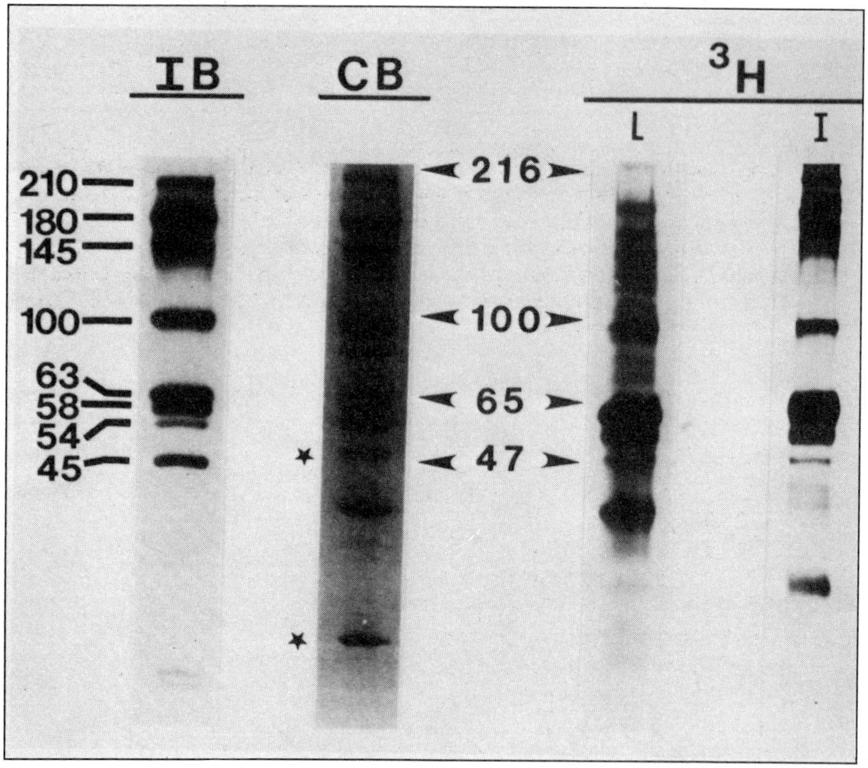

Figure 2

Correspondence between nuclear-pore proteins reactive with specific monoclonal antibodies and proteins which contain *O*-GlcNAc moieties. Nuclear envelope proteins were immunoadsorbed onto a Sepharose column containing a monoclonal antibody directed against nuclear-pore proteins. Some of the immunoadsorbed proteins were then treated with galactosyltransferase, which adds galactose only to free GlcNAc moieties on glycoproteins. The patterns of immunoadsorbed proteins stained with Coomassie blue (lane CB) closely resembles those galactosylated to contain 40 pmol [^{3}H] galactose per 1.5×10^8 nuclear equivalents (lane [H^3] L) or 150 pmol [^{3}H] galactose per 1.5×10^8 nuclear equivalents (lane [H^3] I). Lane IB contains nuclear envelope proteins immunoblotted with ^{125}I-iodinated pore-specific antibody. Stars in lane CB indicate the migration positions of free antibody heavy and light chains. The molecular masses of the pore-specific proteins are given on the left. *Reproduced, with permission, from Ref. 4.*

glycoprotein appears to have multiple sites of *O*-GlcNAc addition (e.g. p62 has 10–12 sites), mostly at serine and to a small extent threonine residues[4]. In support of the contention that nuclear pore *O*-GlcNAc residues are highly immunogenic, many independently isolated nuclear pore monoclonal antibodies also appear to recognize these saccharides[24].

The shared saccharide–protein epitopes recognized by these monoclonal antibodies are highly localized at the cytoplasmic and nucleoplasmic faces of the nuclear pore complexes[23] (Fig. 3). Several laboratories have independently added support for this proposed orientation of *O*-GlcNAc on nuclear pores. A series of careful studies used the latency of mannose-6-phosphatase, an enzyme found in the lumenal spaces of the endoplasmic reticulum and nuclear envelope, to demonstrate the cytoplasmic orientation of *O*-GlcNAc[25]. It was also shown that rhodamine-WGA intensely stains sealed nuclei and that WGA-ferritin selectively localizes to the cytoplasmic and nucleoplasmic faces of the nuclear pore. Pulse–chase experiments on the p62 nuclear pore protein suggest that most of its *O*-GlcNAc moieties are added within five minutes of the protein's synthesis, while the polypeptide is still in the cytoplasm[26]. Together these data suggest that the glycosyltransferase(s) that attach *O*-GlcNAc to protein have their active sites in the cytosol. A specific assay for the polypeptide: *O*-GlcNAc glycosyl transferase has recently been developed (Haltiwanger, Holt and Hart, submitted). Latency studies indeed confirm that the enzyme's active site is within the cytoplasm, but surprisingly the enzyme appears to be membrane associated.

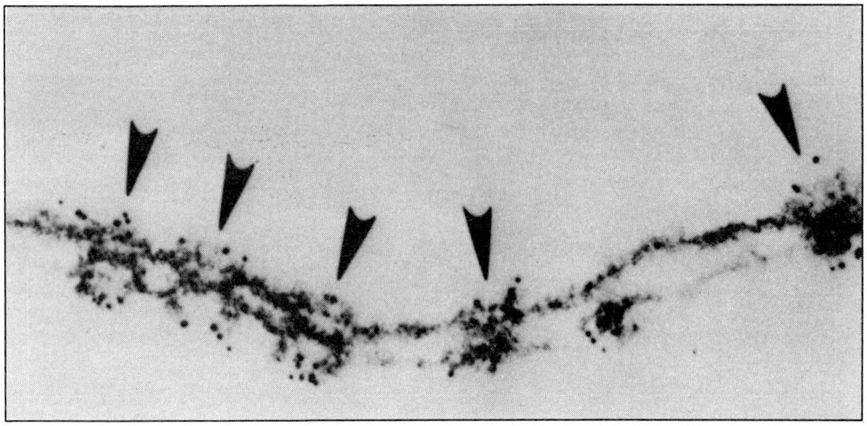

FIGURE 3

Saccharide–protein epitopes recognized by nuclear pore-specific monoclonal antibodies are localized at the nucleoplasmic and cytoplasmic faces of the pore complex. Isolated nuclear envelopes were washed with 0.5 M NaCl, fixed with paraformaldehyde, and incubated for direct immunogold labeling with a nuclear pore-specific monoclonal antibody whose epitope is largely *O*-GlcNAc dependent. The gold particles decorated the nucleoplasmic and cytoplasmic faces of the nuclear pore complexes (indicated by arrows). *Reproduced, with permission, from Ref. 23.*

O-GlcNAc AND NUCLEAR TRANSPORT

As early as 1983 it was demonstrated that an antiserum to a 67 kDa nuclear envelope protein of clam oocytes, as well as the lectin WGA, specifically blocked ATP-dependent RNA transport out of isolated nuclei[5]. More recently, using an *in vitro* transport model comprised of *Xenopus* oocyte extract, rat liver nuclei, and fluorescently-tagged nucleoplasmin (a well characterized *Xenopus* oocyte nuclear protein), it was demonstrated that WGA specifically blocks nucleoplasmin uptake into isolated nuclei[6]. Fluorescent WGA stained the isolated nuclei in a punctate fashion, and electron microscopy indicated that multiple molecules of WGA-ferritin bound at the cytoplasmic faces of nuclear pores. ^{125}I-WGA predominantly bound to a protein of 63–65 kDa, probably the same *O*-GlcNAc-bearing protein discussed above. Similar studies have shown that WGA and a GlcNAc-specific lectin from *Datura stamonium* both block the nuclear uptake of nucleoplasmin and synthetic peptide conjugates containing a nuclear localization signal from simian virus 40 (SV40) large T antigen[7]. However, in these same studies, WGA does not appear to block at least some kinds of RNA export or the passive diffusion of materials into the nucleus. WGA also appeared to have no effect on RNA synthesis[7]. Interestingly, WGA specifically blocks only the translocation, but not the binding steps of nuclear import[9]. In support of these findings, microinjection into *Xenopus* oocytes was used to demonstrate that WGA blocks the uptake of pre-labeled nuclear proteins[8]. WGA was also found to block the transport of fluorescently labeled nucleoplasmin into the nucleus which had been microinjected into the cytoplasm of rat hepatoma cells. The dose response of WGA in this study indicated that, at reduced levels, the lectin slowed nuclear uptake without reducing its final extent[8]. In these studies the uptake of all karyophilic proteins into the nucleus appeared to be similarly affected by WGA. These authors concluded that: (1) WGA does not affect the functional radius of the pore, thus ruling out simple steric blockade as the mechanism; (2) WGA probably blocks the entire spectrum of karyophilic proteins; (3) the intracellular mobility of nucleoplasmin is not affected by WGA; and (4) WGA appears to be able to reduce the rate of uptake of nuclear proteins without affecting its final extent. While these findings are thought provoking, they do not as yet suggest a mechanism for the role of *O*-GlcNAc in the transport of nuclear proteins.

DISTRIBUTION OF *O*-GlcNAc MODIFIED PROTEINS

As mentioned above, *O*-GlcNAc modified proteins appear to be widely distributed throughout various sub-cellular compartments, as well as throughout phylogeny. To date, *O*-GlcNAc-bearing proteins have been found in viruses, yeast, trypanosomes, schistosomes, rodents, frogs and fruit flies, as well as in humans. Thus far, *O*-GlcNAc has not been found in prokaryotes. In studies of human cytomegalovirus, the 149 kDa virion basic phosphoprotein was found to be modified by *O*-GlcNAc addition[27]. The virion basic protein is a tegument protein located between the capsid and viral envelope and accounts for 20% of the viral mass. Thus, even in viruses *O*-GlcNAc residues have localizations unexpected for protein–saccharide moieties. The first published example of *O*-GlcNAc in a parasitic invertebrate organism came from studies of the human blood fluke, *Schistosoma mansoni*[28], in which *O*-GlcNAc is a major form of glycosylation. Studies on human erythrocytes have indicated that the vast majority of proteins bearing terminal GlcNAc residues are localized in the cytoplasmic compartment and largely consist of *O*-GlcNAc[29]. Most of the *O*-GlcNAc in erythrocytes is present on an unidentified 65 kDa protein which is highly soluble and abundant in the cytoplasm. Most interestingly, Band 4.1, a cytoskeletal protein that anchors the underlying cytoskeleton to the plasma membrane, is also modified by *O*-GlcNAc addition. Other major proteins, such as spectrin, actin and hemoglobin contain no traces of *O*-GlcNAc residues. Using sealed vesicles and glucose-6-phosphatase as a marker for latency, several intrinsic membrane proteins of the smooth (SER) and rough endoplasmic reticulum (RER) were shown to have terminal GlcNAc residues facing the cytoplasmic compartment[30]. Linkage analysis of the major 56 kDa protein from SER demonstrated the presence of cytoplasmically oriented *O*-GlcNAc. Interestingly, the SER appeared to contain larger amounts and a different array of proteins with these saccharide structures than the RER.

Functions of Nuclear and Cytoplasmic Glycosylation

Aside from the ostensible roles of *O*-GlcNAc in the transport of macromolecules across the nuclear pore[5–9,22], we know virtually nothing about the possible functions of nuclear or cytoplasmic glycosyl moieties. It has been suggested that *O*-GlcNAc might be important for the proper assembly of multimeric protein complexes, such as those in the nuclear pore complex and those mediated by Band 4.1 in erythrocytes[29]. It also has been postulated that *O*-GlcNAc might reversibly 'block' sites of phosphorylation on proteins, thereby serving a regulatory function[29]. Perhaps there exists a regulated cytoplasmic or nucleoplasmic hexosaminidase that removes *O*-GlcNAc residues in response to various physiologic stimuli? The existence of such an enzyme remains to be investigated. Another suggestion is that clustered *O*-GlcNAc might prevent proteolysis or serve as a targeting signal for nuclear proteins[22]. In the past few months, it has become apparent that numerous DNA-binding proteins contain multiple sites of *O*-GlcNAc addition. Several well-studied transcription factors have been found to be modified by *O*-GlcNAc addition (Jackson and Tjian, pers. commun.), and isolated *Drosophila* polytene chromosomes appear to be remarkably heavily covered with *O*-GlcNAc modified proteins (Kelly and Hart, unpublished). With the availability of *in vitro* transcription and DNA-binding assays, and as more nuclear or cytoplasmic proteins that bear *O*-GlcNAc are identified, it should be only a matter of time before the biological significance of these modifications is at least partially elucidated. Clearly, we are already witnessing an explosive growth in our understanding of this novel form of intracellular protein modification, whose very existence was unknown only a short time ago.

Acknowledgements

We wish to thank Drs Wally Whiteheart, William Kelly and Antonino Passanti for their critical reading of the manuscript. We also gratefully acknowledge grant support from the National Institutes of Health and the Arthritis Foundation.

References

1 Hirschberg, C. B. and Snider, M. D. (1987) *Annu. Rev. Biochem.* 56, 63–87
2 Torres, C-R. and Hart, G. W. (1984) *J. Biol. Chem.* 259, 3308–3317
3 Holt, G. D. and Hart, G. W. (1986) *J. Biol. Chem.* 261, 8049–8057
4 Holt, G. D., Snow, C. M., Senior, A., Haltiwanger, R. S., Gerace, L. and Hart, G. W. (1987) *J. Cell Biol.* 104, 1157–1164
5 Baglia, F. A. and Maul, G. C. (1983) *Proc. Natl Acad. Sci. USA* 80, 2285–2289
6 Finlay, D., Newmayer, D., Price, T. and Forbes, D. (1987) *J. Cell Biol.* 104, 189–200
7 Yoneda, Y., Imamoto-Sonobe, N., Yamaizumi, M. and Uchida, T. (1987) *Exp. Cell Res.* 173, 586–595
8 Dabauvalle, M-C., Schulz, B., Scheer, U. and Peters, R. (1988) *Exp. Cell Res.* 174, 291–296
9 Newmeyer, D. and Forbes, D. (1988) *Cell* 52, 641–653
10 Fransson, L-A. (1987) *Trends Biochem. Sci.* 12, 406–411
11 Stein, G. S., Roberts, R. M., Stein, J. L. and Davis, J. L. (1981) in *The Cell Nucleus Vol. IX* (Busch, H., ed.), pp. 341–356, Academic Press
12 Rizzo, W. and Bustin, M. (1977) *J. Biol. Chem.* 252, 7062–7067
13 Hozier, J. and Furcht, L. (1980) *Cell Biol. Int. Rep.* 4, 1091–1099
14 Kurth, P. D., Bustin, M. and Moudrianakis, E. N. (1979) *Nature* 279, 448–450
15 Glass, W., Briggs, R. and Hnilica, L. (1981) *Anal. Biochem.* 115, 219–224
16 Gerace, L., Ottaviano, Yl and Kondor-Koch, C. (1982) *J. Cell Biol.* 95, 826–837
17 Davis, L. I. and Blobel, G. (1986) *Cell* 45, 699–709
18 Seve, A. P., Hubert, J., Bouvier, D., Masson, C., Geraud, G. and Bouteille, M. (1984) *J. Submicrosc. Cytol.* 16, 631–641
19 Kan, F. and Pinto da Silva, P. (1986) *J. Cell Biol.* 102, 576–586
20 Einck, L. and Bustin, M. (1985) *Exp. Cell Res.* 156, 295–310
21 Reeves, R. and Chang, D. (1983) *J. Biol. Chem.* 258, 679–687
22 Schindler, M., Hogan, M., Miller, R. and De Gaetano, D. (1987) *J. Biol. Chem.* 262, 1254–1260
23 Snow, C. M., Senior, A. and Gerace, L. (1987) *J. Cell Biol.* 104, 1143–1156
24 Park, M., D'Onofrio, M., Willingham, M. and Hanover, J. A. (1987) *Proc. Natl Acad. Sci. USA* 84, 6462–6466
25 Hanover, J., Cohen, C., Willingham, M. and Park, M. (1987) *J. Biol. Chem.* 262, 9887–9894
26 Davis, L. and Blobel, G. (1987) *Proc. Natl Acad. Sci. USA* 84, 7552–7556
27 Benko, D. M., Haltiwanger, R. S., Hart, G. W. and Gibson, W. (1988) *Proc. Natl Acad. Sci. USA* 85, 2573–2577
28 Nyame, K., Cummings, R. D. and Damian, R. T. (1987) *J. Biol. Chem.* 262, 7990–7995
29 Holt, G. D., Haltiwanger, R. S., Torres, C-R. and Hart, G. W. (1987) *J. Biol. Chem.* 262, 14847–14850
30 Abeijon, C. and Hirschberg, C. B. (1988) *Proc. Natl Acad. Sci. USA* 85, 1010–1014

G. W. Hart, G. D. Holt and R. S. Haltiwanger are at the Department of Biological Chemistry, Johns Hopkins University, School of Medicine, 725 N. Wolfe St., Baltimore, MD 21205, USA.

Secretion and Membrane Assembly

William Wickner

Cytoplasmic proteins undergo rapid and stable folding which buries their apolar segments. In contrast, precursors of secreted and membrane proteins have apolar segments which are recognized by chaperones and membrane receptors to distinguish them from soluble proteins.

Proteins begin their life on cytoplasmic polysomes, but are soon sorted to distinct 'compartments', each a membrane or a membrane-limited space with a unique protein composition. As a first step in this process, non-cytosolic proteins are targeted to specific membranes either as nascent chains or soon after their synthesis is complete. Membrane insertion relies on specific regions within a protein, termed leader sequences (alternative names are signal sequence, pre-piece or transit peptide), which are often removed after the protein has crossed the membrane. Energy is required for membrane transit, as high energy phosphate and, in some cases, as an electrochemical potential. This process requires cytoplasmic chaperone proteins (chaperones), membrane-bound receptor systems which couple ATP hydrolysis to protein translocation, and leader peptidases to cleave the proteins which have crossed the membrane. Protein insertion into, and across, biological membranes has recently been reviewed in full elsewhere[1]. This article will focus on the relative roles of conformation, protein : protein recognition, and biophysical partitioning between aqueous and amphiphilic phases in guiding protein transit across membranes. Reviewing these topics provides a basis for updating the membrane trigger hypothesis, first proposed ten years ago[2]. I suggest that cytoplasmic proteins fold rapidly to internalize their apolar residues, while membrane and secreted proteins cannot fold in this manner. Their apolar domains, and especially their apolar leader sequence, remain on the protein surface, but are shielded by special chaperone proteins to prevent aggregation.

A decade ago, discussions of how proteins insert across membranes were dominated by two questions: cotranslational versus post-translational mechanisms, and whether protein transit is through a special proteinaceous pore or directly across the lipid phase of the membrane. The first question was, more precisely, whether proteins could only initiate transfer across a membrane at a particular point in their synthesis, such as a nascent chain just emerging from the ribosome[3]. This was thought to be the case for protein transit into endoplasmic reticulum[4] and was hotly disputed for bacterial cell surface export. In contrast, it was known that protein transit into mitochondria and chloroplasts can clearly occur post-translationally, both *in vivo*[5,6] and *in vitro*[7,8].

In the early 1980s, *in vivo* studies of bacterial export, employing either uncouplers to separate protein synthesis from membrane transit[9,10] or sophisticated two-dimensional peptide maps of nascent chains[11], established that bacterial protein translocation is not coupled to protein synthesis. More recently, it has been established that protein translocation into yeast[12–15] or even mammalian[16,17] endoplasmic reticulum is not obligately coupled to ongoing polypeptide chain growth. It's no longer believed that proteins avoid folding in the cytoplasm by being extruded directly across the membrane as they leave the ribosome. Whether they cross the membrane through a pore or through the lipid remains an open question.

New questions have joined the old ones in this field. Since proteins do not have to cross the membrane as they emerge from the ribosome, what guides their folding to allow a membrane transit later on? The information for translocation clearly resides in the polypeptide sequence of secreted proteins, yet there is no conserved sequence within the leader or mature domains[18]. Are the physical properties of leader peptides, such as their hydrophobicity, helical structure, or charge sufficient to promote membrane transit[19]? Can such non-specific physical properties be recognized by 'helper' proteins that promote translocation? Substantial information has accrued which addresses these questions and provides a basis for revising current models of protein translocation across membranes.

MEMBRANE ASSEMBLY-ACTIVE PROTEIN STRUCTURES

Translocation-competent preproteins have different conformations than the mature, native proteins. Neupert and colleagues have purified two precursor forms of mitochondrial proteins, apocytochrome *c*[20] and water-soluble porin[21]. Each of these is in a very different conformation than its mature counterpart and is active for assembly into mitochondria. Eilers and Schatz[22] examined the mitochondrial uptake of a fusion of the cytochrome oxidase subunit IV prepeptide with dihydrofolate reductase. Stabilizing the mature, folded structure of the dihydrofolate reductase with methotrexate prevented the protein from assembling into the mitochondria. In bacteria, it was shown that leader peptidase which is reversibly blocked from membrane transit by uncoupling the proton-motive force is in a form which is 100 times more susceptible to proteolysis (in lysates) than the mature, transmembrane protein[23]. Randall and Hardy[24] have shown that the form of pre-maltose binding protein which is competent for membrane assembly is sensitive to proteinase digestion, whereas the protein loses assembly competence when it folds into the mature, proteinase-resistant structure. The leader peptide has a direct role in retarding folding and maintaining the translocation-component state[25]. Even for endoplasmic reticulum, there is evidence that prepro-α-factor is more readily digested by protease during interaction with heat shock protein (hsp) 70 and ATP, conditions which promote translocation, than in their absence (B. Koch and R. Schekman, pers. commun.). These data have been loosely interpreted to mean that proteins must remain unfolded to allow membrane transit. However, the term 'unfolded' has also been used loosely, as there is no direct evidence for the unfolded state. Indeed, truly unfolded proteins assume secondary and tertiary structure in milliseconds to seconds after denaturant removal, while protein synthesis and membrane assembly takes seconds to minutes.

While precursor proteins may spontaneously fold to an assembly-competent state, this conformation can be stabilized by soluble proteins termed 'chaperones'[26]. In eukaryotes, heat shock proteins may serve as chaperones. ATP-dependent unfolding of M13 procoat in reticulocyte extracts allows its membrane assembly[27], and can be promoted by hsp 70 (Ref. 28). In yeast, biochemical[29] and genetic[30] approaches have demonstrated that heat shock proteins are involved in protein uptake into the endoplasmic reticulum and mitochondria. Signal recognition particle, which complexes with the N-terminus of presecretory proteins, may function as a chaperone to prevent misfolding and to allow post-translational translocation[31,32]. Even the ribosome itself may be thought of as a chaperone. Lingappa and colleagues[17] have found that full-length nascent chains which are still bound to the ribosome are competent for translocation into dog pancreas microsomes, while release from the ribosome is accompanied by loss of membrane insertion. In bacteria, at least three proteins may function as chaperones: trigger factor[31], groEL[33] and SecB[34,35]. *In vitro* studies have shown that each of these can form a complex with proOmpA, stabilizing that protein for transit across plasma membrane vesicles (S. Lecker, unpublished). The SecB protein has been shown to stabilize pre-maltose binding protein for its membrane insertion, both *in vivo* and *in vitro*[34,35]. Both groEL and trigger factor are dissociable proteins of the bacterial ribosome[36,37], as is eukaryotic signal recognition particle. For recent reviews, see Deshaies *et al.*[38], Hemmingsem *et al.*[26] and Ellis and Hemmingsen[62].

INFORMATION FOR MEMBRANE ASSEMBLY

The leader peptide is clearly essential for a protein to be secreted. Membrane proteins can require additional 'topogenic' elements in addition to the leader sequence. These may function as 'stop transfer sequences'[3] (in that their position in the polypeptide chain specifies the place where the protein will span the membrane[39]) or 'hydrophobic helpers' (apolar regions which do not function as a signal sequence but are necessary for a nearby signal to function[40]). These elements could act as 'signals' to other proteins that catalyse the steps of membrane insertion; alternatively, they may function according to their physical properties, i.e. their basic charge and apolar domain. Of course, these possibilities are by no means mutually exclusive. On balance, the current data, reviewed below, suggest that the major constraint on leader sequence function is simply its basic and hydrophobic character.

Review of leader peptide structure[18,41] has clearly shown that there are no conserved sequence features, except those around the cleavage site (see below). What is conserved is the size (16–26 residues), basic charge at the N-terminus, and an apolar central region of 8–12 residues. Extensive studies of mutant leader peptides have shown that interruption of leader function can be caused by either removing the basic N-terminal[42] residues or by shortening the apolar domain or altering its character by insertion or substitution of polar residues[43,44]. Gierasch and colleagues[19] have shown conclusively that the surface activity of leader peptides (i.e. their ability to insert spontaneously into a lipid monolayer) correlates well with their activity in promoting protein export *in vivo*. Three types of experiments have established the idea that it is strictly the physical character of a leader peptide, rather than its sequence, which promotes translocation. In one approach, pioneered by the late Tom Kaiser, substitution of poly-leucine for the apolar region of a leader sequence was shown still to allow efficient export[45]. In a second approach, it was shown that apolar domains which anchor proteins in the membrane can initiate insertion if placed at the N-terminus and given basic charge[40]. A third type of experiment exploited the power of selection in microbial systems. In these studies, oligonucleotide-directed mutagenesis was used to delete the codons for a normal leader sequence, replacing it with a restriction site. Random DNA, cut with the restriction enzyme, was then cloned into the site and plasmids bearing the chimeric proteins were transformed into cells. The fraction of the encoded fusion protein species which translocated across the membrane was then scored. Using this approach, Baker and Schatz[46] have shown that approximately one in five random peptides will function to support mitochondrial import; sequence evaluation shows that these peptides have the basic charge which typifies leader sequences for that organelle. Kaiser *et al.*[47] performed a similar analysis for invertase, a secreted protein of *Saccharomyces cerevisiae*. They also found that approximately 20% of random sequences support secretion. Sequence analysis showed basic and somewhat apolar character, though with a far broader range than seen with natural leader peptides. Taken together, these diverse experiments establish that, for bacterial and endoplasmic reticulum directed leader sequences, the basic and apolar character is both necessary and sufficient.

What evidence is there for the recognition of leader peptides by proteins? Only a few proteins have been described which could recognize leader regions. These include leader peptidases, signal recognition particle, chaperone proteins and the SecA/SecY 'translocation ATPase' of *E. coli*. All of the information for accurate cleavage by leader peptidase is in the region immediately preceding the cleavage site[48], and is summarized by the '−1, −3 rule'* of von Heijne[18]. However, leader peptidase is not essential for translocation *per se*[49], and mutants in leader peptides which systematically violate the −1, −3 rule block cleavage without affecting translocation[50,51]. Signal recognition particle has been shown to interact selectively with proteins bearing a signal

*G. von Heijne[18] noted that amino acid residues with small side chains are commonly found at positions −1 and −3 with respect to the site of leader peptide cleavage.

sequence[4]; it is not yet clear whether this simply represents recognition of the basic or apolar character of the signals or whether SRP recognizes more subtle structural features. As discussed above and in Deshaies *et al.*[38], chaperones may complex with pre-secretory and organellar proteins to stabilize them for membrane transit. However, there is no evidence that they are specific for leader sequences. Indeed, chaperones may complex with OmpA as well as with proOmpA (S. Lecker and W. W., unpublished) and there is good evidence that SecB protein may recognize the mature region of pre-maltose binding protein more than the leader[34]. We have recently described the direct assay of the *E. coli* 'translocation ATPase', the enzyme responsible for ATP hydrolysis coupled to protein translocation[52]. The assay requires SecA protein, membrane vesicles bearing functional SecY protein, and a translocation-competent protein such as pre-phoE or proOmpA, either freshly diluted from urea or complexed with a chaperone such as SecB, trigger factor, or GroEL. Neither OmpA, nor the leader peptide, will substitute for the full-length proOmpA in supporting this ATPase reaction. However, synthetic leader peptide is a potent inhibitor of this reaction, and its inhibition of ATPase activity is competitive with respect to proOmpA (K. Cunningham and W. Wickner, unpublished). Further studies are clearly needed to determine which aspects of the leader peptide structure allow it to function as a competitive inhibitor.

THE MEMBRANE TRIGGER HYPOTHESIS, 10TH ANNIVERSARY EDITION

I proposed the membrane trigger hypothesis[2] in 1979. The basic features of this model were: (1) Proteins initially fold in a manner compatible with the aqueous cytoplasm, i.e. so as to bury their apolar residues. Interaction with the membrane causes a refolding, due to passage into, or (for secreted proteins) across, the apolar centre of the membrane. (2) Translocation is not coupled to ongoing translation, but is coupled to correct folding and *can* occur post-translationally. (3) Transit would be through the lipid phase *per se* rather than a proteinaceous pore. During the last decade, it has become quite clear that there is a distinction between the initial, translocation-competent form of the protein and the final state. Chaperones may participate in protein stabilization by shielding apolar regions from the cytoplasm prior to secretion. It has also been established that translocation is not mechanistically coupled to ongoing polypeptide chain growth in all well-studied translocation systems: bacteria, chloroplasts, mitochondria, and the endoplasmic reticulum of yeast and mammals. The final point, transit through a pore, still remains a completely open question. The bacterial SecY protein[53] and the mammalian signal sequence receptor[54] are candidates for such a function.

Advances of the last decade fit well with the concept that membranes trigger a refolding of exported proteins. Newly made proteins shield their apolar regions from the cytoplasm, either by folding or by interaction with chaperones. Chaperones may only recognize apolar character, and thus could even bind to cytoplasmic proteins as well as to secreted, organellar and membrane proteins. Chaperones will bind with low affinity, allowing rapid displacement. Cytoplasmic proteins would undergo rapid *intra*molecular folding to a stable form, displacing the chaperones. Non-cytoplasmic proteins, whether secreted from the cell or (in eukaryotes) sequestered in an organelle, form more stable complexes with chaperones. The chaperones are displaced from these complexes by pre-protein interaction with lipids or with membrane receptor systems, such as docking protein, the SecA/SecY translocation ATPase[52], or mitochondrial receptor and general insertion protein[55]. Like the chaperones which they displace, the lipids or receptor proteins only recognize the apolar character of the leader sequences. Soluble proteins may contain such sequences, but internalize them during folding and thereby escape export. Thus, the information for export is encoded not only in the leader region[56] but in the folding pathway of the entire polypeptide chain. In a preliminary sequence survey, Kaiser *et al.*[47] noted that leader-like sequences were systematically excluded from the N-terminus of cytoplasmic proteins. ATP hydrolysis may drive the transfer of apolar domains from receptor association into the membrane lipid phase *per se*. Proton electrochemical potential may assist the transfer of ionizable residues across the membrane. It is widely recognized that proteins insert as loops[57–60]; once initial insertion has been achieved, further transit may not require energy input[61].

There has recently been a major emphasis in this field on the purification and reconstitution of the essential components of translocation. Recent progress in this direction suggests that we will soon leave crude *in vitro* systems behind, and focus on biochemical and genetic tests of the mechanisms of isolated proteins.

References

1 Verner, K. and Schatz, G. (1988) *Science* 241, 1307–1313
2 Wickner, W. (1979) *Annu. Rev. Biochem.* 48, 23–45
3 Blobel, G. (1980) *Proc. Natl Acad. Sci. USA* 77, 1496–1500
4 Walter, P. and Blobel, G. (1981) *J. Cell Biol.* 91, 551–556
5 Hallermayer, G. and Neupert, W. (1976) in *Genetics and Biogenesis of Chloroplasts and Mitochondria* (Buchler, Th., ed.), pp. 807–812, Elsevier
6 Ross, E. and Schatz, G. (1976) *J. Biol. Chem.* 251, 1997–2004
7 Harmey, M. A., Hallermayer, G. and Neupert, W. (1976) in *Genetics and Biogenesis of Chloroplasts and Mitochondria* (Buchler, Th., ed.), pp. 813–818, Elsevier
8 Maccecchini, M. L., Rudin, Y., Blobel, G. and Schatz, G. (1979) *Proc. Natl Acad. Sci. USA* 76, 343–347
9 Date, T., Goodman, J. and Wickner, W. (1980) *Proc. Natl Acad. Sci. USA* 77, 4669–4673
10 Zimmermann, R. and Wickner, W. (1983) *J. Biol. Chem.* 258, 3920–3925
11 Randall, L. L. (1983) *Cell* 33, 231–240
12 Waters, M. G. and Blobel, G. (1986) *J. Cell Biol.* 102, 1543–1550
13 Hansen, W., Garcia, P. D. and Walter, P. (1986) *Cell* 45, 397–406
14 Rothblatt, J. A. and Meyer, D. I. (1986) *EMBO J.* 5, 1031–1036
15 Toyn, J., Hibbs, A. R., Sanz, P., Crowe, J. and Meyer, D. I. (1988) *EMBO J.* 7, 4347–4353
16 Ainger, K. J. and Meyer, D. I. (1986) *EMBO J.* 5, 951–955
17 Perara, E., Rothman, R. E. and Lingappa, V. R. (1986) *Science* 232, 348–352
18 von Heijne, G. (1983) *Eur. J. Biochem.* 133, 17–21
19 Gierasch, L. M. (1989) *Biochemistry* 28, 923–930
20 Korb, H. and Neupert, W. (1978) *Eur. J. Biochem.* 91, 609–620

21 Pfaller, R., Freitag, H., Harmey, M. A., Benz, R. and Neupert, W. (1985) *J. Biol. Chem.* 260, 8188–8195
22 Eilers, M. and Schatz, G. (1986) *Nature* 322, 228–232
23 Wolfe, P. B. and Wickner, W. (1984) *Cell* 36, 1067–1072
24 Randall, L. L. and Hardy, S. J. S. (1986) *Cell* 46, 921–928
25 Park, S., Liu, G., Topping, T. B., Cover, W. H. and Randall, L. L. (1988) *Science* 239, 1033–1035
26 Hemmingsem, S. M., Woolford, C., vanderVries, S. M., Tilly, K., Dennis, D. T., Georgopoulos, C. P., Hendrix, R. and Ellis, R. J. (1988) *Nature* 333, 330–334
27 Wiech, H., Sagstetter, M., Muller, G. and Zimmermann, R. (1987) *EMBO J.* 6, 1011–1016
28 Zimmermann, R., Sagstetter, M., Lewis, M. J. and Pelham, H. R. B. (1988) *EMBO J.* 7, 2875–2880
29 Chirico, W. J., Waters, M. G. and Blobel, G. (1988) *Nature* 332, 805–810
30 Deshaies, R. J., Koch, B. D., Werner-Washburne, M., Craig, E. A. and Schekman, R. (1988) *Nature* 332, 800–805
31 Crooke, E., Guthrie, B., Lecker, S., Lill, R. and Wickner, W. (1988) *Cell* 54, 1003–1011
32 Sanz, P. and Meyer, D. I. (1988) *EMBO J.* 7, 3553–3557
33 Bochkareva, E. S., Lissin, N. M. and Girshovich, A. S. (1988) *Nature* 336, 254–257
34 Collier, D. N., Bankaitis, V. A., Weiss, J. B. and Bassford, P. J., Jr (1988) *Cell* 53, 273–283
35 Gannon, P. M., Li, P. and Kumamoto, C. A. (1989) *J. Bacteriol.* 171, 813–818
36 Neidhardt, F. C., Phillips, T. A., vanBogelen, R. A., Smith, M. W., Georgalis, Y. and Subramanian, A. R. (1981) *J. Bacteriol.* 145, 513–520
37 Lill, R., Crooke, E., Guthrie, B. and Wickner, W. (1988) *Cell* 54, 1013–1018
38 Deshaies, R. J., Koch, B. D. and Schekman, R. (1988) *Trends Biochem. Sci.* 13, 384–388
39 Davis, N. G. and Model, P. (1985) *Cell* 41, 607–614
40 von Heijne, G., Wickner, W. and Dalbey, R. E. (1988) *Proc. Natl Acad. Sci. USA* 85, 3363–3366
41 Perlman, D. and Halvorson, H. O. (1983) *J. Mol. Biol.* 167, 391–409
42 Vlasuk, G. P., Inouye, S., Ito, H., Itakura, K. and Inouye, M. (1983) *J. Biol. Chem.* 258, 7141–7148
43 Emr, S. D. and Silhavy, T. J. (1982) *J. Cell Biol.* 95, 689–696
44 Ryan, J. P., Duncan, M. C., Bankaitis, V. A. and Bassford, P. J., Jr (1986) *J. Biol. Chem.* 261, 3389–3395
45 Kendall, D. A., Bock, S. C. and Kaiser, E. T. (1986) *Nature* 321, 706–708
46 Baker, A. and Schatz, G. (1987) *Proc. Natl Acad. Sci. USA* 84, 3117–3121
47 Kaiser, C. A., Preuss, D., Grisafi, P. and Botstein, D. (1987) *Science* 235, 312–317
48 Dierstein, R. and Wickner, W. (1986) *EMBO J.* 5, 427–431
49 Dalbey, R. and Wickner, W. (1985) *J. Biol. Chem.* 260, 15925–15931
50 Koshland, D., Sauer, R. T. and Botstein, D. (1982) *Cell* 30, 903–914
51 Kuhn, A. and Wickner, W. (1985) *J. Biol. Chem.* 260, 15914–15918
52 Lill, R., Cunningham, K., Brundage, L., Ito, K., Oliver, D. and Wickner, W. (1989) *EMBO J.* 8, 961–966
53 Akiyama, Y. and Ito, K. (1987) *EMBO J.* 6, 3465–3470
54 Wiedmann, M., Kurzchalia, T. V., Hartmann, E. and Rapoport, T. (1987) *Nature* 328, 830–832
55 Pfaller, R., Pfanner, N. and Neupert. W. (1989) *J. Biol. Chem.* 264, 34–39
56 Randall, L. L. and Hardy, S. J. S. (1989) *Science* 243, 1156–1159
57 Halegoua, S. and Inouye, M. (1979) in *Bacterial Outer Membranes: Biogenesis and Functions* (Inouye, M., ed.), pp. 67–114 Wiley
58 Steiner, D. F., Duguid, J. R., Patzelt, C., Chan, S. J., Quinn, P., Labrecque, A. and Hastings, R. (1979) in *Proinsulin, Insulin, and C-peptide* (Baba, S., Kaneko, T. and Yanaihara, N., eds) Excerpta Medica Foundation
59 Kuhn, A. (1987) *Science* 238, 1413–1415
60 Shaw, A. S., Rottier, P. J. M. and Rose, J. K. (1988) *Proc. Natl Acad. Sci. USA* 85, 7592–7596
61 von Heijne, G. and Blomberg, C. (1979) *Eur. J. Biochem.* 97, 175–181
62 Ellis, R. J. and Hemmingsen, S. (1989) *Trends Biochem. Sci.* 14, 339–342

W. Wickner is at the Molecular Biology Institute and Department of Biological Chemistry, University of California, Los Angeles, CA 90024-1570, USA.

A Novel Pathway for Secretory Proteins?

Anne Muesch, Enno Hartmann, Klaus Rohde, Anna Rubartelli, Roberto Sitia and Tom A. Rapoport

In eukaryotes, most proteins which are transported to the extracellular space, into mitochondria or into chloroplasts are synthesized as precursor polypeptides containing cleavable N-terminal signal or targeting sequences. We have searched the literature for proteins that are exported from the cytosol without being proteolytically processed. Some of these proteins contain uncleaved signal or targeting sequences. However, among secretory proteins there is a class that does not possess hydrophobic signal sequences and appears to leave the cell by a secretory pathway clearly distinct from the classical route through the endoplasmic reticulum and Golgi apparatus.

It is generally assumed that secretory proteins in eukaryotic cells are first translocated across the endoplasmic reticulum (ER) membrane, and are then further transported in vesicles through the Golgi apparatus to the plasma membrane, either directly (constitutive secretion) or via secretory granules (regulated secretion)[1].

Translocation of secretory proteins across the ER membrane is triggered by signal sequences which are characterized by a stretch of at least six or seven hydrophobic amino acids. These are usually located at the N-terminus of precursor polypeptides and cleaved off during, or shortly after, transfer of the polypeptide across the membrane. An exception to this scheme is ovalbumin which has an uncleaved internal signal sequence[2]. In contrast to secretory proteins, many membrane proteins which are inserted into the ER have uncleaved signal sequences; these generally represent the most N-terminal membrane-spanning segments[3]. Proteins imported into mitochondria or chloroplasts often have N-terminal cleavable targeting sequences, but unprocessed proteins, like the ADP/ATP-translocator[4], are known to exist.

Given these general considerations, one would expect to detect uncleaved signal or targeting sequences in all unprocessed proteins exported from the cytosolic compartment. We have tested this hypothesis and have found that there exists a class of secretory proteins which lack hydrophobic signal sequences and presumably follow a novel secretory route. Some of them have been noted before as unexplained exceptions to the general rules for secretion[5–7].

Searching for unprocessed proteins exported from the cytosolic compartment

In order to find proteins with internal, uncleaved signal and targeting sequences as well as secretory proteins lacking signal sequences altogether, we have searched for exported polypeptides which remain unprocessed by proteases outside the cytosol. Our approach was based on the consideration stated recently by Bradshaw (Ref. 8; see also p. 183 of this book), that N-terminal acylation of proteins is generally restricted to the cytosolic compartment. Consequently, *N*-acylated proteins found outside the cytosol must have retained their N-terminus and cannot have possessed cleavable signal or targeting peptides.

In order to verify our approach, we have searched the literature for the occurrence of N-terminal acyl groups in proteins with known location in the cell. Among 296 cytosolic proteins, 108 (36%) were found to be acylated. The percentage of acylated proteins in a given cell may, in fact, be significantly higher (60–80%) as suggested by experiments in which proteins were labelled with ^{14}C-acetate[9]. The discrepancy may be due to the fact that our list contains mostly major proteins isolated from different species.

By contrast, among 214 secretory and membrane proteins with cleavable signal peptides, there were only seven cases for which an N-terminal acylation has been reported, four of which appear doubtful. The three remaining exceptions concern cell-specific modifications that are clearly distinct from the cotranslational acylations*.

Likewise, among 73 mitochondrial and chloroplast proteins with cleavable targeting peptides, none was found to be *N*-acylated. Since mitochondria contain acetyl-CoA, the substrate for the acetyltransferase, the lack of acetylation is probably due to the absence of the enzyme.

This analysis thus confirms that N-terminal acylations generally do not occur outside the cytosol. One may assume, however, that the precursor polypeptides are often *N*-acylated *before* the signal or targeting peptide is cleaved off.

INTERNAL SIGNAL AND TARGETING SEQUENCES

Table I lists unprocessed proteins that are exported from the cytosol and which are likely to contain uncleaved signal or targeting sequences. Most of them have been found in our search for *N*-acylated exported proteins, but the non-acylated ones have been collected subsequently in a less systematic way.

The secretory protein ovalbumin has a hydrophobic segment between positions 22 and 41 of the polypeptide chain which has been shown to function as a signal sequence[2]. The related chicken gene Y protein may fall in the same class[10].

The other proteins in Table I are mitochondrial. Cytochrome *c* uses a unique pathway into mitochondria[11], but the others are predicted to have uncleaved mitochondrial targeting sequences. Indeed, sequences fulfilling the criteria of von Heijne[12] (see p. 230) can be found in all of the listed proteins, usually but not always at the N-terminus, with the possible exception of cytochrome oxidase subunit VII. Some of the proteins, e.g. the ADP/ATP-translocator[13], contain several candidate targeting sequences.

*The three cases are the acetylation of β-endorphin and of α-melanotropin in the secretory granules of the intermediate lobe of the pituitary gland[27], and the attachment of glucuronic acid to cutinase from *Fusarium solani pisi*[28].

A NEW CLASS OF SECRETORY PROTEINS

Table II lists secretory proteins that were found in our search for *N*-acylated exported proteins (first seven), and for which the sequence gave no indication of a hydrophobic segment typical for a signal sequence (data not shown). In some cases, the computer programs used (PSIGNAL, SOAP, RAOARGOS) identified weak hydrophobic regions but other evidence indicates that these do not function as signal sequences (see below). Also, in most cases the existence of a cleavable signal peptide can be formally excluded since the sequence has been determined both from the protein and from the nucleotide sequence of the corresponding cDNA.

Of course, secretory proteins lacking a signal sequence may also be expected among non-acylated proteins. Indeed, we have found five non-acylated proteins that appear to fall in the same class (Table II, last five). The bacterial enzyme chloramphenicol acetyltransferase which lacks a hydrophobic signal sequence has also been reported to be secreted from eukaryotic cells[14].

TABLE I
Proteins with presumed uncleaved signal or targeting sequences[a]

Protein[b]	Location	N-terminal blocking group
Ovalbumin (chicken)	extracellular	acetyl
Gene Y protein (chicken)	extracellular	?
Cytochrome *c*	mitochondrial	acetyl
ADP/ATP-translocator (bovine)	mitochondrial	acetyl
Cytochrome oxidase VII (bovine)	mitochondrial	acetyl
Ubiquinol-cytochrome *c* reductase 8 kDa protein (bovine)	mitochondrial	acetyl
Ubiquinone-binding protein QP-C (bovine)	mitochondrial	acetyl
3-Oxoacyl-CoA thiolase (rat)	mitochondrial	blocked
Ubiquinol-cytochrome *c* reductase 14 kDa protein (yeast)	mitochondrial	blocked?
Cytochrome oxidase VIc (rat)	mitochondrial	?
Cytochrome oxidase Vc (potato)	mitochondrial	none
Cytochrome oxidase VII (yeast)	mitochondrial	none
α-Isopropyl-malate-synthase (yeast)	mitochondrial	?
tRNA (guanine-N^2,N^2) dimethyl-transferase (yeast)	mitochondrial	?

[a]Membrane proteins with uncleaved signal sequence inserted into the ER membrane are numerous and are not included in the table. Also, outer membrane proteins of mitochondria are generally unprocessed and have been omitted.
[b]A full list of references will be supplied by the authors on request.

CHARACTERISTICS OF THE NEW CLASS OF SECRETORY PROTEINS

Although the pathway of secretion is not yet known for the proteins listed in Table II, the absence of a signal sequence indicates that it does not involve the classical route through the ER and Golgi apparatus. Indeed, interleukin-1 could not be translocated *in vitro* into canine microsomes[15] and cannot be detected in the ER in cells which secrete it[16]. A number of drugs which inhibit the classical secretory pathway in fact stimulate the release of interleukin-1β from monocytes (Rubartelli, Cozzolino, Talio and Sitia, submitted). Despite the fact that many proteins in Table II have potential Asn-glycosylation sites (some of them contain as many as six), none of them is glycosylated. This

Table II
Secretory proteins without classical signal sequence[a]

Protein[b]	Source	N-terminal blocking group
Coagulation factor XIII a-chain (human)	blood platelets and plasma; other tissues (placenta, uterus, spleen, prostate, macrophage)	acetyl
Prothymosin (α1-thymosin) (rat)	thymus	acetyl
Parathymosin (rat)	thymus	acetyl
β4-Thymosine (bovine)	thymus	acetyl
Transglutaminase (pig)	liver	acetyl
Mammary-derived growth inhibitor (MDGI) (bovine)	mammary gland	acetyl
Lipocortin (human)	monocytes, neutrophils, renal medulla	blocked
Interleukin-1 (human)	many cell types including macrophages	none
Platelet derived growth factor (PD-ECGF) (human)	platelets	none
Basic fibroblast growth factor (bovine)	macrophages, many tissues (brain, hypothalamus, pituitary, retina, adrenal gland, kidney)	none
Acidic fibroblast growth factor (endothelial cell growth factor) (bovine)	neuronal tissue	none
Adult T-cell leukemia (ATL)-derived factor (human)[c]	T cells	none

[a]The human plasminogen activator inhibitor-2 may also belong to this class, but has not been included in the table because the absence of Asn-glycosylation has not been firmly established.
[b]A full list of references will be supplied by the authors on request.
[c]The factor is identical with thioredoxin (A. Holmgren, pers. commun.).

argument is particularly important for proteins with internal weak hydrophobic sequence segments. Glycosylation of interleukin-1β does occur if a cleavable signal sequence is attached to its N-terminus[17]. The presence of cysteines but absence of disulfide bridges in some of the proteins (e.g. factor XIII and transglutaminase) also shows that chemical modifications restricted to the lumen of the endoplasmic reticulum do not occur.

All proteins listed in Table II appear to be present in high concentrations inside the cell, largely in the cytosol. This raises the possibility that they are liberated by cell lysis. However, we consider this possibility unlikely. The secretory proteins are synthesized in a great variety of cells (see Table II) and for some of them an extracellular function has been demonstrated. In human monocytes secreting interleukin-1, cytosolic proteins could not be detected in the medium, nor was there any indication of cell death[18] (Rubartelli *et al.* op. cit.). The major extracellular form of interleukin-1β, the 17 kDa C-terminal portion, is not found in cell lysates[18] (Rubartelli *et al.* op. cit.). Interleukin-1α is secreted from the same cells at a much slower rate than interleukin-1β[18] (Rubartelli *et al.* op. cit.). Generally, many of the proteins listed in Table II appear to leave the cell only slowly with half-times of several hours.

Some of the proteins listed in Table II are not secreted by certain cells, even if synthesized in large amounts[19–21]. It therefore appears that there is some cell specificity as to whether secretion occurs.

What is the mechanism of secretion?

Several mechanisms of secretion may be envisaged and it is possible that different pathways are used by the proteins listed in Table II. Intracellular vesicles which fuse with the plasma membrane may be involved, as suggested by the occurrence of trypsin-protected vesicular interleukin-1β in secreting but not in non-secreting cells (Rubartelli *et al.* op. cit.). Alternatively, the proteins may directly traverse the plasma membrane.

An interesting possibility is suggested by the results of studies on the secretion of hemolysin by hemolytic bacteria[22,23] and of the **a**-factor from *Saccharomyces cerevisiae*[24]. In both cases, classical hydrophobic signal sequences are absent. Secretion of hemolysin and of **a**-factor requires the products of the genes *HlyB* and *STE6*, respectively, which are homologous to the mammalian P-glycoprotein associated with the phenomenon of multidrug resistance[25]. The homology includes the presence of cytosolic ATP-binding domains and of membrane-spanning regions. These proteins appear to belong to a superfamily of ATP-driven transport ATPases. This superfamily includes (among others) the products of the cystic fibrosis gene, that of the *white* gene of *Drosophila* and several periplasmic solute transport systems of Gram-negative bacteria[26]. Since homologues of the P-glycoprotein participate in known cases of nonclassical secretion of (poly)peptides, it is tempting to speculate that member(s) of this family are generally involved in the translocation of proteins lacking hydrophobic signal sequences (see also Ref. 29). A similar mechanism may hold for the translocation of peptides into the ER for antigen presentation by the major histocompatibility complex.

References

1 Palade, G. (1975) *Science* 189, 347–358
2 Tabe, L., Krieg, P., Strachan, R., Jackson, D., Wallis, E. and Colman, A. (1984) *J. Mol. Biol.* 180, 645–666
3 Hartmann, E., Rapoport, T. A. and Lodish, H. (1989) *Proc. Natl Acad. Sci. USA* 86, 5786–5790
4 Pfanner, N. and Neupert, W. (1987) *Curr. Top. Bioenerg.* 15, 177–219
5 Jaye, M., Howk, R., Burgess, W., Ricca, G. A., Chui, J-M., Ravera, M. W., O'Briess, S. J., Modi, W. S., Maciag, T. and Dohan, W. N. (1986) *Science* 233, 541–545
6 Esch, F., Baird, A., Ling, N., Ulno, N., Hill, F., Denoroy, L., Klepper, R., Gospodarowicz, D., Boehlen, P. and Guillemin, R. (1985) *Proc. Natl Acad. Sci. USA* 82, 6507–6511
7 Furutani, Y., Notake, M., Yamayoshi, M., Yamayoshi, J. J., Nomura, H., Mayumi, O., Furata, R., Fukui, T., Yamada, M. and Nakamura, S. (1985) *Nucleic Acids Res.* 13, 5869–5882
8 Bradshaw, R. A. (1989) *Trends Biochem. Sci.* 14, 276–279; see also p. 183 of this book
9 Brown, J. L. (1979) *J. Biol. Chem.* 254, 1447–1449
10 Heilig, R., Muraskowsky, R., Kloepfer, C. and Mandel, J. L. (1982) *Nucleic Acids Res.* 10, 4363–4382
11 Nicholson, D. W., Koehler, H. and Neupert, W. (1987) *Eur. J. Biochem.* 164, 147–157
12 von Heijne, G., Steppuhn, J. and Herrmann, R. G. (1989) *Eur. J. Biochem.* 180, 535–545
13 Hartl, F-U., Pfanner, N., Nicholson, D. W. and Neupert, W. (1989) *Biochim. Biophys. Acta* 988, 1–45
14 Bunker, C. A. and Moore, D. D. (1988) *Gene* 67, 279–286
15 Moore, B., Zhang, Y., Adesnik, M. and Sabatini, D. D. (1988) *J. Cell Biol.* 107, 699a
16 Singer, I. I., Scott, S., Hall, G. I., Limjuco, G., Chin, J. and Schmidt, J. A. (1988) *J. Exp. Med.* 167, 389–407
17 Baldari, C., Marray, J. A. H., Chiara, P., Cesareni, G. and Gallotti, C. L. (1987) *EMBO J.* 6, 229–234
18 Hazuda, D. J., Lee, J. C. and Young, P. R. (1988) *J. Biol. Chem.* 263, 8473–8479
19 Young, P. R., Hazuda, D. J. and Simon, P. L. (1988) *J. Cell Biol.* 107, 447–456
20 Cozzolino, F., Rubartelli, A., Aldinucci, D., Sitia, R., Torcia, M., Shaw, A. and Di Guglielmo, R. (1989) *Proc. Natl Acad. Sci. USA* 86, 2369–2373

21 Fuhlbrigge, R. C., Fine, S. M., Unanue, E. R. and Chaplin, D. D. (1988) *Proc. Natl Acad. Sci. USA* 85, 5649–5653

22 Mackman, N., Baker, K., Gray, L., Haigh, R., Vicaud, J-M. and Holland, I. B. (1987) *EMBO J.* 6, 2835–2841

23 Felmlee, T., Pellett, S. and Welck, R. A. (1985) *J. Bacteriol.* 163, 94–105

24 McGrath, J. P. and Varshavsky, A. (1989) *Nature* 340, 400–404

25 Endicott, J. A. and Ling, V. (1989) *Annu. Rev. Biochem.* 58, 137–171

26 Riordan, J. R., Rommens, J. M., Kerem, B-S., Alon, N., Rozmahel, R., Grzelczak, Z., Zielenski, J., Lok, S., Plavsic, N., Chou, J-L., Drumm, M. L., Iannuzzi, M. C., Collins, F. S. and Tsui, L-C. (1989) *Science* 245, 1066–1073

27 Glembowski, C. C. (1982) *J. Biol. Chem.* 257, 10501–10509

28 Kolattukudy, P. E. (1984) *Methods Enzymol.* 106, 210–217

29 Featherstone, G. (1990) *Trends Biochem. Sci.* 15, 169–170

A. Muesch, E. Hartmann, K. Rohde and T. A. Rapoport are at the Zentralinstitut für Molekularbiologie der AdW der DDR, 1115 Berlin-Buch, GDR. A. Rubartelli and R. Sitia are at the Instituto Nazionale per la Ricerca sul Cancro, 16132 Genova, Italy.

The Regulation of Protein Polymerization

Carl Frieden

Reversible polymerization reactions are essential for many cellular functions. This review briefly discusses some of the mechanisms for controlling the rate and extent of these processes.

The ability of certain proteins to undergo reversible association–dissociation reactions has been investigated for many years, both with respect to the possible physiological role of the polymerization and to the physical characterization of polymerization processes. While it is possible to examine many of the characteristics of these polymerizing systems using current sophisticated physical methods, their study still presents some interesting and difficult problems. Many earlier studies were concerned with the role of polymerization as related to the kinetic properties of enzymes, while recent attention has been focused on non-enzymic proteins that are equally as important in controlling certain cellular functions. It has become clear in recent years that such proteins undergo reversible, possibly rapid, association and dissociation and that these reactions are subject to metabolic control. This review will deal with some of the mechanisms for the control and regulation of polymerization processes for both enzymic and non-enzymic systems and thus consider those factors which control either the rate of association or dissociation or the extent of the polymerization.

Polymerization mechanisms

Mechanisms for polymerization fall into two general classes which represent extreme cases of a very general mechanism. The two extremes are linear and nucleated polymerization[1]. One major difference between these two types is the relative rate constants for association and dissociation of monomers from the polymer species of various molecular weights (see Fig. 1). Linear polymerization, in most cases, can be considered in terms of a self-association process[1]: monomers add continuously to a growing chain essentially independently of the length of the chain (Fig. 1a). If the association constant for monomer addition is absolutely independent of polymer length (i.e. an isodesmic association), there is no predominant species formed and no cooperative polymerization behavior. For the most part, linear polymerization processes appear to be indefinite, i.e. the extent of polymerization continues to increase as the protein concentration increases. However, a special case of linear polymerization would be its polymerization to form a species of discrete molecular weight, e.g. a monomer–tetramer polymerization (Fig. 1b), where the structure of the tetramer is such that no surfaces are available for further monomer addition. Linear polymerization systems have usually been considered in terms of their characterization under equilibrium conditions[1,2]. In these systems, ligand binding to a polymer species of

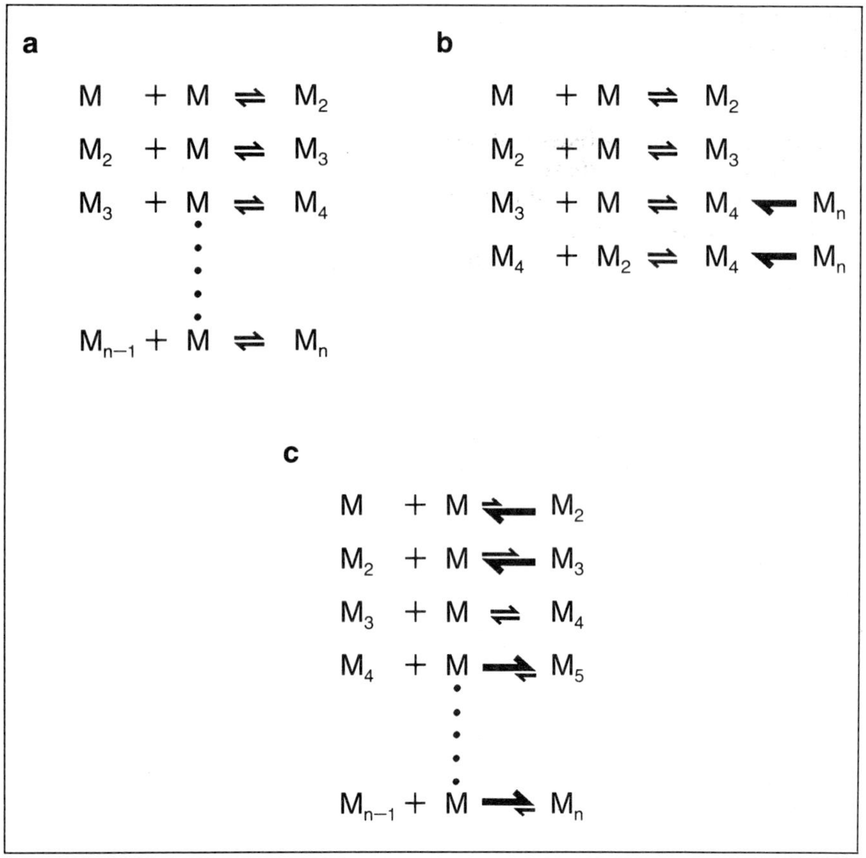

Figure 1

General mechanisms for polymerization of identical monomers (M). (**a**) Indefinite polymerization with identical association rate constants for each step. (**b**) Example of a discrete polymerization to a tetramer. (**c**) Nucleated polymerization. Size and thickness of arrows reflect relative values for association and dissociation rate constants.

specific molecular weight can influence the molecular weight distribution of polymers formed and can be a mechanism for controlling the extent of polymerization as well as the rate. If we wish to be concerned with the control of polymerization processes, then it is necessary to take into account ligand binding as well as the rates of the association and dissociation processes.

There are several interesting exceptions to the concept of indefinite polymerization. For example, in contrast to small oligomers such as tetramers, α-crystallin, the major protein in the lens of the eye, may polymerize into a micellar-type structure[3]. Clathrin, which forms the protein basket located in coated pits, is another such example[4]. This protein, which has an essential role in endocytosis, assembles into polyhedral cages under the appropriate conditions. Although much work has been done on the properties of this interesting and unusual protein[5], the mechanism of the polymerization process has not been worked out in detail.

Nucleated polymerization describes the situation in which there is an unfavorable association of several monomers to form a nucleus that serves as the point of further growth (elongation) of the polymer (Fig. 1c). When nucleation is characterized by a rather poor association process but there is a rather good association constant in the subsequent elongation process, then the process can be highly cooperative with very low concentrations of polymers of intermediate molecular weight. As discussed below, the rates of the polymerization or depolymerization for these processes have been an important consideration in determining their cellular function.

Some polymerizing systems, specifically those of viral coat proteins, are extraordinarily complex and since little is known about the control of the assembly process, these systems will not be discussed here. Although in the following discussion I have divided the various mechanisms for control into separate categories it should be understood that there may be no clear separation between them.

CONTROL INHERENT IN THE POLYMERIZATION PROCESS

Regardless of mechanism, polymerizing systems are at least second order in the association process and usually first order in dissociation. Thus, both the rate of association and the extent of polymerization are affected by the protein concentration. For linear polymerization processes, as defined above, these properties strictly follow laws of mass action. For an isodesmic polymerization (Fig. 1a) the total equilibrium concentration of polymer, C_t, is given by

$$C_t = C_1/(1-K_I C_1)^2$$

where K_I is the intrinsic association constant and C_1 is the monomer concentration. The rate of polymer formation (dC_t/dt) will be dependent on the monomer concentration although in a rather complex way. Since the distribution between different molecular weight species seems usually to be adjusted rapidly not many studies have been directed towards determining rate constants for the association and dissociation process for this type of polymerization. In such studies, however, it is frequently observed that the association rate constant, while clearly dependent on the molecular weight, is close to diffusion control limits values (10^7–10^8 s^{-1} mol^{-1}; Ref. 6) and in the absence of other factors, such as indicated below, this type of polymerization is probably neither a highly regulated process nor one that is kinetically important.

The situation is quite different for the nucleated polymerization mechanism in which there are several factors that control the polymerization process and polymerization or depolymerization processes can take seconds or minutes to be completed and perhaps longer to come to equilibrium. These reactions usually involve a time-dependent lag in the association which depends on the extent of forming nuclei from the monomer relative to the rate at which elongation occurs from this nucleus[7]. In addition, because the process is highly cooperative, there are very low concentrations of intermediate species and it is these concentrations which determine the rate of elongation. Ultimately, there is a given critical concentration of monomer in equilibrium with the polymer, below which essentially no polymerization can occur. Thus the intrinsic rate constants for nuclei formation and dissociation, for elongation and for the binding of monomer to the ends of the polymer, control the rate and extent of the polymerization process. These rate constants can be affected by many factors including temperature, pH and ionic strength. In general, the rates and extent are very sensitive to pH and this is undoubtedly an important parameter *in vivo*. For example, some of the kinetic parameters involving nucleation have been determined for actin polymerization. At pH 8, the dissociation constant for dimer formation is very unfavorable (8×10^5 μM). Trimer formation is about 10^5 times more favorable and tetramer formation

30-fold more favorable than trimer formation[8]. Of interest is the observation that the major contributing factor to the pH and temperature dependence of the polymerization is the ability to form the dimer. For example, dimer formation is 300–500 times more favorable at pH 6 relative to pH 8 (Ref. 9). There are, of course, other proteins which show the characteristics of nucleated polymerization and, in addition to actin, the recent literature is dominated by studies of Hb S (Ref. 10) and tubulin[11].

Nucleating proteins

Although systems that follow a nucleated polymerization mechanism can polymerize in the absence of added nuclei, polymerization obviously proceeds better in the presence of some nucleating substance since the addition of a nucleus will bypass any unfavorable nucleation step. Actin, tubulin or flagellin, for example, will polymerize more readily in the presence of short polymers created by fragmenting longer polymers. There are also proteins that serve to nucleate the polymerization. For actin, a well studied nucleating protein is gelsolin, which forms a tight complex with actin. In the presence of Ca^{2+}, the actin : gelsolin ratio is 2 : 1 (Refs 12, 13). This 2 : 1 complex is probably analogous to the naturally formed actin nucleus and in its presence the rate of the elongation process controls the overall rate of polymerization. The Ca^{2+} dependence of the formation of the gelsolin : actin complex and other effects[14] add other layers of possible control mechanisms to polymerization processes.

Ligand binding

If a ligand binds preferentially to a polymer species of specific molecular weight, then the presence of that ligand will increase the concentration of that species. Such ligand binding may be particularly important in enzyme systems where the kinetic properties of the polymer differ from those of the monomer. A number of enzyme systems have been examined[15] and two examples of such systems are acetyl CoA carboxylase and phosphofructokinase. In the former case, citrate (an activator) induces polymerization and activation of the enzyme[16]. In the latter case, a reversible dimer–tetramer reaction (a limited linear polymerization case) is influenced by substrates and allosteric effectors.The extent of dimer–tetramer reaction not only influences the allosteric properties of the enzyme but also the activity since the dimer slowly isomerizes to a form that is inactive[17]. However, as indicated in the previous section, ligand binding may affect polymerization indirectly via ligands interacting with other proteins that in turn influence the polymerization rate or extent.

Nucleotide hydrolysis or exchange

The systems which dominate the literature in this area are actin and tubulin[18]. As is well known, both proteins in their monomeric form (a single polypeptide chain for actin and a dimer for tubulin) contain a tightly bound nucleoside triphosphate (ATP in actin and GTP in tubulin) that is hydrolysed to the corresponding diphosphate during polymerization. The properties of the monomeric species differ depending on the state of the bound nucleotide in such a way as to be a controlling factor in the polymerization process. For example, the critical concentration is much larger and therefore the extent of polymerization less when the bound nucleotide (in both the polymer and monomer) is the diphosphate. This difference in critical concentration may be in part responsible for the dramatic effects observed in microtubule depolymerization. Microtubules undergo a drastic and rapid depolymerization (almost certainly related to the role of microtubules in the mitotic spindle) which can actually be visualized in video imaging experiments[19,20]. Amazingly, this process can occur in the same solution where other microtubules continue to polymerize (referred to as dynamic instability[21]). The process seems primarily related to the fact that the end of the microtubule polymer includes a cap of monomers in which the GTP has not been hydrolysed. Hydrolysis of the GTP in this cap appears to be a major factor in the catastrophic depolymerization but the process obviously depends on the rate of elongation relative to the rate of hydrolysis and rates of association and dissociation of monomers containing either GTP or GDP. Since the depolymerization process produces monomer that contains the hydrolysed nucleotide, the rate at which the nucleoside triphosphate displaces the diphosphate can also be an important factor in controlling the rate and extent of polymerization. A similar effect has not been observed for actin polymers, perhaps because the cap is much smaller. However, actin monomers containing ADP may also dissociate from the polymer and the rate of exchange of this ADP by ATP may still be an important factor in controlling the rate and extent of polymerization.

INTERACTIONS WITH OTHER PROTEINS

It would appear that accessory proteins that affect polymerization mostly interact with nucleated polymerization systems. I have already mentioned one protein, gelsolin, which enhances the rate of actin polymerization as a consequence of bypassing the nucleation step. Many other actin binding proteins exist[22] that can affect the polymerization or depolymerization process. For example, there are proteins which bind specifically to the actin monomer, others which bind specifically to one end of the polymer and others which may serve to crosslink polymers. These latter proteins probably influence depolymerization rates. As indicated above, the effect of some of these proteins may depend on other factors such as the Ca^{2+} concentration, other small molecules or the pH. Actin is not alone in this regard: accessory proteins have been found to affect the assembly of clathrin into polyhedral structures[23], there are a number of microtubule associated proteins (MAPs) which affect the rate and extent of tubulin polymerization[24] and other proteins are associated with intermediate filaments[25].

Phosphorylation–dephosphorylation

Certainly one of the dramatic developments in regulation is the role of phosphorylation and dephosphorylation in all aspects of metabolism. There have been recent reports that indicate that phosphorylation–dephosphorylation reactions may serve an important regulatory role in polymerizing systems as well, especially for that class of proteins called intermediate filaments. These fibrous proteins constitute an extremely heterogeneous multigene family which appear to have similar structures[25,26]. In general, these proteins were considered to be reasonably insoluble and rapid depolymerization at neutral pH had not been observed *in vitro*. There is now evidence, however, that phosphorylation can result in such a depolymerization for certain species of desmin[27], vimentin[28] and others of the intermediate filament family, although it is not clear yet whether this will be a general mechanism for regulating the depolymerization process.

In addition to phosphorylation directly affecting polymerization, there are other proteins whose interaction with polymerizing proteins is affected by phosphorylation. The extent of phosphorylation of some MAPs or actin binding proteins, for example, influences their interaction with microtubules or actin. These and many other systems remain to be characterized in detail.

Conclusion

In recent years it has become clear that freely reversible polymerization–depolymerization reactions, far from being just interesting systems for physical–chemical studies, are essential components of cellular functions. The most startling observation and impressive indication of this statement is the dynamic instability of microtubules. Currently, the mechanisms for the control of the polymerization processes are recognized as being important areas of investigation. Of these mechanisms, the role of protein phosphorylation and dephosphorylation is exceedingly important and will probably be the subject of the most intense study over the next few years.

Acknowledgement

Work of the author was supported by Grant DK 13332 from the National Institutes of Health.

References

1 Nichol, L. W. (1981) in *Protein–Protein Interactions* (Frieden, C. and Nichol, L., eds), pp. 2–17, Wiley

2 Jeffrey, P. D. (1981) in *Protein–Protein Interactions* (Frieden, C. and Nichol, L., eds), pp. 215–256, Wiley

3 Augusteyn, R. C. and Koretz, J. F. (1987) *FEBS Lett.* 222, 1–5

4 Brodsky, F. M. (1988) *Science* 242, 1396–1402

5 Pearse, B. M. F. and Crowther, R. A. (1987) *Annu. Rev. Biophys. Biophys. Chem.* 16, 49–68

6 Koren, R. and Hammes, G. G. (1976) *Biochemistry* 15, 1165–1170

7 Frieden, C. and Goddette, D. W. (1983) *Biochemistry* 22, 5836–5943

8 Frieden, C. (1983) *Proc. Natl Acad. Sci. USA* 80, 6513–6517

9 Zimmerle, C. T. and Frieden, C. (1988) *Biochemistry* 27, 7766–7772

10 Noguchi, C. T. and Schecter, A. N. (1985) *Annu. Rev. Biophys. Biophys. Chem.* 14, 239–263

11 Timasheff, S. N. and Grisham, L. M. (1980) *Annu. Rev. Biochem.* 49, 565–591
12 Bryan, J. and Kurth, M. C. (1984) *J. Biol. Chem.* 259, 7480–7487
13 Doi, Y. and Frieden, C. (1984) *J. Biol. Chem.* 259, 11868–11875
14 Yin, H. L., Hartwig, J. H., Maruyama, K. and Stossel, T. P. (1981) *J. Biol. Chem.* 256, 9693–9697
15 Frieden, C. (1971) *Annu. Rev. Biochem.* 40, 653–696
16 Vagelos, P. R., Alberts, A. W. and Martin, D. B. (1963) *J. Biol. Chem.* 238, 533–543
17 Pettigrew, D. W. and Frieden, C. (1979) *J. Biol. Chem.* 254, 1887–1895
18 Carlier, M. F. (1988) *Cell Biophysics* 12, 105–117
19 Schulze, E. and Kirschner, M. (1988) *Nature* 334, 356–359
20 Walker, R. A., O'Brien, E. T., Pryer, N. K., Soboeiro, M. F., Voter, W. A., Erickson, H. P. and Salmon, E. D. (1988) *J. Cell Biol.* 107, 1437–1448
21 Mitchison, T. J. and Kirschner, M. W. (1984) *Nature* 312, 232–237
22 Pollard, T. D. and Cooper, J. A. (1986) *Annu. Rev. Biochem.* 55, 987–1035
23 Keen, J. H., Willingham, M. C. and Pastan, I. H. (1979) *Cell* 16, 303–312
24 Olmsted, J. B. (1986) *Annu. Rev. Cell. Biol.* 2, 421–457
25 Steiner, P. M. and Roop, D. R. (1988) *Annu. Rev. Biochem.* 57, 593–625
26 Lazarides, E. (1982) *Annu. Rev. Biochem.* 51, 219–250
27 Inagaki, M., Gonda, Y., Matsuyama, M., Nishizawa, K., Nishi, Y. and Sato, C. (1988) *J. Biol. Chem.* 263, 5970–5978
28 Inagaki, M., Nishi, Y., Nishizawa, K., Matsuyama, M. and Sato, C. (1987) *Nature* 328, 649–652

C. Frieden is at the Department of Biochemistry and Molecular Biophysics, Washington University School of Medicine, St. Louis, MO 63110, USA.

Structural Basis of DNA – Protein Recognition

Richard G. Brennan and Brian W. Matthews

Recent structure determinations of several repressor–operator complexes have shown how proteins can recognize specific binding sites on DNA. Although each of these repressor proteins belongs to the 'helix-turn-helix' class of DNA-binding proteins, they do not use a simple code for recognition.

Protein recognition of specific DNA-binding sites is critical for the biological function of all cells. Transcription, replication and restriction are but a few of the cellular activities that are under the control of DNA-binding proteins. Over the past several years substantial progress has been made in elucidating the structural and molecular mechanisms by which proteins recognize and bind specific DNA sequences. In light of the recent structure determinations of several DNA-binding proteins and their nucleic acid complexes, it is timely to review the field (for other recent reviews see Brennan and Matthews[1], Schlief[2], Pabo and Sauer[3]). Although structures of proteins involved in replication (Klenow fragment[4,5]), restriction (*Eco*RI[6]) and DNA superstructure (DNA-binding protein II[7], histone octamer[8], histone complex[9]) are known, discussion here will be limited to proteins that are integral to the control of gene transcription.

Transcriptional regulatory proteins

Detailed structural insight into one of the mechanisms by which gene repressors and activators carry out their transcription regulatory roles was first provided by the X-ray structure determinations of the λ phage cro repressor protein[10] and the *E. coli* catabolite gene activator protein–cAMP complex[11]. These proteins are both active as dimers *in vivo*. Although very different in their overall three-dimensional structure, each protein displayed a very striking structural feature, namely, a pair of twofold related α-helices which protruded from the protein's surface. These helices were tilted and had a center-to-center separation of approximately 34 Å. Combining this structural information with the available biochemical data on cro–DNA binding, a model for the cro–O_R3 operator complex was proposed[10] in which the cro dimer was docked to B-form DNA with the twofold axis of the protein coincident with the twofold axis of the operator DNA. In this way the two α_3-helices of the cro dimer were located in successive major grooves and could thereby utilize their amino acid side chains for site-specific recognition and binding. The twofold related α_2-helices also were located near the DNA so that the N-terminal main chain and side chains were able to make nonspecific contacts with the phosphodiester backbone (Fig. 1). It was observed that the α_E and α_F helices of the CAP structure were strikingly similar to the α_2 and α_3 helices of cro[12] and subsequently proposed that these bihelical motifs play analogous roles in the two structures[13]. The functionality of this

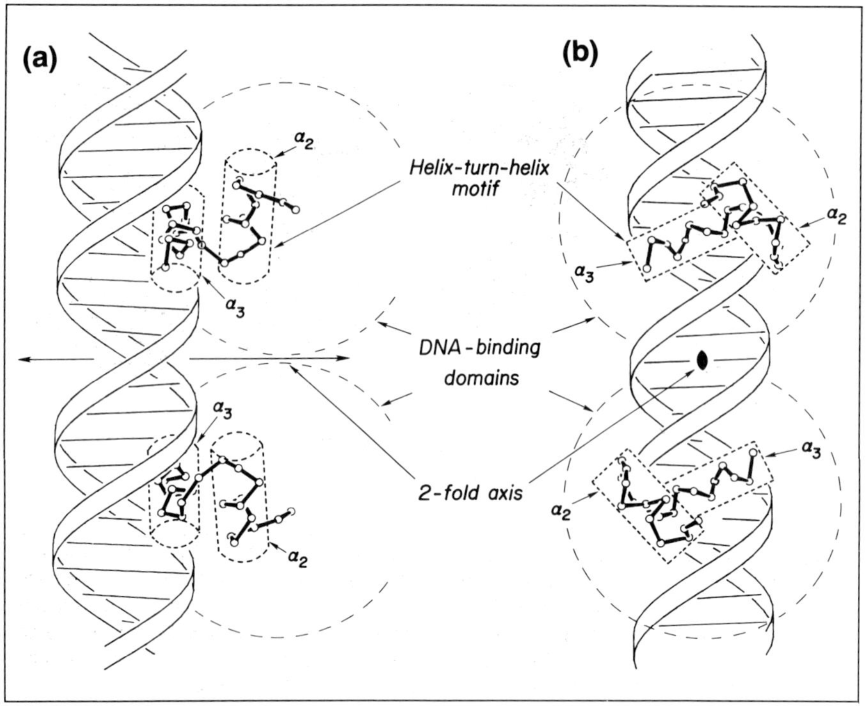

Figure 1

The figure illustrates the general nature of the interaction now known to occur in many DNA-regulatory proteins between a common α_2–α_3 or 'helix-turn-helix' helical unit and right-handed B-form DNA; (**a**) is a 'side view' with the twofold axis of symmetry (arrowed) extending from left to right, and (**b**) shows the view 'face on' (after Ref. 23).

DNA-binding motif was further corroborated by the structure determination of the λ repressor headpiece[14,15].

The common DNA-binding structure is now known as the helix-turn-helix motif (Fig. 2). A matrix of contacts between the helix-turn-helix and the DNA, including hydrogen bonds, salt bridges and van der Waals contacts has been proposed to account for site recognition and binding specificity for each of the above complexes[13,14,16,17]. Additional nonspecific interactions for cro and λ repressor have been attributed to floppy C- (residues 62–66) and N- (residues 1–5) terminal arms, respectively, which in the cro–O_R3 model complex are proposed to interact with the phosphodiester backbone of the minor groove[16] and in the λ repressor–O_R1 model complex to wrap around the DNA and make contacts in the major groove 'behind' the helix-turn-helix binding site[14]. Further, modeling studies and electrostatic calculations[16,18] indicated that cro could make a tighter complex with its recognition site if the DNA were slightly bent (Fig. 3) or if the protein underwent 'hinge-bending' motion.

That the helix-turn-helix DNA-binding motif is common among many gene regulatory proteins has been borne out by additional structural and amino acid sequence homology studies[1,3,19–26]. In particular, the crystal structure determination of the *E. coli* trp repressor complexed with its corepressor L-tryptophan[27] and the cro repressor protein[28] and repressor headpiece[29] from bacteriophage 434 have revealed DNA-binding substructures, as predicted[19,21,23], as has the NMR structure determination of the *lac* repressor headpiece[20,30]. A recent NMR study of the homeodomain from *Antennapedia* has also suggested a similar motif[31]. The close structural correspondence between the helix-turn-helix motifs in different DNA-binding proteins is highlighted by calculating the root-mean-square (rms) discrepancy between the 21 α-carbons in the helix-turn-helix of the subject protein and that of cro. None shows an rms difference of more than 0.9 Å (Table I)[1,32]. It should be emphasized that the presence of a helix-turn-helix motif in a given protein structure is no guarantee that it is employed in DNA binding. A case in point is the non-specific DNA-binding protein, DNA-binding protein II from *Bacillus stearothermophilus*, the structure of which reveals a helix-turn-helix motif (Table I). Yet a different DNA-binding mechanism, involving the β-sheet structure of this protein, has been invoked[7]. Backbone structures that correspond quite well to the helix-turn-helix motif in cro have also been located in cytochrome *c* peroxidase and the L7/L12 ribosomal protein (Table I)[1,32]. However, none of the helix-turn-helix motifs of these latter three proteins agrees as well with cro as the helix-turn-helix units from the proteins that are known to be involved in transcriptional regulation (Table I). Furthermore, their amino acid sequences do not correspond clearly to those observed in known DNA-binding proteins. It is quite possible that small structural differences may be significant in determining the functional role of the motif. Additionally, the function of the helix-turn-helix may be affected profoundly by the presence or absence of effector molecules. Sigler's studies of *trp* repressor have shown that the active, ligand-bound form of the repressor dimer retains its helix-turn-helix motifs in the correct alignment for DNA binding. On removal of ligand, however, the relative positions of the twofold-related helix-turn-helix motifs are changed in such a way that they can no longer match the shape of the DNA. The L-tryptophan co-repressor stabilizes the conformation that allows the 'flexible reading heads' to bind to their DNA recognition sites[33]. The crystal structures of additional ligand binding, gene regulatory proteins, such as the biotin operon repressor, *birA*[34] and met repressor[35], will be of great interest to see if similar ligand induced conformational changes are a common mechanism in the regulation of DNA binding.

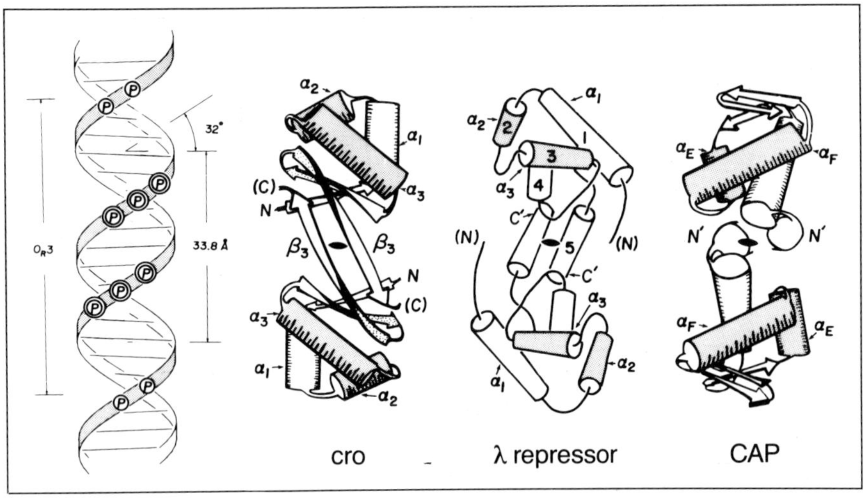

FIGURE 2

Schematic drawing of a segment of Watson–Crick B-form DNA together with dimers of uncomplexed cro, λ-repressor N-terminal domains, and CAP C-terminal domains viewed down their respective twofold symmetry axes. The corresponding α_2–α_3 (or α_E–α_F) helices are labeled. DNA phosphates whose ethylation interferes with binding of both λ-repressor and cro are indicated by the letter P within a double circle. Phosphates whose ethylation effects λ-repressor binding, but not cro, are indicated by a P in a single circle (after Ref. 17).

TABLE I

Structural correspondence between helix-turn-helix motifs

Protein	Residues	Root mean square discrepancy[a] (Å)
λ cro	16–36	–
Phage 434 repressor	17–37	0.43
Phage 434 cro	19–39	0.56
λ Repressor	33–53	0.68
Trp repressor	64–84	0.89
CAP	169–189	0.90
Cytochrome *c* peroxidase	132–152	1.39
DNA-binding protein II	7–27	1.43
L7/L12 ribosomal protein	16–36	1.61

[a]The table gives the root mean square discrepancy between 21 α-carbons in the helix-turn-helix of the subject protein and that of cro repressor.

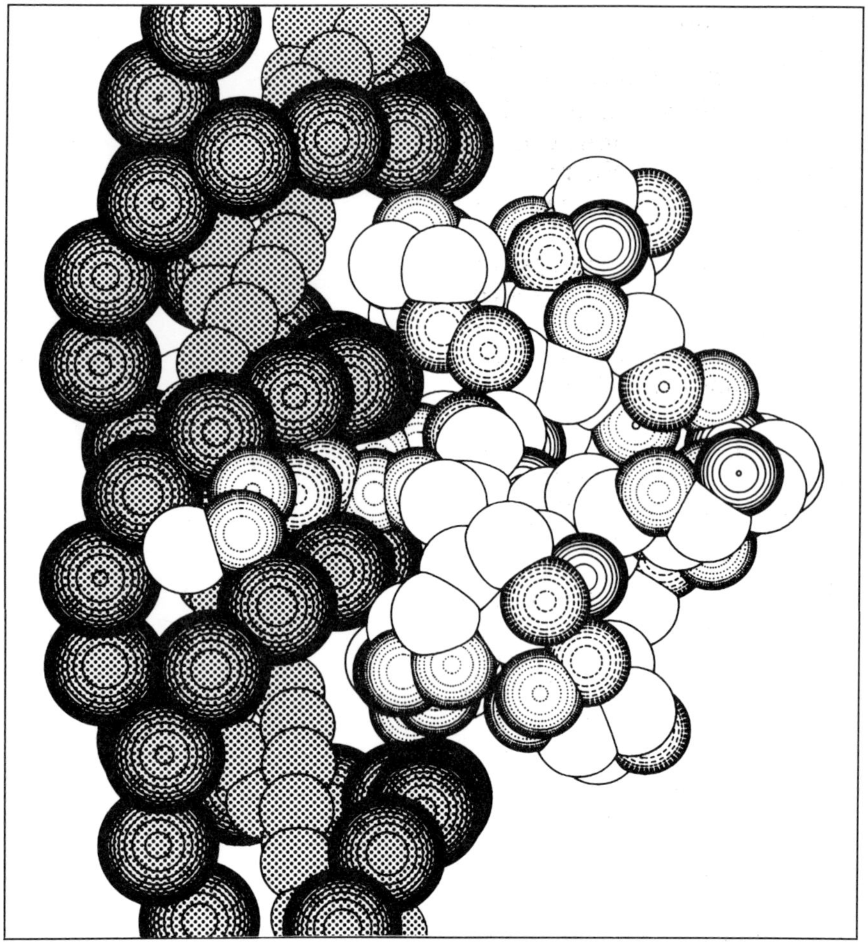

FIGURE 3

Stylized drawing showing the complementarity between the structure of cro-repressor protein and DNA. The DNA is represented by large dotted spheres centered at the phosphate positions, and small dotted spheres that follow the bottom of the major groove. In the protein, each residue is represented by a single sphere. Acidic residues have solid concentric circle shading, basic residues have broken circle shading, uncharged hydrophilic residues have dotted circle shading and hydrophobic residues have no shading. This drawing was based on model building using, as a basis, the structure of cro repressor[10,16]. The overall features of the complex have recently been confirmed by the crystal structure determination of a cro–operator complex.

PROTEIN–DNA COMPLEXES

Model building, based on the known structures of transcriptional regulatory proteins, suggested that DNA–protein recognition could be attributed to specific interactions between the amino acid side chains of the recognition helix of protein and the exposed major groove functional groups of the DNA recognition site. However, to ascertain the accuracy of these proposed interactions it was clear that the structures of appropriate protein–DNA complexes had to be obtained. To date, the crystal structures of five transcriptional regulatory protein–DNA complexes[36–40] and one NMR-determined complex[41,42] have been solved. In general, these structures strongly support the features that were anticipated in the proposed protein–DNA model complexes. (1) The twofold symmetry axis of each protein is coincident with the twofold axis of its DNA operator fragment. (2) The operator fragment assumes an essentially B-DNA conformation, although deformation is observed in some of the complexes. (3) The recognition helices of the proteins are located in consecutive major grooves of the DNA. Some highlights of each complex follow.

The first repressor–operator complex to be determined was that of the 434 repressor headpiece bound to an O_R1 operator fragment. A moderate resolution structure was obtained using a 'blunt-end' 14 base pair fragment[36] and a higher resolution structure resulted from the use of a 'sticky-end' 20-mer peptide[37]. Since there are no major conformational differences between the two structures, discussion will be confined to the higher resolution structure. The repressor protein undergoes very little conformational change upon binding to the operator. Recognition and binding is the result of direct interactions between the amino acid side chains of the α_3-helix and the base pairs in the major groove. Specific contacts are made by the side chains of Glu28, -29 and -33, which hydrogen bond to base pairs A·T (no. 1), G·C (no. 2) and A·T (no. 4), respectively (see Ref. 37 for numbering scheme), and by Thr27 and Glu29, which are involved in van der Waals interactions with the methyl group of base pair A·T (no. 3). Most striking are the repressor induced effects on the DNA. These include bending to a radius of curvature of 65 Å, and the overwinding of the middle four base pairs and compression of the minor groove in the center of the operator fragment such that the phosphate to phosphate distance is reduced from 11.5 Å for canonical B-DNA to 8.8 Å in the complex. As a further result, the middle A·T base pairs participate in non-Watson–Crick bifurcated hydrogen bonds, which facilitate binding of the 434 repressor to the O_R1 site.

In contrast to these results, the DNA in the 434 cro repressor–14 base pair operator complex is straight, even though the sequence is identical to that of the 434 repressor–14-mer complex[38]. Nevertheless, the series of specific protein–DNA contacts to the first four base pairs (ACAA) are identical to those made by the 434 repressor headpiece. The three-dimensional backbone structures of the 434 cro and 434 repressor headpiece monomers, both of which are composed of five α-helices, are essentially identical and can be superimposed with an rms discrepancy of 0.77 Å. Like the 434 repressor, the 434 cro monomer does not change its conformation upon binding. In order to maintain the same sequence-specific contacts the 434 cro *dimer* assumes a different relative orientation from that taken by the *dimeric* 434 repressor headpiece. This difference arises from the different monomer–monomer contacts. The 434 cro dimer is stabilized by hydrophobic contacts made between the bulky side chains of Phe46 (helix α_4) and Tyr61 (helix $\alpha_{5'}$). The 434 repressor dimer, on the other hand, is more tightly packed as Pro46 (helix α_4) to Leu61 (helix $\alpha_{5'}$) contacts allow a closer approach of these helices.

The structure determination of the complex formed by the λ repressor headpiece and a 20 base pair O_L1 operator fragment also reveals a similar 'direct readout' recognition mechanism[39]. Here again, the protein structure does not undergo significant conformational changes upon binding to DNA. In the complex the DNA is essentially straight B-form without bifurcated bonds. The contacts made between the α_3 recognition helix and the base pairs are, for the most part, as predicted by model building studies[3,15], although there are interesting side chain–side chain and side chain–DNA interactions that were not foreseen. In recognizing base pair A·T (no. 2), Gln44 forms two hydrogen bonds with the adenine base. Unexpectedly, this contact is further stabilized by a hydrogen bond between the side chains of Gln44 and Gln33, which in turn is positionally anchored by hydrogen bonding to a nearby phosphate oxygen. Additional multidentate hydrogen bonds, such as those from Asn55 and Lys4 to base pair G·C (no. 6), demonstrate the need for high-resolution structures in order to define the interactions of a given DNA–protein complex with rigor. An additional feature of the λ repressor–O_R1 complex is the nonspecific contacts made by the N-terminal arms. As predicted these extensions wrap around the DNA and form hydrogen bonds with the major groove bases on the opposite side of the DNA helix.

Recently, the structure of a λ cro-17 base pair consensus operator fragment has been solved to moderate resolution (Brennan, R. G., Roderick, S. L., Takeda, Y. and Matthews, B. W., unpublished). In common with the above repressor–operator complexes, the recognition helix (α_3) lies within the major groove, and the side chains of this helix contact the exposed edges of the DNA bases. Unlike some of the other repressor–operator complexes, but as seen in the 434 repressor headpiece–O_R1 and *trp* repressor–18-mer complexes (see below), the blunt-end 17 base pair operator fragment is bent. Furthermore, the DNA in the middle is overwound and the minor groove somewhat compressed. These latter characteristics were unanticipated since the middle seven base pairs are all 'less flexible' C·G pairs. Particularly striking in the λ cro–17-mer complex is the conformational change that the cro dimer undergoes upon binding DNA. The cro dimer is held together by hydrophobic contacts between the two symmetry related β_3-strands of each monomer (Fig. 2). It is the only repressor complex of known structure that contains any β-sheet structure. Upon binding DNA, the β_3-strands twist such that the monomers, including the helix-turn-helix substructures are rotated by more than 35° with respect to each other. Although this quaternary change is dramatic, it should be emphasized that for the most part, the three-dimensional structures of the complexed and uncomplexed *monomers* are the same.

Unexpected binding modes of the helix-turn-helix

The most unexpected structure of a DNA–protein complex described to date is that of the *trp* repressor bound to an 18 base-pair *trp* operator consensus sequence[40]. In this complex, which has been refined to very high resolution, the repressor structure is relatively unchanged from its unbound state. The largest shifts are located in the N-terminus and the α-helices D and E, the helices involved in DNA binding. However, these shifts are all less than 2 Å when compared with the uncomplexed repressor. The N-terminal arms (residues 2–14), which have been shown to enhance *trp* repressor's binding to both specific and nonspecific DNA[43], are poorly ordered in the structure. The DNA fragment displays only Watson–Crick hydrogen bonding and is bent towards its ends, the cumulative effect of small local distortions. The complex displays an extensive protein–DNA contact surface in which 2900 $Å^2$ is inaccessible to bulk solvent at this interface. Furthermore, there is extensive hydrogen bonding, which includes both direct side chain–DNA and water mediated protein–DNA bonds. What is so unusual about this complex is the complete *lack of direct* interactions between the side chains of the recognition helix and the base pairs in the major groove of the DNA.

Sigler *et al.*[40] have argued that the operator sequence is read 'indirectly' by a combination of protein–water–nucleic acid hydrogen bonds and sequence-dependent conformational effects on the DNA phosphate backbone. The water molecules can be considered as an 'extension' of the hydrogen-bonding surface of the protein and, in this sense, the proposed model of recognition could be regarded as a variation on that adopted by the λ- and 434-cro and repressor proteins. On the other hand, the conformation changes in the DNA structure are not large, and it is not clear that they would be sufficient to explain sequence-specific recognition by *trp* repressor.

One possibility that needs to be considered is that the structure is that of a nonspecific complex. The conditions of crystallization, i.e. low salt and high alcohol, might favor the formation of a nonspecific complex. Also, the observed complex seems to be at variance with *in vivo* binding studies of mutant *trp* repressors by Youderian *et al.*[44] which implicate residue Thr81 as key to trp repressor's specific recognition of base pairs A·T (no. 5) and G·C (no. 6). In the observed complex, Thr81 makes a hydrogen bond to a phosphate group and does not appear to be able to play such a role. Clearly, additional biochemical, genetic and structural studies are desirable to clarify the binding mechanism of this repressor–operator complex.

The structures of repressor–operator complexes already determined by X-ray crystallography make it clear that the helix-turn-helix motif is used in somewhat different ways by different DNA-binding proteins. Recent ^{1}H NMR studies by Kaptein *et al.*[41,42] on complexes between *lac* repressor headpiece and 14 and 22 base pair operators have suggested even more dramatic variations. In this case the orientation of the recognition helix appears to be *reversed* in the major groove as compared to the recognition helices of the aforementioned repressors. That is to say, the N-terminus of the recognition helix is closer to the operator's center of dyad symmetry. These authors suggest that several other gene regulatory proteins, including galR, deoR and CAP, bind in a similar reverse manner.

NON-SPECIFIC PROTEIN–DNA CONTACTS

Although the emphasis in this review has been on specific protein–DNA contacts, nonspecific contacts are energetically critical and cannot be ignored (see reviews by von Hippel[45,46]). In a number of cases, these non-sequence-specific contacts appear to be mediated by flexible 'arms' and loops of the repressor proteins[39,40]. One such contact, however, deserves special attention as it is common to all five crystallographic complexes. In these repressors, the first helix of the helix-turn-helix begins with a glutamine. In each case the *peptide* nitrogen of this glutamine is found to make a hydrogen bond to the first phosphate group of the recognition sequence. Furthermore, each of these helices is aligned such that its N-terminus is pointing toward the negative DNA phosphate backbone, thereby satisfying the positive charge of the α-helix dipole. Such 'nonspecific' contacts not only contribute to the binding energy of the specific complex but presumably also play a role in the alignment of the recognition helix in the major groove.

FUTURE PROSPECTS

Although the determination of several sequence-specific DNA–protein complexes has clearly demonstrated how proteins can recognize sequence-specific sites on DNA, there are many questions that remain unanswered. Many transcriptional regulatory proteins are now known to contain similar helix-turn-helix DNA-binding motifs, but do not use their helix-turn-helix motifs in exactly the same way. This is true even for proteins that recognize the same DNA operator. Thus, there is no simple recognition code whereby a given base pair on the DNA is recognized by a given type or combination of amino acids on the protein. Rather, DNA–protein recognition is seen to involve a complex network of interdependent hydrogen-bonding, van der Waals and ionic interactions. This complexity makes it difficult to contemplate the synthesis of novel repressor proteins designed to recognize specified DNA target sequences.

While much of the structural data to date relates to the helix-turn-helix class, details of other novel DNA-binding motifs should be forthcoming shortly. The phage P22 ARC repressor, which has been shown to use its N-terminal amino acids for binding, and seems not to contain a helix-turn-helix motif, is now under structural study[47,48]. Also other DNA-recognition motifs including 'zinc fingers'[49,50] and 'leucine zippers'[51,52] are being actively pursued (see *TIBS* 14, 137–140 for a recent review). Knowledge of the modes of DNA-recognition that are used by these other classes of DNA-binding proteins are eagerly awaited.

REFERENCES

1 Brennan, R. G. and Matthews, B. W. (1989) *J. Biol. Chem.* 264, 1903–1906
2 Schlief, R. (1988) *Science* 241, 1182–1187
3 Pabo, C. O. and Sauer, R. T. (1984) *Annu. Rev. Biochem.* 53, 293–321
4 Ollis, D. L., Brick, P., Hamlin, R., Xuong, N. G. and Steitz, T. A. (1985) *Nature* 313, 762–766
5 Freemont, P. S., Friedman, J. M., Beese, L. S., Sanderson, M. R. and Steitz, T. A. (1988) *Proc. Natl Acad. Sci. USA* 85, 8924–8928
6 Frederick, C. A., Grable, J., Melin, M., Sanozi, C., Jen-Jacobson, L., Wang, B. C., Greene, P., Boyer, H. W. and Rosenberg, J. M. (1984) *Nature* 309, 327–331
7 Tanaka, I., Appelt, K., Dijk, J., White, S. W. and Wilson, K. S. (1984) *Nature* 310, 376–381
8 Richmond, T. J., Finch, J. T., Rushton, B., Rhodes, D. and Klug, A. (1984) *Nature* 311, 532–537
9 Burlingame, R. W., Love, W. E., Wang, B-C., Hamlin, R., Xuong, N-H. and Moundrianakis, E. N. (1985) *Science* 228, 546–553
10 Anderson, W. F., Ohlendorf, D. H., Takeda, Y. and Matthews, B. W. (1981) *Nature* 290, 754–758
11 McKay, D. B. and Steitz, T. A. (1981) *Nature* 290, 744–749
12 Steitz, T. A., Ohlendorf, D. H., McKay, D. B., Anderson, W. F. and Matthews, B. W. (1982) *Proc. Natl Acad. Sci. USA* 79, 3097–3100
13 Weber, I. T. and Steitz, T. A. (1984) *Proc. Natl Acad. Sci. USA* 81, 3973–3977
14 Pabo, C. O. and Lewis, M. (1982) *Nature* 298, 443–447
15 Lewis, M., Jeffrey, A., Wang, J., Ladner, R., Ptashne, M. and Pabo, C. O. (1983) *Cold Spring Harbor Symp. Quant. Biol.* 47, 435–440
16 Ohlendorf, D. H., Anderson, W. F., Fisher, R. G., Takeda, Y. and Matthews, B. W. (1982) *Nature* 298, 718–723
17 Takeda, Y., Ohlendorf, D. H., Anderson, W. F. and Matthews, B. W. (1983) *Science* 221, 1020–1026

18 Ohlendorf, D. H. and Matthew, J. B. (1985) *J. Biol. Chem.* 260, 5860–5862
19 Anderson, W. F., Takeda, Y., Ohlendorf, D. H. and Matthews, B. W. (1982) *J. Mol. Biol.* 159, 745–751
20 Matthews, B. W., Ohlendorf, D. H., Anderson, W. F. and Takeda, Y. (1982) *Proc. Natl Acad. Sci. USA* 79, 1428–1432
21 Sauer, R. T., Yocum, R. R., Doolittle, R. F., Lewis, M. and Pabo, C. O. (1982) *Nature* 298, 447–451
22 Weber, I. T., McKay, D. B. and Steitz, T. A. (1982) *Nucleic Acids Res.* 10, 5085–5102
23 Ohlendorf, D. H., Anderson, W. F. and Matthews, B. W. (1983) *J. Mol. Evol.* 19, 109–114
24 Brennan, R. G., Weaver, L. H. and Matthews, B. W. (1986) *Chemica Scripta* 26B, 251–255
25 Dodd, I. B. and Egan, J. B. (1987) *J. Mol. Biol.* 194, 557–564
26 White, S. W. (1987) *Protein Eng.* 1, 373–376
27 Schevitz, R. W., Otwinowski, Z., Joachimiak, A., Lawson, C. L. and Sigler, P. B. (1985) *Nature* 317, 782–786
28 Mondragon, A., Wolberger, C. and Harrison, S. C. (1989) *J. Mol. Biol.* 205, 179–188
29 Mondragon, A., Subbiah, S., Almo, S. C., Drottar, M. and Harrison, S. C. (1989) *J. Mol. Biol.* 205, 189–200
30 Zuiderweg, E. R. P., Kaptein, R. and Wüthrich, K. (1983) *Proc. Natl Acad. Sci. USA* 80, 5837–5841
31 Otting, G., Qian, Y-Q., Müller, M., Affolter, M., Gehring, W. and Wüthrich, K. (1988) *EMBO J.* 7, 4305–4309
32 Richardson, J. S. and Richardson, D. C. (1988) *Proteins: Structure, Function and Genetics* 4, 229–239
33 Zhang, R-G., Joachimiak, A., Lawson, C. L., Shevitz, R. W., Otwinowski, Z. and Sigler, P. B. (1987) *Nature* 327, 591–597
34 Brennan, R. G., Sanjay, V., Matthews, B. W. and Otsuka, A. O. (1989) *J. Biol. Chem.* 264, 5
35 Rafferty, J. B., Phillips, S. E. V., Rojas, C., Tsoulot, G., Saints-Girons, I., Guillon, Y. and Cohen, G. N. (1988) *J. Mol. Biol.* 200, 217–219
36 Anderson, J. E., Ptashne, M. and Harrison, S. C. (1987) *Nature* 326, 846–852
37 Aggarwal, A. K., Rodgers, D., Drottar, M., Ptashne, M. and Harrison, S. C. (1988) *Science* 242, 899–907
38 Wolberger, C., Dong, Y., Ptashne, M. and Harrison, S. C. (1988) *Nature* 335, 789–795
39 Jordan, S. R. and Pabo, C. O. (1988) *Science* 242, 893–899
40 Otwinowski, R. W., Schevitz, R. W., Zhang, R-G., Lawson, C. L., Joachimiak, A., Marmorstein, R. Q., Luisi, B. F. and Sigler, P. B. (1988) *Nature* 335, 321–329
41 Boelens, R., Scheek, R. M., van Boom, J. H. and Kaptein, R. (1987) *J. Mol. Biol.* 193, 213–216
42 Lamerichs, R. M. J. N., Boelens, R., van der Marel, G. A., van Boom, J. H., Kaptein, R., Buck, F., Fera, B. and Rüterjans, H. (1989) *Biochemistry* 28, 2985–2991
43 Carey, J. (1989) *J. Biol. Chem.* 264, 1941–1945
44 Bass, S., Sorrells, V. and Youderian, P. (1988) *Science* 242, 240–245
45 von Hippel, P. H. and Berg, O. G. (1987) in *DNA–Ligand Interactions* (Guschlbauer, W. and Saenger, W., eds), pp. 159–171, Plenum Press
46 Berg, O. G. and von Hippel, P. H. (1988) *Trends Biochem. Sci.* 13, 207–211
47 Knight, K. L. and Sauer, R. T. (1989) *Proc. Natl Acad. Sci. USA* 86, 797–801
48 Jordan, S. R., Pabo, C. O., Vershon, A. K. and Sauer, R. T. (1985) *J. Mol. Biol.* 185, 445–446
49 Brown, R. S., Sander, C. and Argos, P. (1985) *FEBS Lett.* 186, 271–274
50 Berg, J. M. (1988) *Proc. Natl Acad. Sci. USA* 85, 99–102
51 Landshulz, W. H., Johnson, P. F. and McKnight, S. L. (1988) *Science* 240, 1759–1764
52 O'Shea, E. K., Rutkowski, R. and Kim, P. S. (1989) *Science* 243, 538–542

R. G. Brennan and B. W. Matthews are at the Institute of Molecular Biology, University of Oregon, Eugene, OR 97403, USA.

Helix-turn-helix, Zinc-finger, and Leucine-zipper Motifs for Eukaryotic Transcriptional Regulatory Proteins

Kevin Struhl

Four distinct structural motifs have been proposed for the DNA-binding domains of eukaryotic transcriptional regulatory proteins; the helix-turn-helix, two kinds of zinc finger, and the leucine zipper. Within each structural motif, there are often families of related proteins that recognize similar DNA sequences and are conserved throughout the eukaryotic kingdom. However, the processes of transcriptional activation and repression appear to be independent of the specific type of protein–DNA interaction.

Eukaryotic genes are regulated differentially in response to a complex set of environmental and developmental cues. In terms of transcriptional regulation, eukaryotic promoters are large, complex arrangements of short DNA sequences that are recognized by a wide variety of specific DNA-binding proteins that activate or repress transcription. The distinct transcriptional regulatory patterns of individual genes or sets of genes are determined primarily by the specific protein–DNA interactions that occur at the promoters.

Structural and functional analyses of eukaryotic DNA-binding proteins indicate that small autonomous domains containing less than 100 amino acid residues are sufficient for specific DNA-binding activity. Although detailed structural information is not yet available, the primary sequences and biochemical properties of various DNA-binding domains are suggestive of at least four basic structural motifs. These motifs are probably important for the overall structure of the DNA-binding domain rather than being directly involved in the specific contacts between protein and DNA, because different proteins containing a particular structural motif can recognize a variety of DNA sequences.

The helix-turn-helix

The first and by far the best characterized structural motif for a DNA-binding domain is the helix-turn-helix found in prokaryotic activator and repressor proteins (reviewed in Ref. 1). X-ray crystallographic analysis of several proteins and protein–DNA complexes have elucidated the structures in detail, and functionally important interactions have been uncovered by analysing mutant proteins and target DNA sites. As is obvious from the name, the crucial structure in this class of proteins contains two α-helices that are separated by a β-turn. Despite considerable sequence variability, the helix-turn-helix motifs have a highly conserved geometry. Amino acids within one of these helices, the so-called 'recognition helix', directly contact bases exposed in the major groove of the target DNA; the other α-helix lies across the major groove and makes some non-specific contacts to DNA (Fig. 1). The overall structural homology among helix-turn-helix motifs

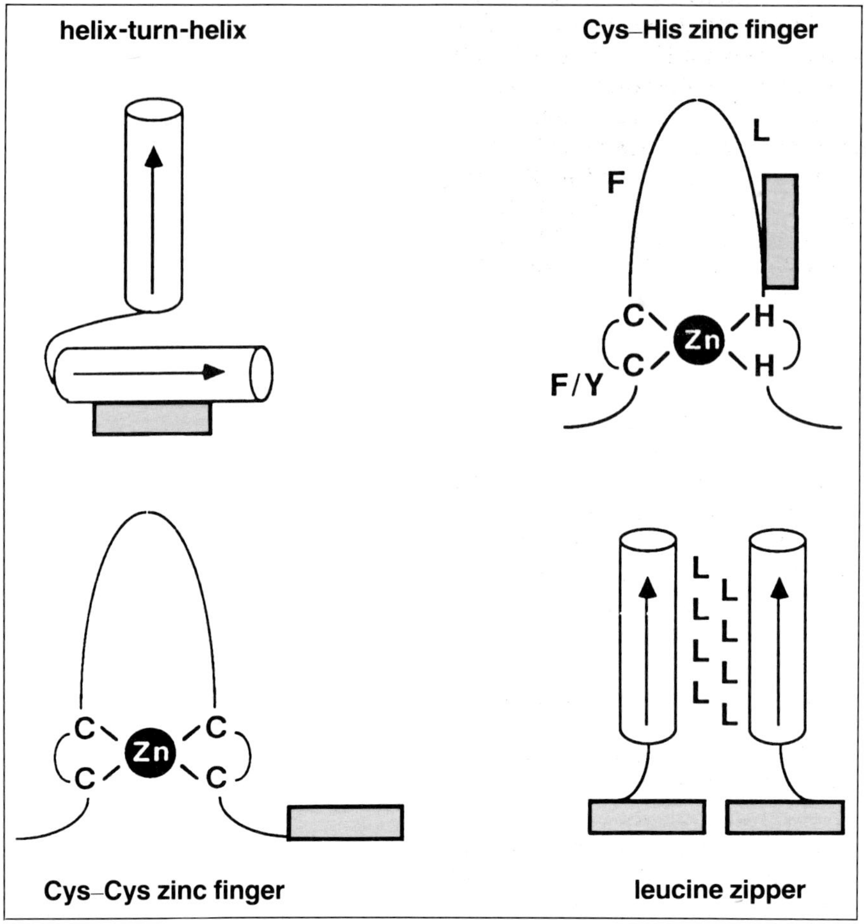

Figure 1

Schematic representation of the four structural motifs described in the text. α-Helices are represented by cylinders with arrows indicating directionality; conserved amino acid residues are shown in one-letter code, and zinc ions shown as black circles. The shaded boxes indicate the regions of the proteins proposed to be involved in specific contacts to DNA.

makes it possible in some cases to change the specificity of the protein–DNA interaction by altering or swapping recognition helices[2]. The prokaryotic helix-turn-helix proteins bind as dimers to DNA sequences that have dyad symmetric character.

In eukaryotic organisms, the helix-turn-helix motif was first invoked for a family of *Drosophila* proteins that control many of the key decisions in early development (reviewed in Refs 3, 4). These proteins contain a highly conserved 'homeodomain' about 60 amino acids in length within which lies a region with sequences common to many prokaryotic helix-turn-helix proteins. Homeodomains have now been found in a wide variety of eukaryotic organisms ranging from yeast to human, and more than 80 such proteins have already been identified. With the exception of the more diverged yeast proteins, different homeodomains show 40–90% amino acid sequence identity and can be classified into distinct subtypes based on the extent of homology. The subtypes often differ in their putative recognition helices, suggesting that they recognize different DNA sequences. Although there is no direct evidence that homeodomains contain helix-turn-helix motifs, it has been shown in several cases that the proteins specifically bind DNA *in vitro*. In cases where the functions of these proteins are known or inferred, homeodomains are frequently associated with proteins that control cell fate during development. (See *TIBS* 14, 52–56 [1989] for a review of vertebrate homeodomain proteins.)

The yeast MATα2 protein, which regulates cell type by binding to operator sequences and repressing transcription, is the best characterized eukaryotic protein with a putative helix-turn-helix motif[5]. The unusually large (30 bp) α2 operators have highly conserved sequences at both ends with an approximate twofold symmetry, but lack sequence similarities in the middle. In accord with this unusual arrangement, α2 dimers directly interact with half-sites at each end of the operator but do not contact the center of the operator. The protein (α2) is remarkably flexible as it can bind to an operator with the central 13 bp deleted such that the half-sites are immediately adjacent and located on the opposite side of the DNA helix. Like the prokaryotic proteins, α2 contains two domains; a C-terminal homeodomain that interacts specifically with operator DNA although with a reduced affinity compared to the full length protein, and an N-terminal region that does not bind DNA but facilitates dimerization. Amino acid substitutions within the α2 homeo-like region abolish function, presumably by virtue of a loss of DNA-binding activity[6].

THE ZINC FINGER

The unusual sequence of TFIIIA, a protein required for transcription of the 5S RNA genes by RNA polymerase III, led to the proposal of a distinct DNA-binding motif, the zinc finger[7]. TFIIIA contains 7–11 zinc atoms per molecule, and it is composed of nine repeating units of approximately 30 residues. Each unit contains two invariant pairs of cysteines and histidines as well as other conserved amino acids. In the original zinc-finger model, the cysteine and histidine pairs serve as a tetrahedral coordination site for a single zinc ion, and the amino acids between these coordination sites project out as fingers[7] (Fig. 1). (See also *TIBS* 12, 464–469 [1987] for review.) A different model, which invokes an antiparallel β-sheet and α-helix between the zinc coordination sites, has been proposed by analogy with structures of other metalloproteins[8]. Deletion analysis of both TFIIIA protein and its large target site within the 5S RNA structural gene indicate that the zinc fingers contribute to specific DNA binding in a modular fashion, and can be viewed as being arranged in a linear array along the target sequence[9,10]. However, individual fingers do not interact precisely with adjacent nucleotides, but rather clusters of fingers bind to the three functionally important regions of the binding site.

DNA sequence analysis of cloned genes indicates that more than 15 proteins from a variety of eukaryotic organisms from yeast to man contain regions that strongly resemble the putative zinc-finger motif of TFIIIA (reviewed in Ref. 11). Some of these proteins are known to be transcriptional activators, while others have presumptive roles in development or in sex determination. Depending on the protein, the number of potential zinc fingers ranges from two to more than ten. In several cases, it has been demonstrated that these proteins bind specifically to DNA, and that both the zinc-finger region and zinc are necessary for binding. In both models, specific DNA contacts are proposed to occur with residues in the finger regions, possibly the putative α-helix[7,8]. However, a synthetic peptide containing the two zinc fingers of yeast ADR1 is unable to recognize its target site, even though it forms a discrete structure in the presence of zinc that can interact non-specifically with DNA[12]. This suggests that the zinc finger has an essential structural role for a functional DNA-binding domain, but may not always be directly involved in protein–DNA interactions.

There is a related, but distinct, class of specific DNA-binding proteins, exemplified by the yeast GAL4 transcriptional activator and mammalian steroid receptors, in which putative zinc fingers are formed by coordination with four cysteine residues; replacement of one cysteine pair by a histidine pair abolishes DNA-binding activity (reviewed in Ref. 11) (Fig. 1). In the case of GAL4, an important role for zinc is inferred from mutants that function only at high zinc concentrations *in vitro* and *in vivo*[13]. Finger swapping experiments suggest that the all cysteine zinc fingers, though essential for DNA-binding, are not important for specificity. For example, the yeast PPR1 zinc finger can functionally replace the GAL4 zinc finger, but the resulting chimeric protein still binds to GAL4 recognition sites (S. Johnston, pers. commun.). Similarly, binding specificities of the glucocorticoid and thyroid receptors can be interchanged by replacing short regions between the two zinc fingers (R. Evans, pers. commun.). Thus, as may be the case for the TFIIIA-like proteins, it is likely that the zinc finger is important for maintaining overall structure and perhaps non-specific interactions to DNA, whereas distinct sequences are involved in direct contact to DNA.

The leucine zipper

Recently, a new class of proteins has been identified, and it has been proposed that they utilize a novel motif for DNA binding, the leucine zipper[14]. These proteins, which include the yeast GCN4 transcriptional activator, the jun, fos and myc oncoproteins, and the C/EBP enhancer binding protein, all contain four or five leucine residues that are spaced exactly seven residues apart and hence could be viewed as being repeated every two turns of an α-helix. In the initial structural model, it was proposed that the leucine residues are important for interdigitating two α-helices, one from each monomer unit, that provide the structural basis for the dimer formation[14] (Fig. 1).

Consistent with this model, GCN4, C/EBP, and jun bind as dimers to their target sites, and fos and jun can form DNA-binding heterodimers[14–18]; in the case of GCN4, a 60-residue region containing the putative leucine zipper is fully competent for dimer formation and specific DNA-binding. Moreover, a synthetic peptide corresponding to the GCN4 leucine zipper forms stable dimers that are essentially 100% α-helical[19]. Some problems with the original leucine zipper model are that the α-helices are parallel[19], and that a variety of mutations of the conserved leucines in GCN4 and fos have little if any functional effect[18] (Sellers and Struhl, unpublished). It has been suggested that the leucine zipper may indeed be a more conventional coiled coil structure[19].

Although the leucine zipper is probably important for dimer formation, it is likely that sequences outside this region are involved in DNA interactions[14]. (1) By analogy with zinc-finger proteins, the various leucine zippers show modest conservation beyond the leucine residues, and the proteins bind different DNA sequences. (2) GCN4 and jun are structurally related[20] and bind essentially identical DNA sequences but are unable to form heterodimers[21], suggesting that the residues involved in direct contacts to DNA are located within an adjacent 30-residue stretch that is most highly conserved. This adjacent region is also found in fos, which binds to the identical DNA sequence when complexed with jun as a heterodimer. (3) Leucine-zipper motifs have recently been observed in proteins that also contain zinc fingers or homeodomains; perhaps the leucine zipper acts as an independent dimerization region for different kinds of DNA-binding domains.

RELATED FAMILIES OF PROTEINS RECOGNIZING SIMILAR DNA SEQUENCES

Although different eukaryotic DNA-binding proteins containing a particular structural motif often recognize unrelated DNA sequences, one fundamental aspect of eukaryotic organisms is that they contain families of proteins that interact with similar DNA sequences. In contrast, prokaryotic helix-turn-helix proteins generally have unique DNA sequence recognition properties; i.e. each protein recognizes a distinct set of target sites (the binding of the bacteriophage λ repressor and cro proteins to common operators is a notable exception). Related protein families have been identified for all the structural classes. Six different *Drosophila* homeodomains bind to a common AT-rich consensus sequence even though their protein sequences vary, and the mammalian OCT-1, OCT-2 and Pit-1 proteins (which have similar homeodomains and an additional POU domain[22]) recognize related sites[3]. For the zinc-finger proteins, the steroid receptors show considerable amino acid sequence homology and they bind related DNA sequences[11]. The retinoic acid and thyroid receptors have nearly identical specificities, and the consensus sequences for binding by the glucocorticoid receptor are similar at six out of eight positions. In the case of the putative leucine-zipper motif, mammalian cells have a variety of proteins (jun, junB, junD, fos, fra and others) that interact with a common sequence usually known as an AP-1 site[16–18]. Moreover, multiple leucine-zipper proteins recognizing a common target sequence are found even in a unicellular eukaryote; e.g. the yeast GCN4, AP-1 and probably other proteins[23,24].

The existence of multiple proteins that recognize related sequences increases the precision and flexibility for coordinately and independently regulating genes, particularly those involved in processes of fundamental importance such as cell growth. For example, the bacteriophage λ repressor and cro proteins recognize similar but not identical sequences that control the developmental decision between lysis and lysogeny[25]. In multicellular organisms, the related protein families might provide a mechanism for controlling a core group of genes in a variety of cell types and developmental stages (e.g. the *Drosophila* homeodomain proteins) or in response to extracellular signals (e.g. the steroid receptors). Precision and flexibility could be achieved by subtle differences in sequence recognition properties or DNA-binding affinities such that individual promoters would be strongly affected by particular proteins (or combination of proteins). In this regard, the dimeric nature of the leucine-zipper proteins (and probably the zinc-finger and helix-turn-helix proteins) suggests an additional mode of flexibility involving heterodimers between different proteins. Such heterodimers might yield new proteins with distinct recognition properties, or they might make it possible to influence gene expression only when two specific physiological conditions occur.

EVOLUTIONARY CONSERVATION OF EUKARYOTIC DNA-BINDING PROTEINS

It is now clear that the mechanism of transcription is remarkably conserved throughout the eukaryotic kingdom. Yeast upstream activator proteins function in a variety of eukaryotic organisms, vertebrate transcription factors function in yeast cells, and the yeast and mammalian TATA factors are functionally interchangeable for transcription *in vitro*[26–28]. Such functional conservation undoubtedly indicates that the basic mechanism of transcriptional initiation has existed since the first eukaryotic organisms.

In addition to the conservation of the basic transcriptional activation mechanism, eukaryotic cells from yeast to human contain structurally similar and functionally analogous transcription factors that recognize essentially identical sequences. Some early examples include the yeast GCN4 protein and the vertebrate jun oncoprotein[20,21], the yeast PHO2 and *Drosophila* engrailed protein[29], the human CCAAT-binding protein CP-1 and the yeast HAP2,3 proteins[30]. Moreover, both CP-1 and HAP2,3 bind as heteromeric complexes involving two distinct proteins, and the yeast and human subunits can be functionally exchanged[30]. It seems likely that these examples represent the tip of the iceberg, and that striking similarities throughout the eukaryotic kingdom will soon become the rule rather than the exception.

Although eukaryotic cells have related transcription factors that recognize similar DNA sequences, the homologues often perform different functions in their respective organisms. For example, GCN4 and HAP2,3 activate the amino acid biosynthetic and oxygen regulated genes respectively in yeast, whereas their evolutionary counterparts, jun and CP-1, activate a variety of genes whose functions appear unrelated. This phenomenon may reflect the fact that for transcription factors that bind to multiple promoters, it would be difficult to alter the sequence recognition properties of the regulatory protein at any point in evolution without affecting the transcription of many genes. However, the functions of such transcription factors could diverge in individual organisms. By analogy, the genetic code is essentially universal and eukaryotic TATA elements and prokaryotic −10 sequences are similar even though transcriptional and translational mechanisms are quite different.

ROLE OF DIFFERENT STRUCTURAL MOTIFS IN TRANSCRIPTIONAL REGULATION

Eukaryotic transcription factors contain at least two essential functions, a DNA-binding domain and a transcriptional activation region, that are generally located in separate parts of the protein. Yeast transcriptional activation functions are defined by short acidic regions with minimal primary sequence-homology regions (reviewed in Refs 31, 32). Different acidic regions of GCN4 and GAL4 are equally capable of activating transcription even though their primary sequences are dissimilar. Moreover, acidic regions are observed in other yeast activator proteins, acidic character is the common feature of functional transcriptional activation regions selected from short *E. coli* DNA segments, and amino acid substitutions that alter the level of activation are usually associated with a change in net negative charge. Despite the importance of acidic character, functional transcriptional activation regions have additional structural features beyond negative charge. High resolution deletion analysis and proteolytic mapping suggest that activation regions are repeating structures composed of small units that might be α-helices with amphipathic character. It has been proposed that acidic activation regions are surfaces used for interactions with other proteins of the transcription machinery.

A crucial feature of acidic activation regions is that they function autonomously when fused to different DNA-binding domains (reviewed in Refs 31, 32). For example, functional transcriptional activator proteins can be produced by fusing the DNA-binding domain of the *E. coli* LexA repressor (a helix-turn-helix protein) to the activation regions of proteins containing zinc fingers (GAL4, HAP1), leucine zippers (GCN4, jun, fos, myc), and homeodomains (bicoid). Moreover, the distance and orientation of the GCN4 and GAL4 activation regions with respect to their DNA-binding domains is functionally unimportant.

Acidic regions of DNA-binding proteins are likely to be commonly employed for transcriptional activation in all eukaryotic organisms. GAL4 activates transcription from appropriate target promoters in mammalian cells, and the acidic activation region is required[33,34]. Conversely, transcriptional activation in yeast cells by the jun oncoprotein and the glucocorticoid receptor requires acidic sequences in addition to the DNA-binding domains[35,36]. Thus, acidic activation regions probably contact some part of the transcription machinery that is conserved functionally throughout the eukaryotic kingdom such as a TATA-binding protein or RNA polymerase II.

In this view, the DNA-binding domain has two important roles in transcriptional regulation. (1) It brings the protein to the DNA such that it can easily interact (through the activation region) with other components to form a functional transcriptional initiation complex. (2) The high specificity of the protein–DNA interactions provides the major mechanism by which genes are differentially expressed. For either of these roles, the distinct structural motifs are equivalent. Thus, the helix-turn-helix, zinc-finger, and leucine-zipper motifs utilized for recognizing specific DNA sequences represent different structural solutions to a common function. Did these structural motifs arise independently during evolution, or does one of them represent the primordial DNA-binding domain?

References

1 Pabo, C. O. and Sauer, R. T. (1984) *Annu. Rev. Biochem.* 53, 293–321
2 Wharton, R. P. and Ptashne, M. (1985) *Nature* 316, 601–605
3 Levine, M. and Hoey, T. (1988) *Cell* 55, 537–540
4 Scott, M. P., Tamkun, J. W. and Hartzell, G. W., III (1989) *Biochim. Biophys. Acta Reviews on Cancer* 989, 25–48
5 Sauer, R. T., Smith, D. L. and Johnson, A. D. (1988) *Genes Dev.* 2, 807–816
6 Porter, S. D. and Smith, M. (1986) *Nature* 320, 766–768
7 Miller, J., McLachlan, A. D. and Klug, A. (1985) *EMBO J.* 4, 1609–1614
8 Berg, J. (1988) *Proc. Natl Acad. Sci. USA* 85, 99–102
9 Vrana, K. E., Churchill, M. E. A., Tullius, T. D. and Brown, D. D. (1988) *Mol. Cell. Biol.* 8, 1684–1696
10 Pieler, T., Hamm, J. and Roeder, R. G. (1987) *Cell* 48, 91–100
11 Evans, R. M. and Hollenberg, S. M. (1988) *Cell* 52, 1–3
12 Parraga, G., Horvath, S. J., Eisen, A., Taylor, W. E., Hood, L., Young, E. T. and Klevit, R. E. (1988) *Science* 241, 1489–1492
13 Johnston, M. (1987) *Nature* 328, 353–355
14 Landschulz, W. H., Johnson, P. F. and McKnight, S. L. (1988) *Science* 240, 1759–1764
15 Hope, I. A. and Struhl, K. (1987) *EMBO J.* 6, 2781–2784
16 Nakabeppu, Y., Ryder, K. and Nathans, D. (1988) *Cell* 55, 907–915
17 Halazonetis, T. D., Georgopoulos, K., Greenberg, M. E. and Leder, P. (1988) *Cell* 55, 917–924
18 Kouzarides, T. and Ziff, E. (1988) *Nature* 336, 646–651
19 O'Shea, E. K., Rutkowski, R. and Kim, P. (1989) *Science* 243, 538–542
20 Vogt, P. K., Bos, T. J. and Doolittle, R. F. (1987) *Proc. Natl Acad. Sci. USA* 84, 3316–3319
21 Struhl, K. (1987) *Cell* 50, 841–846
22 Herr, W., Storm, R. A., Clerc, R. G., Corcoran, L. M., Baltimore, D., Sharp, P. A., Ingraham, H. A., Rosenfeld, M. G., Finney, M., Ruvkun, G. and Horvitz, H. R. (1988) *Genes Dev.* 2, 1513–1516
23 Harshman, K. D., Moye-Rowley, W. S. and Parker, C. S. (1988) *Cell* 53, 321–330
24 Jones, R. H., Moreno, S., Nurse, P. and Jones, N. C. (1988) *Cell* 53, 659–667
25 Johnson, A. D., Poteete, A. R., Lauer, G., Sauer, R. T., Ackers, G. R. and Ptashne, M. (1981) *Nature* 294, 217–223
26 Guarente, L. (1988) *Cell* 52, 303–305
27 Buratowski, S., Hahn, S., Sharp, P. A. and Guarente, L. (1988) *Nature* 334, 37–42
28 Cavallini, B., Huet, J., Plassat, J-L., Sentenac, A., Egly, J-M. and Chambon, P. (1988) *Nature* 334, 77–80
29 Burglin, T. R. (1988) *Cell* 53, 339–340
30 Chodosh, L. A., Olesen, J., Hahn, S., Baldwin, A. S., Guarente, L. and Sharp, P. A. (1988) *Cell* 53, 25–35
31 Struhl, K. (1987) *Cell* 49, 295–297
32 Ptashne, M. (1988) *Nature* 335, 683–689
33 Kakidani, H. and Ptashne, M. (1988) *Cell* 52, 161–167
34 Webster, N., Jin, J. R., Green, S., Hollis, M. and Chambon, P. (1988) *Cell* 52, 169–178
35 Struhl, K. (1988) *Nature* 332, 649–650
36 Schena, M. and Yamamoto, K. R. (1988) *Science* 241, 965–967

K. Struhl is at the Department of Biological Chemistry, Harvard Medical School, Boston, MA 02115, USA.

INDEX